PRINCIPLES
OF CODE
ENFORCEMENT

James M. Foley

PEARSON

Boston   Columbus   Indianapolis   New York   San Francisco   Upper Saddle River
Amsterdam   Cape Town   Dubai   London   Madrid   Milan   Munich   Paris   Montreal   Toronto
Delhi   Mexico City   São Paulo   Sydney   Hong Kong   Seoul   Singapore   Taipei   Tokyo

Publisher: Julie Levin Alexander
Publisher's Assistant: Regina Bruno
Executive Editor: Marlene McHugh Pratt
Senior Acquisitions Editor: Stephen Smith
Associate Editor: Monica Moosang
Editorial Assistant: Samantha Romano
Director of Marketing: David Gesell
Marketing Manager: Brian Hoehl
Marketing Specialist: Michael Sirinides
Managing Production Editor: Patrick Walsh

Production Project Manager: Debbie Ryan
Production Editor: Julie Boddorf
Media Product Manager: Lorena Cerisano
Senior Design Coordinator: Christopher Weigand
Cover Designer: Karen Salzbach
Composition: iEnergizer/Aptara®, Inc.
Full-Service Project Management: Rashmi Tickyani,
    Aptara®, Inc.
Printing and Binding: Courier Kendallville
Cover Printer: Lehigh-Phoenix Color/Hagerstown

Library of Congress Cataloging-in-Publication Data
Foley, James M.
    Principles of code enforcement / James M. Foley.
        pages cm
    Includes bibliographical references.
    ISBN-13: 978-0-13-262591-3
    ISBN-10: 0-13-262591-1
    1. Fire prevention—Inspection—United States.   2. Fire prevention—Standards—United States.   3. Fire investigation—United States—Case studies.   4. Fires—United States—History.   I. Title.
    TH9176.F65 2014
    363.37'60973—dc23
                                                                                    2013002149

10 9 8 7 6 5 4 3 2 1

PEARSON

ISBN 13: 978-0-13-262591-3
ISBN 10:   0-13-262591-1

DEDICATION

This book would not have been possible without the love and understanding of my wife Patricia and my two sons, Brian and Brandon. I thank you for your support, love, and understanding with the countless hours I had to spend writing this book.

I also dedicate this book to my fellow co-workers at the Atlantic City Fire Department: Chiefs Scott McKnight and Robert Wojcik; Fire Captains Joe Haney, Mike Ruley, Gordon Pherribo, Bill Brooks, Leon Fontville, Anthony Ingenito, Carnice Lambert, Allen Huggins, Ledford Mack, and Mohammed Hatafi; and secretaries Nancy, Debbie, and Sybil whose assistance and cooperation made this project easier. You all have dedicated your careers to making people safer from fire, and I and the public owe you all a debt of gratitude.

I also would like to dedicate this book to my former mentors: the late Chief Thomas Gienger (GFVFD), Battalion Chief (Ret.) Jim McMasters (CFD), the late Battalion Chief John Duffy, Deputy Chief (Ret.) Anthony Ingenito, and Fire Official (Ret.) Joseph Goukler, Jr., of the ACFD. Your leadership, belief in me, and encouragement made me pursue a higher level of education in the fire service and enabled me to write this book. The opportunities and encouragement you provided will help shape future fire service leaders to focus greater attention on fire prevention; I thank you all.

Last, I would like to thank two very good friends: Battalion Chief (Ret.) Joseph Ward, whom I have known since high school and college days and who encouraged me to take my first assignment in the fire prevention bureau; and my friend, Fire Inspector Luke Schlachter, who died while on duty. Your silent leadership and willingness to always take the most difficult jobs encouraged others to be better in the fire prevention field. You set a great example for me, and you are greatly missed by your brothers.

CONTENTS

Chapter 2 The Code Enforcement System   17

Chapter 3 Understanding Statistics and Community Risk Reduction   36

Chapter 6   Violation Correction Orders and the Legal Process   107

Chapter 7   Identifying Common Fire Code Violations   133

Chapter 8   Inspection of Hazardous Materials   160

Chapter 9 **Means of Egress   195**

Chapter 10 Passive Fire Protection 223

Chapter 11 Fire Alarm Systems   245

Chapter 12 Water-Based Fire Protection Systems   281

Chapter 14 Fire Protection Plan Review   350

Answers to End of Chapter Review Questions

The Need for Public Safety Education

Fire Investigations

PREFACE

It is my privilege to be involved in this educational project by my good friend, James Foley. As an instructor and course developer at the national, state, and local levels, I have had the honor to be associated with Jim for nearly a decade. His new textbook, *Principles of Code Enforcement*, is packed with practical information. Chief Foley is a highly respected fire service leader, and an educator and emergency responder who knows the complex challenges of crawling down dark hallways and standing up to the political influences of the Atlantic City gaming industry. Over many years, he has built a solid reputation as an advocate in public safety and risk reduction for first responders. The valuable information contained within these pages, when properly applied, will directly reduce fires and increase safety.

Today, emergency responders face enormous difficulties, from the dangers in the fire stations to the dangers of the fire ground. This text provides tangible tips to outline the sustainable steps needed to ensure that the public and first responders have not forgotten the large fire losses of the past, and it will become the cornerstone for creating a fire-safe community.

Firefighting operations have changed since my first run back in January of 1973. As a firefighter, company commander and chief officer, I quickly learned that situational awareness is an essential element. Fire attack, ventilation, search, and rapid intervention crews (RIC) all hinge on having useable information. Building construction, fire load, limits of bunker gear, and the need for wearing an SCBA combined with smaller staffing has altered the mind-set of the modern incident commander. Reliable technology and familiarity with the fire protection features of various occupancy groups are vital. Citizens depend on early warning in order to avoid harm, and first responders require immediate information to be effective. Valuable knowledge of historical events, the evolution of codes, and changes in consensus standards and adoption of regulations found within this text will increase the likelihood of better life safety and incident stabilization decisions.

All stakeholders from the onset of the design of a structure to the response phase need technical knowledge of the auxiliary appliances needed during an emergency. Proper prevention planning and the courage to apply the fire code to all structures is the single opportunity to make a significant difference. Jim Foley's text captures critical case studies that will help today's new generation of fire service leaders correctly interpret and understand the purpose of a strong fire code. Code enforcement officers need to develop a fair process so that elected officials and business entrepreneurs can fully support enforcement procedures. First responders have an obligation to gather background information on active and passive protective systems in their advance-planning phase. This text is the best available tool for gaining insight into practical approaches to improving response and recovery efforts in any community.

First responders, at the strategic, tactical, and task levels, have to make many decisions during the first few moments on the fire ground. You can increase the odds of stopping the crisis, properly allocating resources, and ensuring that "everyone goes home" by investing your time in comprehending the intent of this text. This intent is a simple one: to identify a pipeline of potential problems from past practices, cost, culture, resources, and risk. Then the text finds the straightforward solution—enforcement of the fire code during the construction and occupancy phase!

I found this text to be thorough, time-tested, and insightful. Having a "baseline" on the importance of built-in fire protection features is the only logical way to prepare for the challenges of command during a rapidly escalating emergency.

As you use your knowledge of prevention projects, you will be supporting the overall mission of the founding fathers of the fire service—preventing devastating disasters! It is my privilege to continue as a strong supporter of quality fire prevention and risk reduction endeavors.

William Shouldis
Deputy Fire Chief (ret)
Philadelphia (PA) Fire Department

Using This Book

This book contains 14 chapters, and it can be easily adapted to a fire prevention college-level course in a 15-week semester. The text covers the recommended requirements for higher education established by both FESHE and NFPA 1031, *Professional Qualifications for Fire Inspectors,* and NFPA 1037, *Professional Qualifications for Fire Marshals.* The text also identifies several of the 16 life safety initiatives specified for firefighter safety and survival in "Everybody Goes Home" from the National Fallen Firefighters Foundation.

Chapter 1: Builds the history and background of fire prevention by examining major fires and events that helped to shape the codes of today. It explores the very foundations of the American fire services approach to fire prevention.

Chapter 2: Discusses the basics of fire prevention organization including how the model codes developed, training standards for fire inspectors, prescriptive versus performance codes, and strategies for code development as well as an overview of budgeting for fire marshals and fire officials.

Chapter 3: Identifies the analysis necessary for community risk assessment. This chapter provides statistical analysis in a simplified format to turn data into useful information. This helps identify useful measurements for the fire marshal or fire official to use in support of fire prevention programs.

Chapter 4: Develops an understanding of fire prevention's mission and the vision and leadership necessary to deal with fire department change. Fire prevention inspectors must learn how to become effective managers and good leaders in their organizations in a changing code enforcement and building construction environment.

Chapter 5: Takes the fire inspector through the entire fire inspection process from the review of the building file to the closing interview with the owner explaining each key element in the process.

Chapter 6: Identifies the after-inspection activities of preparing and writing fire inspection reports and preparing notices of violation. This chapter reviews the process for board of appeals and court hearings and helps the fire inspector prepare for giving testimony in fire code enforcement cases.

Chapter 7: Describes the commonplace fire code violations that occur in most building use groups. The fire inspector will learn how to identify and correct these common fire hazards.

Chapter 8: Addresses the inspection of facilities containing large quantities of hazardous materials. The chapter reviews the background of federal regulations that are the foundation of the fire code and helps the inspector identify proper handling, storage, and use requirements.

Chapter 9: Explores the elements of the means of egress both from a building code design perspective and the fire code maintenance requirements. This chapter illustrates key egress concepts that all fire inspectors must understand to deal with emergency exiting in all types on buildings.

Chapter 10: Identifies the elements of passive fire protection employed in building construction, including the five types of construction and the important role of fire test conformance. This section helps the fire inspector identify the passive fire protection features of the building.

Chapter 11: Describes the importance of fire alarm systems and provides the fire inspector with the technical knowledge of how they operate and how they should be tested. This chapter helps the fire inspector identify deficiencies in fire alarm coverage or equipment performance.

Chapter 12: Reviews the applicable NFPA standards and requirements for water-based fire protection systems and helps the fire inspector identify key inspection points for automatic sprinklers, fire standpipes, fire pumps, and other specialized fire protection systems.

Chapter 13: Explores the fire inspector's role in community planning related to the establishment of effective water supply for the fire department. This chapter explores planning and zoning requirements as well as evaluating both urban and rural water supply systems.

Chapter 14: Offers fire inspectors additional insight into the plan review process as it relates to fire protection system evaluation. The chapter identifies the different types of architectural and engineering drawings that are reviewed and the types of fire protection information that are required to be provided on submitted plans.

While many fire inspector texts may cover these areas of subject matter, this text is unique in that it provides in-depth technical knowledge coupled with practical experience and presents it in a simple-to-understand manner. This text will not only be useful to new fire inspectors in training but will provide a reference resource of information to seasoned fire inspectors, fire marshals, fire officials, and fire chiefs for years to come.

Ancillary Materials

Through Resource Central, this text invites students to enrich their classroom learning with quizzes, web links, and additional content related to the topic of Code Enforcement. Instructors will also find here a full complement of supplemental teaching materials such as test banks, PowerPoint lectures, and an instructor's manual to facilitate the teaching process.

ACKNOWLEDGMENTS

Deputy Chief (Ret.) James Smith, Philadelphia Fire Department – I want to thank Chief Smith for recommending me for this project and for his support in this and other endeavors.

Deputy Chief (Ret.) Bill Shouldis, Philadelphia Fire Department – I want to thank Bill for writing the Preface for this book and for his friendship, expertise, and support over the years in the many projects we have worked on together.

Fire Captain Michael Ruley, Atlantic City Fire Department – Thanks, Mike, for assisting me with the photographs.

Fire Captain Joseph Haney, Atlantic City Fire Department – Thanks, Joe, for your support and encouragement and the input from your EFO project.

Fire Chief Dennis J. Brooks, Atlantic City Fire Department – Thanks, Chief, for your continued support and for allowing me access to fire prevention after my retirement.

Fire Chief (ret.) John J. Bereheiko and Deputy Fire Chief (Ret.) Victor "Rick" Francesco – Your leadership, support, and friendship over the years allowed me to expand my horizons. Without that ability, I could not have completed this text.

Special thanks is extended to the following individuals who reviewed this textbook and who made many valuable suggestions for change. Their contributions are reflected throughout the text.

Tim Henshaw
Fire Protection Engineer
City of Raleigh Fire Marshal's Office

Ernie Misewicz
Adjunct Instructor
University of Alaska Fairbanks

Jon Napier
Division Chief/Fire Marshal
Kent Fire Department Regional Authority

Steve Beumer
Assistant Fire Marshal
Aurora Fire Department/Community College of Aurora

Jeff Huber
Professor of Fire Science
Lansing Community College

Gary Edwards
Fire Science Program Director
Montana State University

Michael Garcia
Deputy Fire Chief / Fire Marshal
Long Beach Fire Department

George Hettenbach, BBA, MS
Adjunct Instructor/Captain (Retired)
Upper Darby Fire Department, PA

Charles F. Jenkins
Fire Marshal
Glendale Fire Department, Glendale, Arizona

Nathan Sivils
Director of Fire Science
Blinn College, Bryan, Texas

Martin Walsh
Professor
San Diego Miramar College, San Diego, CA

David McClean
Fire Science Faculty
Southern Maine Community College, South Portland, Maine

Scott Alderman
Fire Chief
Lewisville Fire Department, Lewisville, NC

Danny W. Roach
Assistant Fire Chief/Training Chief
Whitfield County Fire Department
Dalton, Georgia

Gregory Aymot
Fire Protection Supervisor
State University of New York at Albany

Anthony Gianantonio
Lieutenant – FO
Palm Bay Fire Rescue, Palm Bay, FL

JAMES M. FOLEY has been a firefighter since the age of 18 and has a career spanning 42 years in the fire department and fire prevention services. Chief Foley began his career in the Green-Fields Village Volunteer Fire Department in 1970 and became a career firefighter in the City of Camden New Jersey in 1974, where he was assigned to Engine Company #6 in North Camden. He later transferred to the Atlantic City New Jersey Fire Department in 1976 and was assigned to Ladder Company # 3 and later Engine Company #1 of the Atlantic City Fire Department. In 1980 he was promoted to Fire Inspector and was assigned to the Fire Prevention Division, Plans Review Unit during the development and building of the Atlantic City casinos. In 1985 he was promoted to the rank of Assistant Chief

Fire Inspector and was assigned to supervise the fire inspection unit of the Fire Prevention Division. The fire chief also tasked Chief Foley with the development of the hazardous materials response team and the fire department heavy rescue unit. Chief Foley was a founding member and Task Force Leader with New Jersey Task Force 1, the State of New Jersey Urban Search and Rescue Team. Chief Foley was one of the first responders to the World Trade Center disaster on September 11, 2001, with NJ-TF1 on a ten-day deployment. Chief Foley was again promoted in 1998 to the position of Fire Official at a Deputy Chief Rank, where he directed the fire prevention inspection, fire investigation, and public fire safety education efforts of the fire department until his retirement in 2010 with 34 years service in the ACFD. Chief Foley has written many articles in fire service trade journals on fire prevention and building construction fire safety issues and is also currently an adjunct professor at Camden County College in the Fire Science Department. He instructs college classes in fire science chemistry, hazardous materials, and fire protection systems. Chief Foley is also an instructor at Kean University and also instructs with the New Jersey Division of Fire Safety in the State Fire Inspector and Fire Official recertification program.

Chief Foley holds a Bachelor of Science degree in fire protection technology and fire administration from the University of Maryland and a Bachelor of Arts degree in biology life sciences from Rowan University. He is a licensed Fire Subcode Official and HHS Fire Protection Inspector, a certified Fire Official /Fire Inspector, and a certified Level II Fire Service Instructor and Drill Ground Instructor as well as a Hazardous Materials Technician Instructor. Chief Foley is a member of the New Jersey Deputy Fire Chiefs Association, the NJ-IAAI, and a former member of the New Jersey Master Planning Advisory Council to the New Jersey Department of Community Affairs.

Chief Foley can be contacted at: jimfoley2@yahoo.com

FIRE AND EMERGENCY SERVICES HIGHER EDUCATION (FESHE) GRID

Course Description:

This course provides students with the fundamental knowledge of the role of code enforcement in a comprehensive fire prevention program.

The following grid outlines Principles of Code Enforcement course requirements and where specific content can be located within this text:

FESHE Course Requirements	1	2	3	4	5	6	7	8	9	10	11	12	13	14
1. Explain the code enforcement system and the fire inspector's role		X	X	X	X	X	X	X	X	X	X	X	X	X
2. Describe the code and standards development and adoption process	X	X			X									X
3. Describe the difference between prescriptive and performance codes		X								X				X
4. Describe the legal authority and limitations relevant to fire code inspections		X		X										X
5. Describe the importance of documentation					X	X					X	X		X
6. Recognize the ethical practices of code enforcement officers				X	X									
7. Explain the application and interrelationship of codes, standards, recommended practices, and guides		X		X	X									X
8. Describe the differences of how codes apply to new and existing buildings		X		X										X
9. Identify appropriate codes and their relationship to other requirements in the built environment		X		X	X		X	X	X	X	X	X		
10. Describe the political, business, and other interests that influence the code enforcement process	X			X										
11. Identify the professional development process for code enforcement practitioners		X		X										X

1

Understanding the Historical Significance of Fires and Their Impact on Fire Codes

Courtesy of J. Foley

KEY TERMS

fire experience, *p. 2*

fire inspector, *p. 2*

fire prevention code, *p. 2*

fire warden, *p. 2*

National Board of Fire Underwriters, *p. 5*

National Fire Protection Association, *p. 5*

OBJECTIVES

Upon completing this chapter, the reader should be able to:

- Understand the relationship between firefighting and the mission of fire prevention.
- Understand the history that shaped modern fire prevention methods.
- Understand the impact of tragic fires and their effects on fire code enforcement and code change.

Resource**Central** For additional review and practice tests, visit www.bradybooks.com and click on Resource Central to access text-specific resources for this text. To access Resource Central, follow the directions on the Student Access Card provided with this text. If there is no card, go to www.bradybooks.com and follow the Resource Central link to buy access from there.

Early Fire Codes

fire inspector ▪ A person actively employed in the inspection, application, and enforcement of fire prevention codes. Fire inspectors may be nationally certified or may be state certified for compliance with NFPA 1031, *Standard for Professional Qualifications of Fire Inspector and Plan Examiner*.

fire prevention code ▪ A legally adopted set of fire prevention rules and regulations to be enforced as law within a municipality or state jurisdiction. Fire prevention codes usually are adaptations of model codes such as the Uniform Fire Code by the International Codes Conference or NFPA 1 from the National Fire Protection Association. Some jurisdictions may also adopt local code requirements or may amend adopted model code based on local conditions.

fire experience ▪ Statistical analysis of fire data within a jurisdiction or state based on occupancy type, fire loss, fire injuries, and fire deaths as well as other contributing factors. Fire experience lends credence to fire code regulations to prevent future occurrences of similar types of fires by modifying ignition sources, fuel loads, or human behavior.

A **fire inspector** is expected to enforce fire code requirements identified in a **fire prevention code** or related regulation. These code requirements address some area of hazard mitigation by prohibition, safety distance, passive or active protection methods, or changes in the behavior of building occupants. Fire code requirements usually are based on some history of **fire experience** and often on lessons learned from great fire tragedies. Enforcement of fire prevention codes traditionally has been entrusted to the fire service. This is because of its congruency with the fire department's mission to protect life and property.

Building codes can be traced back to 1772 BC in Babylon during the reign of King Hammurabi. Hammurabi's code is best known for its "eye for an eye, tooth for a tooth" harshness. Hammurabi established many laws and rules, including a building code provision that any builder who built a home that collapsed and killed the owner would be put to death; and if the owner's son were killed, then the builder's son would also be put to death. These types of regulations would certainly encourage builders to perform their craft properly.[1] Fire prevention code enforcement historically began in 300 BC, in ancient Rome. The Emperor formed a band of slaves called the *familia publica* who were stationed around the walls of Rome to watch for uncontrolled fires. When Caesar Augustus became emperor of the Roman Empire, he developed a more professional fire service called the *Vigiles*, which was composed of 7,000 centurions who acted as firefighters, code enforcers, and policemen. They had the authority to act as judge and jury, inflicting corporal punishment or even death upon any Roman who had caused an uncontrolled fire. The Vigiles existed for approximately 500 years[2] until the fall of the Roman Empire.

Fire prevention codes were first documented in England under King Alfred (890 AD) and were reinforced later by William the Conqueror (1066 AD) with the enactment of the *Curfew Law*, which meant "cover the fire." Residents of England had to extinguish fires at night at the sounding of a bell. The curfew law was abandoned later by King Henry I, who required that all homeowners have a bucket of water ready and bring it to any fire that would occur in their neighborhood.[3]

The Great Fire of London in 1666 burned for five days and destroyed over 13,000 homes, churches, and buildings. In December of that year the City Council passed its first building code requiring that all buildings be made of brick and stone and forbidding hazardous occupations in the courtyards.[4] The London fire also gave rise to the development of fire insurance companies in the British Isles.

Fire Prevention in Colonial America

As the New World began to be colonized by the British, Spanish, and French, fire was a necessity for the colonists' survival in that it provided heat and cooked food; however, it also threatened the colonies, as forts and homes were constructed of only combustible materials such as wood and grasses. The hazard of uncontrolled fire was a threat to the colonists' very existence.

The need for fire prevention became evident when Jamestown, Virginia, was settled in 1607 by the London Company under the direction of Captain John Smith. Jamestown suffered a devastating fire in 1608 that destroyed most of the shelters and provisions of the colonists. In fact, over its history, Jamestown would have four major fires, eventually leading to Williamsburg's becoming Virginia's state capitol after the last fire in 1699 (Figure 1.1).[5]

fire warden ▪ An official assigned to prevent or fight fires, as in a forest, logging operation, camp, or town.

As the American colonies grew, so did the need for fire prevention and fire control. In 1648, the City of New York, then called New Amsterdam, enacted a law under Director General Peter Stuyvesant to hire four **fire wardens** to inspect all chimneys in the city and issue fines for those in need of cleaning or disrepair.[6] Later, eight citizens were appointed as "Rattle Watchers." They had wooden rattles that they would spin to make noise and

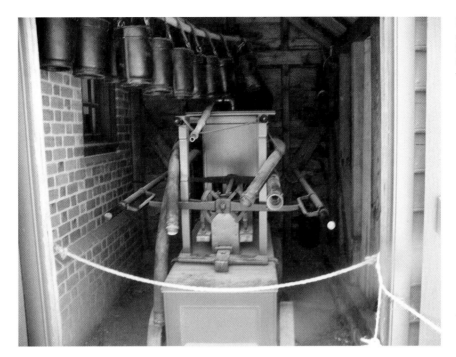

FIGURE 1.1 Early American fire engine in Williamsburg, VA. *Courtesy of J. Foley*

direct citizens with buckets to the location of any fire. This was a distant forerunner to the central fire alarm station. New York was not alone in passing fire prevention laws. In 1631, Governor John Winthrop of Boston passed a law forbidding thatched roofs and wooden chimneys in the city.

In Philadelphia in 1736, one of the most notable founding fathers of this nation, Benjamin Franklin, founded the first volunteer fire company in America (see Figure 1.2). The Union Fire Company, established at Second and Quarry streets in Philadelphia, later became Engine Company #8 of the Philadelphia Fire Department in 1871 and the Philadelphia Fire Museum in 1976. Franklin played a major role in the development of the American fire service. He was a proponent of fire prevention and authored many papers

FIGURE 1.2 Benjamin Franklin was the first American fire chief. Pictured is his portrait at Engine 8 and Ladder 2 in Philadelphia, PA. *Courtesy of J. Foley*

FIGURE 1.3 Pictured is a fire mark that shows that this home was insured by the "Fire Association of Philadelphia." It is located on the oldest street in America, Elfreth's Alley in Philadelphia, PA. *Courtesy of J. Foley*

on fire safety, the most notable being "An ounce of prevention is worth a pound of cure." This paper dealt with the hazard of carrying hot coals in a shovel rather than in a bed warmer in the winter and emphasized the need for clearance to combustible bedding.[7] In 1751, after many disastrous fires in Philadelphia and elsewhere, Franklin and members of the Union Fire Company began the first fire insurance company in America, The Philadelphia Contributorship. A citizen would buy fire insurance, and the property would undergo inspection before a policy would be issued. The Contributorship paid its first fire loss in 1753 for a fire in a house on Water Street, which Franklin reported in his Gazette. Franklin stated that the house was quickly repaired at no cost to the owner due to fire insurance which promoted insurance purchases. Houses in the city that were covered by insurance would have a placard called a "fire mark" placed on them for the local fire company to see when they responded (see Figure 1.3). This ensured prompt fire suppression efforts by the firefighters. The Philadelphia Contributorship is still in the fire insurance business today.

History of Fire Prevention

Fire prevention has always been a vital part of the fire service's mission in the United States, but traditionally it has been underfunded as compared to fire suppression at the local level. The United States continues to lag behind other modern nations in the approach it takes to fire prevention. It is not necessarily that the wrong things are being done; rather, the right things are not done aggressively enough. The fire problem has significantly improved in the United States; however, it still has a long away to go. What spurred the American fire service into undertaking better and improved methods of code enforcement was a series of fire prevention conferences on the subject at the turn of the twentieth century. Fire prevention inspectors and fire officials need to examine the lessons of these conferences so that they can better understand how fire prevention methods and organizations came to be as they are.

FIRST NATIONAL FIRE PREVENTION CONVENTION: 1913

The first American National Fire Prevention Convention was held in October 1913 in Philadelphia (see Figure 1.4). This conference set the pattern for most fire prevention bureaus as they exist today. The Convention was attended by both public and private sector professionals who had an interest in fire safety from across the United States. Powell Evans, Chairman of the Philadelphia Fire Prevention Commission and Director of Public Safety, and J. S. Mallory, the acting Fire Marshal for the City of Philadelphia, chaired the event and presented a landmark report on public fire protection management, which states the following:[8]

> *Fire departments, as a rule, all over this country, up to this time have been selected and managed too much on old "Rule of Thumb" lines, and not sufficiently*

FIGURE 1.4 Independence hall in Philadelphia, PA. The first National Fire Prevention Convention was held in Philadelphia in 1913. The Conference established the path for fire prevention bureaus as they exist today. *Courtesy of J. Foley*

instructed in or brought up to modern methods. Perhaps when the public pays this important body of employees as well as it pays the police, and demands better service, it will get it. After emphasizing the need for fire fighting schools, we pass on to our particular subject, the use of firemen for fire prevention purposes; which provides better occupation for the men, familiarizes them with the properties under their control, and enables them to perform their duties in emergencies with greater speed and safety than otherwise. The system has been employed in a few cities as long as fifteen years and is spreading rapidly. It has worked to advantage whenever used.

Their report proceeded with an inventory of fire prevention practices of major American cities. It notes that most major fire departments were conducting fire inspections using fire department personnel. It identifies many cities that used firefighters and fire officers to conduct these inspections, which made them more prepared in the event of a fire. The 1913 convention also spoke about educating the public in fire safety and encouraged states to take a more active role in fire prevention and safety education. Many cities across the country, encouraged after this convention, established fire prevention bureaus in the fire departments as recommended by the **National Board of Fire Underwriters** and the **National Fire Protection Association**.

PRESIDENT'S CONFERENCE ON FIRE PREVENTION: 1947

After the end of World War II, in May 1947, President Harry Truman appointed Major General Phillip B. Fleming, the administrator of the Federal Works Agency, as chairman to convene the President's Conference on Fire Prevention. This conference assembled the leading minds in fire prevention, public education, and building and fire codes of that time. The United States was seeing increased fire loss and death tolls after the war. It was estimated that these losses could exceed a trillion dollars in property loss by 1952 if nothing was done. President Truman was so concerned with this problem that he wanted a national effort to suppress the effects of fire nationwide. The outcome of this conference was the establishment of the three "E's" of fire prevention—Enforcement, Engineering,

National Board of Fire Underwriters ▪ The National Board of Fire Underwriters (NBFU) was formed in 1866 after 75 separate insurance underwriting companies joined to establish four common goals—to balance insurance rates, balance broker fees, fight arson, and provide measures for protection of common interests. The organization later became the American Insurance Association (AIA) and published the National Building Code.

National Fire Protection Association ▪ The National Fire Protection Association (NFPA) was founded in 1896 by a group of individuals from the insurance underwriters and sprinkler industries. Their goal was to develop safety standards for proper installation of automatic sprinklers. The first standard was produced in 1897 on automatic sprinkler systems and later became known as pamphlet 13. Today, the NFPA is one of the largest consensus standard-making organizations in the field of fire science in the world.

and Education—and they became the foundation of fire prevention bureaus' missions across the country. Most modern fire prevention bureaus today align all of these efforts with their mission of public safety and organizational structure.

WINGSPREAD I–V CONFERENCES: 1966 TO 2006

The Johnson Wax Company established a series of conferences beginning in 1966 (Wingspread I) to gather fire service and industry leaders to help identify problems in the fire service and industry related to fire safety and education. These conferences helped to establish goals and objectives to address these safety issues over the next 10 years. Wingspread conferences currently are held every 10 years, and recommendations from past conferences are reviewed as to whether the issues are resolved or need further recommendations. The latest Wingspread Conference in 2006 identified twenty concerns for the fire service, three of which are of critical interest to fire prevention professionals:

1. *The Fire Problem in the United States*—The fire problem in the United States is a political problem, not a technological problem. It will not be solved without participation in the political process. Fire chiefs and fire service organizations need to more fully participate in the political process on a local, state, regional, and national level.
2. *Professional Development*—Significant strides have been made in fire service professional development, but improvement is still needed. The fire service needs to continue to evolve as a profession as have other governmental entities that operate in the environments where we work as well as other governmental organizations and the private sector. These skills are as important in the volunteer and combination fire services as they are in the career fire service.
3. *Fire Prevention and Public Fire Education*—All aspects of fire prevention have become core components of effective fire service delivery. All fire departments, regardless of size, should value and strive to provide the full range of fire prevention and life safety education services. Increasingly, fire departments are being expected to take on all-hazard / all-risk messaging in addition to traditional fire safety efforts.[9]

WILLIAMSBURG CONFERENCE: 1970

The National Fire Protection Association conducted the Williamsburg Conference of 1970. The purpose of this conference was to establish a forum for discussion of the nation's fire problem. Two representatives from every major fire organization were selected to form the Joint Council on National Fire Service Organizations. The Council's goal was to allow fire service organizations to meet informally and exchange ideas, and to improve cooperation between organizations and establish national goals on fire service issues.

AMERICA BURNING: 1974

America Burning was the turning point in the modernization of fire prevention methods in the United States. The America Burning report was an outgrowth of the Wingspread and Williamsburg Conferences, when the Joint Council of Fire Service Organizations called upon President Richard M. Nixon to reestablish the National Commission on Fire Prevention and Control (see Figure 1.5). The America Burning report examined in detail the fire problem in America and determined that 12,000 Americans were dying annually and over 300,000 were being injured in fires each year. The report pointed out that for every U.S. dollar spent in the fire service, 95 cents was spent suppressing the fire, and only 5 cents was spent trying to prevent the fire from starting.[10] America Burning focused the nation's fire service on fire prevention, building code enforcement, and arson investigation. This report was also responsible for the establishment of a national fire focus by establishing the U.S. Fire Administration and the National Fire Academy to improve professional knowledge of the nation's fire service. The report encouraged master planning by the fire service in the evaluation of community risk versus tolerable loss. A vital component of that was risk reduction through better fire inspection, code enforcement, and public fire safety education programs. The federal government in 1987, and again in 2002, revisited America Burning to measure progress. The recommendations and the process of defining the country's fire service will continue to evolve as the service confronts new issues posed by an ever-changing and complex society.

Tragedy and Code Development

Students of fire prevention and protection must understand that all building and fire codes are political documents. They are the products of advocate negotiations and the consensus of the stakeholders, including firefighters, fire protection industry representatives, builders and architects, insurance representatives, government, and business representatives. At the federal level, the government has established its own safety codes for the military, Department of the Interior, and other federal fire agencies. These codes often include references to NFPA standards. The states generally adopt model building and fire codes from the ICC or NFPA and amend them to the particular state requirements. Local jurisdictions may also incorporate local fire safety requirements in ordinances as long as state law provides that power to local government and these local requirements do not conflict with state or federal laws. Fire and building codes often are driven by tragic events that have caused great loss of life or large property and economic loss. The struggles to adopt regulations often are political and sometimes lead to positive code changes; at other times, competing interests may negate such changes for reasons of economics or fear of government overreach. As an example, after the Meridian One high-rise fire on February 23, 1991, in which three firefighters died, Philadelphia City Fire Chief Rodger Ulshafer called for the retrofit of all high-rise buildings with automatic sprinklers. By December, Mayor Wilson Goode signed into law a requirement for all commercial high-rise buildings over 75 feet to have automatic sprinklers installed within eight years. This was a positive change brought about by the magnitude and impact of the Meridian One fire.

In 2009, a majority of the states voted to adopt automatic sprinkler protection for one- and two-family dwellings into the International Building Code. The requirement was passed and is now an IBC requirement. In several states, however, the requirement will be postponed or not be adopted because of the perceived economic impact on the home building industry due to a soft economy. This delay has occurred despite significant statistical fire data that show 84 percent of all fire deaths occur in residential dwellings. One reason why this issue is such a difficult sell for the fire service is that people die in home fires in small numbers, and so these incidents generally do not draw attention at a state or national level like a major high-rise fire does.

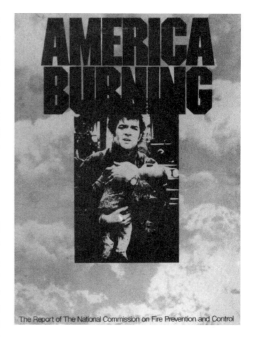

The Report of The National Commission on Fire Prevention and Control

FIGURE 1.5 The 1974 report to President Richard Nixon, "America Burning," began the national dawning of a renewed effort on the importance of fire prevention, fire code enforcement, and fire safety education. *Courtesy of the Federal Emergency Management Agency*

Significant Fires Impacting Code Change

Benjamin Franklin once stated that "those who do not learn the lessons of history are doomed to repeat them"; therefore, the lessons learned in each major fire incident must be remembered by fire prevention professionals, or they will eventually be repeated. Tragedies often can spark innovation and improvement to fire codes and fire prevention techniques. Let us review these major disasters to better understand how tragedy can create change when the lessons on what caused them are learned.

THE GREAT CHICAGO AND PESHTIGO FIRES: 1871

Most people when they hear of the Chicago Fire of 1871 think of Mrs. O'Leary's cow kicking over a lantern at 137 DeKoven Street. The story, however, is only a great myth that made good reading in the newspapers at the time. While it is true the fire did start in O'Leary's barn, the actual cause was more likely the work of two local characters, Daniel "Peg Leg" Sullivan and Dennis Reagan. The fire burned for three days, from October 8 to October 10, and when it was finally extinguished, over 2,000 acres of the city were destroyed, over 17,500 buildings were burned down, and 90,000 people were left homeless.[11] The Chicago Fire killed some 200 to 300 individuals in the course of the three days. An irony of the Chicago fire story was that simultaneously, in the small Wisconsin town of Peshtigo, a forest fire swept through on October 8, 1871, which killed 1,500 to 2,400 people—a far greater human loss than in Chicago, and the entire town was wiped off the map.[12] Forty years later, the Fire Marshals Association of North America, the oldest members section of the NFPA, pushed to have "The Great Chicago Fire" commemorated with a week of fire prevention and home fire drills. President Calvin Coolidge later proclaimed in 1925, on the anniversary of the fire, the establishing of "National Fire Prevention Week," which is commemorated every year by fire departments across the nation during the month of October. Fire Prevention Week is an opportunity to focus the public's attention on the fire prevention safety issues of the community. Fire departments should use this time as a vehicle for public support. When fire marshals and fire officials gain media attention, they can help shape the public's opinion on fire safety issues and positive code change.

THE IROQUOIS THEATER FIRE—CHICAGO, ILLINOIS: 1903

The Iroquois Theater burned in 1903, claiming 602 victims in a fire that lasted only several minutes. The Iroquois, a new theater located at 24–28 West Randolph Street in Chicago, was described as "fireproof" by its builders. There were over 30 fire exits in the auditorium, but most were not well marked or were obstructed by heavy drapery. The theater was occupied by over 1,700 persons attending a matinee of "Mr. Blackbeard" starring Eddie Foy, a famous actor of the period. During the performance, a piece of gauze in the fly loft fell upon a carbon arc lamp and burst into flames. Stagehands quickly tried to lower the asbestos fire curtain to protect the proscenium arch and contain the fire, but it tangled and hung up two-thirds of the way down. The flames leaped under the curtain and out into the audience. As the audience began to move away from the fire they found that several of the exits in the rear of the auditorium were locked. People then rushed toward the main entrance where they had originally entered. Many of the victims fell or were suffocated as smoke rapidly obscured the exits and filled the auditorium. The majority of occupants, mostly women and children, were either crushed or trampled to death in the panic to reach safety at the main entrance of the theater. This fire forced better fire safety inspections and stricter regulations in the city of Chicago for places of public assembly. The fire also motivated the NFPA to examine exiting requirements in places of public assembly. Some of the results of this fire included marking of emergency exits and increasing the capacity of the main exit to accommodate a major portion of the occupants.[13] The Iroquois Theater was considered a state-of-the-art building in its day; the result of the fire was better fire code enforcement. Many fire code violations in the building, including locked exits, were overlooked by inspectors. Resulting code requirements began to address the marking of exits and the capacity of the main exit. The asbestos curtain also was cutting-edge technology at the time, but it failed because it was not properly inspected and tested.

TRIANGLE SHIRTWAIST FIRE—NEW YORK CITY: 1911

The Triangle Shirtwaist Company occupied the upper floors of the Asch Building, a nine-story commercial high-rise building in New York City. This building also was described as one of the new "fireproof" buildings in the city. Triangle Shirtwaist was a sweat shop that employed immigrant workers in the garment manufacturing industry. Working conditions for the mostly female employees were less than safe or humane by today's standards. On March 25, 1911, a fire broke out on the ninth floor of the building, and within a short period of time 146 of the 500 immigrant workers at Triangle Shirtwaist lay dead in the city streets below. The main route of escape was a stairwell that had a locked door to prevent employees from stealing materials for personal use. An exterior fire escape was used by many, but this collapsed and crashed to the ground under the weight of the escaping workers. With no possible means of escape and fire department ladders too short to reach the workers, women lined up at windows and jumped to their deaths one after another rather than suffer burning to death. The Triangle Shirtwaist fire was significant as it gave impetus to the labor union movement to improve employee working conditions. It also highlighted the need for better fire inspections and enforcement of commercial building regulations in New York City and throughout the country. By October 1911, the public outcry was so significant that the American Society of Safety Engineers (ASSE) was formed in New York. The ASSE is dedicated to the safety and health of working men and women everywhere, and it celebrated its one-hundredth anniversary in 2011. This fire brought to light the need for safer exit systems in tall buildings.

THE COCOANUT GROVE FIRE—BOSTON, MASSACHUSETTS: 1942

The Cocoanut Grove Fire occurred in Boston in 1942. The building originally was an auto garage. Barnett "Barney" Welansky, the future club's owner, purchased the building in 1933 and turned it into the biggest nightclub attraction in Boston. Many Hollywood stars, such as Bob Hope and Bette Davis, visited the club regularly. "Buck Jones," the singing cowboy, who died in the fire, also frequented the club on a regular basis. On the night of the fire, the Cocoanut Grove became crowded at the conclusion of the Holy Cross–Boston College football game. The owner ordered that more tables be added to accommodate the crowd. An estimated 1,000 patrons were in the nightclub, which was licensed to hold only 460 occupants. The Cocoanut Grove had makeshift lighting and papier mâché palm trees to make the interior look like Hawaii or the South Seas. The interior finishes were lacquered and cheap and were not fire retardant. The fire began when a busboy was directed to replace a light bulb in the ceiling of the Melody lounge. The young man could not see clearly, so he struck a match and accidentally ignited the paper décor on the ceiling. Many occupants died in their attempt to escape through a revolving door at the main entrance, while other patrons escaped through basement windows. In all, 492 persons died in this fire. In the days after, the City of Boston examined all nightclubs for similar fire code hazards. Barney Welansky was charged with negligence and reckless homicide and cited for failure to obtain building permits and for allowing substandard electrical work. Welansky was found guilty and was sentenced to 12 years in prison, where he died of cancer.[14] The Cocoanut Grove fire spurred Underwriters Laboratories to begin further investigations into the flammability of interior finishes and the NFPA to begin developing a standard called "The Building Exit Code," which later would evolve into NFPA 101, *Life Safety Code®*. As an interesting side note, in 1990, a fire researcher, Charles Kenny, discovered that methyl chloride, a highly flammable gas, had been used in the club's refrigeration system. Kenny suggested that perhaps there was a leak in the refrigeration system, releasing the flammable methyl chloride gas, which could have been ignited when the busboy struck the match, contributing to the rapid flashover and flame spread in the building.[15] The Cocoanut Grove fire led to many building code changes related to use of revolving doors as means of egress, as well as the development of NFPA 701 for fire resistance of decorative materials.

WINECOFF HOTEL FIRE—ATLANTA, GEORGIA: 1946

The Hotel Winecoff was built in 1913 in Atlanta, Georgia. Its owners deemed this fifteen-story luxury hotel "absolutely fireproof," even though it had no fire alarm or automatic fire sprinkler system.[16] The building was of a European design with a central core stairway and elevator shaft. The stair doors often were left open and were not protected by fire-rated doors. The fire occurred on December 7, 1946, at 3:00 a.m., with over 280 guests in the hotel. Of these, 119 were killed in the fire, as their escape was cut off down the main stairway by smoke. Thirty of the victims were high school seniors on a field trip from Rome, Georgia, and most of them died that night. The Hotel Winecoff fire was not the only hotel fire to take lives in 1946; in June, three fires took 77 lives at the LaSalle, Canfield, and Baker hotels. These hotel fires led to a nationwide effort for better code enforcement in the hotel industry and made hotels a major fire inspection target for communities around the nation.[17]

OUR LADY OF THE ANGELS SCHOOL—CHICAGO: 1958

The Our Lady of the Angels School fire in Chicago, Illinois, in December 1958 claimed the lives of 92 elementary school students and three nuns. This was not the first major fire to occur in a school in the United States. The 1937 New London School gas explosion in New London, Texas, killed 298 children, and the Collingswood School fire in Cleveland, Ohio, in 1908 killed 174 students. The fire began in the front stairwell at Our Lady of the Angels school and blocked the egress of students from the building. Many had to jump from the second floor, dying as a result of fall injuries because the interior second-floor stairways did not have fire doors, and smoke quickly filled the corridors and classrooms. After the fire, Mr. Percy Bugbee, then president of National Fire Protection Association, was interviewed and made the following comment: "There are no new lessons to be learned from this fire; only old lessons that tragically went unheeded."[18] This fire changed laws in Chicago regarding the installation of fire alarm boxes and fire alarm systems within school buildings. The fire also caused a nationwide increase in school safety inspections and fire drills conducted to safely evacuate students.

THE BEVERLY HILLS SUPPER CLUB FIRE—SOUTHGATE, KENTUCKY: 1977

The Beverly Hills Supper Club fire occurred in 1977 in Southgate, Kentucky. The nightclub was located in a rural area just 30 miles from Cincinnati. The original club was small, but as business grew, several additions were constructed without proper building and fire inspections and permits from the state of Kentucky. As these additions were made, existing exits were lost or emptied into new rooms with inadequate exit capacity. The building was 54,000 square feet and had neither automatic fire sprinklers nor any automatic or manual fire alarm system. When the fire broke out in an electrical closet by the Zebra room, the flames quickly spread up a monumental stairway and down the exit access corridor toward the Cabaret Room, where 134 victims died. Flammable wood paneling and flammable interior finishes accelerated the fire's spread, making evacuation untenable fairly quickly. The club had about 2,400–2,800 patrons at the time of the fire, and 194 lost their lives. A code analysis of the building by the National Fire Protection Association demonstrated that the building was overoccupied and had insufficient exterior exits for the patrons to safely escape.[19]

History from the Iroquois Theater and Cocoanut Grove fires had repeated itself once again. This fire demonstrated the need for building and fire code enforcement and active regular fire inspections in places of public assembly.

MGM GRAND FIRE—LAS VEGAS, NEVADA 1980

The MGM Grand was the showplace of Las Vegas and a major casino hotel facility. In 1980, it suffered a major fire in the casino, resulting in 84 persons losing their lives. The fire began in a deli pie case that had an electrical ground short circuit. The fire spread to nontreated combustible foam materials in the cabinet and quickly developed into a raging inferno, flashing over the 60,000-square-foot casino in 12 seconds according to the NFPA fire investigation report. The fire was fueled by burning plastics and furnishings, and smoke spread to all areas of the building via unprotected vertical openings and seismic joints. Most of the victims died from smoke inhalation on the top two floors of the hotel.[20] This fire brought to light the need for automatic fire suppression and fire alarm systems in hotels and high-rise buildings with large undivided areas like casinos. The MGM Grand fire and several other hotel fires in the 1980s set in motion an effort to promote sprinklers in hotels throughout the country. The Marriott Hotel Corporation teamed up with the International Association of Fire Chiefs and Federal Emergency Management Agency (FEMA) to conduct test fires with residential automatic sprinklers in a program called "Operation San Francisco" in 1983. This program later became "Operation Life Safety" and implemented the installation and development of residential sprinkler technology in hotels throughout the country.

THE GREAT ADVENTURE HAUNTED CASTLE FIRE—JACKSON, NEW JERSEY: 1984

The fire at Great Adventure in Jackson, New Jersey, claimed the lives of eight teenagers in a walk-through dark ride called the "Haunted Castle." The Castle was constructed of 17 box trailers interconnected by plywood and covered on the exterior with a façade of a castle. There were 29 persons in the ride at the time of the fire. Allegedly, the fire was started by a teenager operating a cigarette lighter in the dark part of the walk-through haunted house, igniting the plastic foam on the walls and ceiling. The open flame quickly ignited the combustible finish and spread thick smoke and heat throughout the trailers. This fire led to the enactment of code change in the New Jersey State Uniform Fire Prevention and Construction Codes for dark ride amusements. The regulations were added to the IBC and now apply throughout the country. Dark ride amusements now require the retroactive installation of fire alarms, exit and emergency lighting, and automatic sprinkler protection in New Jersey and other parts of the country.

UNION CARBIDE CHEMICAL RELEASE—BHOPAL, INDIA: 1984

Although the Bhopal Union Carbide incident occurred in India, it had a far-reaching effect on building and fire code changes in the United States. The enactment of environmental regulations and community right-to-know laws, dealing with dangerous chemicals within communities, spurred changes in the hazardous materials regulations of the international fire code and the NFPA codes. The Bhopal incident was caused by the release of methyl isocyanate, or MIC. This material is an immediate chemical compound used in the manufacture of carbaryl, or Sevin, a pesticide still widely used today. Somehow, water entered a 42-ton tank containing MIC and began an exothermic chemical reaction. The tank overpressurized and vented, releasing toxic gases over the town of Bhopal. Over 3,700 deaths were related to the release, and another 3,000 died within weeks. The Indian government has reported over 8,000 deaths from chemical exposure related to this incident and over 558,000 injuries resulting from the MIC release. Union Carbide conducted its own investigation and concluded that a disgruntled worker placed water into the tank and sabotaged the chemical plant. This theory, however, was never proved or disproved by the Indian government. Union Carbide also pointed to many safety violations in the plant including poor maintenance, storing MIC in large tanks, and filling them beyond recommended safety levels. The investigation also concluded that the switching off of many safety systems to save money was a cause of this incident.[21] The Bhopal incident was a driving force behind the enactment of the Superfund Amendment and Reauthorization Act, or SARA Title III, in 1986 by Congress. Eventually, the regulation created changes in the later additions of the model building and fire prevention codes.

ONE MERIDIAN PLAZA—PHILADELPHIA, PENNSYLVANIA: 1991

The One Meridian Plaza fire was one of the premiere high-rise fires of the twentieth century, occurring on February 23, 1991. The fire was caused by improperly disposed paint finishing rags that spontaneously ignited on the twenty-second floor of the building. The fire occurred on a Saturday night, leading to the deaths of three Philadelphia fire fighters from Engine Company 11. There were many building and fire code issues at One Meridian, including the improper setting of pressure-reducing standpipe valves. This problem with PRV valves had been experienced before in Los Angeles at the 1988 Interstate Bank Fire in Wiltshire. The fire's aggressive development also attacked the primary and secondary emergency power feeds on the twenty-second floor, causing the fire pump and emergency generator system to completely fail. Flames spread rapidly through auto-extension on the exterior perimeter curtain walls, and firefighters could not get adequate water to the fire floors to mount an effective attack. Eventually, the firefighters stretched 5-inch hose up two stairs to the twenty-second floor, but the fire had gained too much headway for it to be effective. During the course of the fire, another 11 firefighters almost became trapped in a mechanical room near the roof while searching for Engine 11's crew. They could not find the exits from the room because the exit markings were poor. The fire raged for more than 18 hours, and eventually all the firefighters had to evacuate the building due to the potential for total structural collapse. The firefighters watched One Meridian burn from a safe location for several hours. When the fire progressed to the twenty-ninth floor, which had automatic sprinklers installed, the nine operating sprinkler heads suppressed the fire. This fire caused changes in the Philadelphia Fire Code, requiring the installation of automatic fire sprinklers within 8 years in all commercial high-rise buildings.[22]

THE STATION FIRE—WARWICK, RHODE ISLAND: 2003

The Station nightclub fire occurred as a result of fire code violations, and it led to the loss of 94 lives. The media concentrated on pyrotechnics as the cause of the fire, but the real contributing factor was the egg-shaped soundproofing foam plastic installed on the wall behind the stage. While the use of pyrotechnics was certainly a code violation, the fireworks acted only as the match. When the pyrotechnics were discharged, the foam immediately ignited, rapidly spreading flames and dense, black smoke throughout the small venue. The main exit, where over 40 victims died, had been restricted to about 3 feet clear width by a podium obstructing the double-doors exit. This fire changed the fire prevention code in the state of Rhode Island, causing a nationwide review of small nightclubs for fire safety violations. Of particular interest is that a week before this fire, in the city of Chicago at the E2 nightclub, 21 people died in a crushing incident, and 50 people were injured when pepper spray was discharged in a crowd on the second floor. Thinking it was a terrorist strike, people ran down an exit stairway to a locked door and were crushed. The E2, like the Station, was a relatively small nightclub. Both of these incidents demonstrated the need for constant inspection of egress facilities at late-night establishments.[23]

FIGURE 1.6 Front of the Sofa Super Store furniture showroom where nine Charleston firefighters died. The fire demonstrated the need for active fire inspection programs and coordination with the fire department. *Courtesy of Adam K. Thiel*

THE SOFA SUPER STORE FIRE—CHARLESTON, SOUTH CAROLINA: 2007

The Sofa Super Store fire in Charleston, South Carolina, again demonstrated the need for active fire code enforcement (see Figure 1.6). A trash fire started at the rear of the Sofa Super Store building in a loading dock area at 1807 Savannah Highway. Within the first 30 minutes, nine firefighters lost their lives. The area of fire origin was a 2,200-square-foot addition erected with no building permits or fire inspections. The store itself was 42,000 square feet and was attached by a metal corridor to a 17,000-square-foot warehouse in the rear. The loading dock was constructed between these two buildings. The store originally was a supermarket, built in the 1960s. Upon purchasing the store, Sofa Super Store added two additional wings to the showroom. These additions increased the total square footage above 12,000 square feet, which would require the installation of automatic sprinklers under the building code. The store owners instead separated these wings by fire separation walls and roll-down fire-rated doors, keeping the total square footage below the sprinkler requirements, and thus these structures were grandfathered. In 1987, the authority to conduct fire inspections in commercial buildings was transferred from the Charleston Fire Department to the Department of Public Services. The Public Service Department had the authority to conduct annual fire inspections, but had last inspected Sofa Super Store in 1998, 9 years earlier.[24] The Sofa Super Store fire demonstrated the need for fire departments to be actively engaged in the inspection and enforcement of fire codes. Good inspection practices would have detected the illegal additions, and automatic sprinklers would have been required as well as proper maintenance of the fire doors that already existed in the building. Fire prevention inspections could have detected and corrected these code violations, and the brave firefighters most likely would not have died on that fateful day.

FIGURE 1.7 The attack on the World Trade Center in 2001 changed the way building and fire prevention codes address fire protection and emergency egress within buildings. This event opened an entire new paradigm for safety issues to be confronted by building designers and code enforcement personnel in the future, and that is terrorism. *Courtesy of Federal Emergency Management Agency*

THE WORLD TRADE CENTER—NEW YORK CITY: 9/11/2001

The World Trade Center attack was one of the most horrific fires and building collapses in modern history (see Figure 1.7). The lessons learned from this incident will impact fire prevention services for years to come. The World Trade Center disaster is perhaps the most investigated incident to be learned from in modern times. The National Institute of Standards and Technology (NIST) already has drawn many conclusions with regard to improving building and fire codes to address evacuation in tall buildings. NIST is making suggestions on the failure of steel fireproofing and the effective size of evacuation stairways and is examining redundancy in building systems and communications systems to improve the escape of people trapped in burning high-rise buildings. The future of fire prevention and the role that fire inspectors will play in detecting and preventing terrorism is a developing part of every fire prevention organization's mission. As everyone learns to "expect the unexpected," fire inspectors will need to be more aware of the potential for terrorism and how to prevent or reduce its harm in the building environment. The World Trade Center attack also has provided lessons on how people evacuate and the need for building management buy-in during evacuation planning and fire drills. The companies at the World Trade Center that evacuated upper-level management also had most of their employees survive, while those companies whose bosses did not evacuate lost many of their employees.[25] The incident demonstrated how company loyalty and ordinary daily routines can effect emergency evacuations. It will drive fire code change into the future as we gain real-world knowledge and experience from one of the most tragic days in American history.

Summary

In conclusion, history demonstrates the need for fire prevention in every civilized society all the way back to ancient Rome. Fire prevention codes generally were an outgrowth or response to fire tragedies that have occurred throughout human history. As fire prevention codes developed in the early part of the twentieth century, major fires that killed many people in large numbers helped to advance changes to improve safety. These code requirements generally applied to ignition sources, exit facilities, flammable interior finishes, fuel fires, appliances, smoking materials, chimney maintenance, general housekeeping, and changing peoples' attitudes and behaviors.

It is clear that fire prevention is, and always has been, an integral part of the fire department's mission and should always remain so. The ability of firefighters to have a say in the environment they are expected to work in is critical to their safety. Safety has been a paramount issue in the fire service, and it has been the theme of most of the national conferences held on the subject since 1913. Fire prevention code development and performance are functions of proper code application and active code enforcement programs. The best codes in the world are of little use if they are not enforced properly.

Fire codes must be stringent enough to prevent the fire problems of the past, yet flexible enough to allow new technologies in building materials and construction. Fire codes have changed from prescriptive codes that are specification oriented to fire performance codes based on fire experience, anticipated fuel packages, and scientific measurement. Computer and engineering technology in the future will allow risk assessment of buildings to be performed based on analytic fire models to develop safety parameters and strategies. The key element, however, will always be the fire inspector's enforcement of the regulations to ensure proper application and performance. It is the proper application and enforcement of fire codes that has the largest impact on fire losses and life safety. Fire inspectors need to be concerned with both the building occupants and the firefighters who answer the call in the event of a fire. Proper fire code enforcement makes fires more controllable with less property damage and loss of life. Fire inspectors must be keenly aware that if attention is not paid to installation details, fire growth and development potential increases, having a negative impact on fire control. Review of fire incidents and active interaction with fire service incident commanders is necessary if we are to stop repeating history. It is critical that fire inspectors examine past fires and learn their lessons so that these tragedies do not reoccur. It is equally important that fire inspectors examine the present fire environment and where we are heading in the future. Fire inspectors must have a firm foundation in fire science, building construction, risk analysis, material safety, and fire investigation to continue to stay ahead of the ever-evolving fire problem. Changes in building materials from natural substances to synthetic composites present new fire challenges that have not been experienced before. Fire inspectors must stay current in the industry as the move toward more performance-based building standards continues. The fire inspectors of the future will need to stay current with technology and will require far greater knowledge in the fire science field than did their predecessors.

Review Questions

1. List and describe three regulations from colonial America that were enacted to prevent fires.
2. Many fire prevention conferences were held from 1913 to 2006. Describe some of the major accomplishments of these conferences and how they affected fire prevention in the nation.
3. In 1947, the President's Conference on Fire Prevention initiated the three "E's" of fire prevention. What are they?
4. List and describe three major fires that impacted code change. Describe the underlying fire prevention violations and how they could have been prevented.
5. In 1974, the America Burning report changed the direction of most fire departments in the field of fire prevention. What were the major outcomes of this report on America's fire problem?

Suggested Readings

Bugbee, Percy. 1971. *Men Against Fire: The Story of the National Fire Protection Association 1896–1971.* Boston, MA: NFPA.

Cotes, Arthur, and Percy Bugbee. 1988. *Principles of Fire Protection.* Boston, MA: NFPA.

The National Commission on Fire Prevention and Control. 1974. *America Burning.* Washington, DC: U.S. Government Printing Office.

Endnotes

1. http://avalon.law.edu/ancient/hamframe.asp
2. http://www.btinternet.com/~graeme.kirkwood/GenHis.htm
3. Ibid.
4. Ibid.
5. http://www.history.org/history/teaching/enewsletter/volume7/sept08/firefighting.cfm
6. http://www.wwdmag.com/Firefighting-in-America-article3822
7. http://www.ushistory.org/franklin/philadelphia/insurance.htm
8. Powell Evans and J. C. Mallory, "Public Fire Protection Management," *Report of the First American National Fire Prevention Convention*, 150–151.
9. Wingspread V, *Statements of National Significance to the Fire Service and Those Who Serve* (Atlanta, Georgia, 2006), 1–3.
10. *America Burning* (National Commission on Fire Prevention and Control Report, 1974), 7.
11. http://www.thechicagofire.com/cause.php
12. http://www.peshtigofire.info/causes.html
13. http://www.chicagotribune.com/news/politics/chi-chicagodays-iroquoisfire-story,0,6395565.story
14. http://www.withthecommand.com/2003-Aug/MD-tom-public1.html
15. http://www.celebrateboston.com/disasters/cocoanut-grove-fire.htm
16. http://winecoffhotelfire.com/
17. http://www.dailykos.com/story/2009/11/01/799261/-How-Regulation-came-to-be-The-Hotel-Fires-of-1946-Part-II
18. Hal Bruno, "Old Lessons Continue to Go Unheeded," *Firehouse Magazine*, August 2003.
19. National Fire Protection Association, *Beverly Hills Supper Club, Southgate, KY, May 28, 1977* (Quincy, MA: NFPA, 1978).
20. Richard Best and David P. Demers, *Investigation Report on the MGM Grand Hotel Fire, Las Vegas Nevada, November 21,1980* (Quincy, MA: NFPA, 1982).
21. Ashok Kalelkar and Arthur D. Little, *Investigation of Large-Magnitude Incidents: Bhopal as a Case Study* (London, England: The Institute of Chemical Engineers Conference on Preventing Major Chemical Accidents, May 1988).
22. J. Gordon Routley, Charles Jennings, and Mark Chubb, *Highrise Office Building Fire, One Meridian Plaza, Philadelphia, Pennsylvania* (Washington, DC: National Fire Data Center , 1991).
23. http://www.crowdsafe.com/new.asp?ID=1265
24. http://www.cdc.gov/niosh/fire/reports/face200718.html
25. E. R. Galea and S. Blake, *Collection and Analysis of Human Behaviour Data Appearing in the Mass Media Relating to the Evacuation of the World Trade Centre Towers of 11 September 2001* (London: Fire Safety Engineering Group, University of Greenwich, 2004).

2

The Code Enforcement System

Courtesy of J. Foley

KEY TERMS

Commercial Rating Schedule, *p. 19*

Factory Mutual, *p. 19*

FIRESCOPE, *p. 21*

Fire Suppression Rating Schedule, *p. 19*

formal technical opinion, *p. 28*

line item budget, *p. 31*

National Board of Fire Underwriters, *p. 19*

performance code, *p. 26*

prescriptive code, *p. 26*

severability clause, *p. 28*

Underwriters Laboratories, *p. 19*

OBJECTIVES

Upon completing this chapter, the reader should be able to:

- Understand how the code enforcement system developed and the fire inspector's role in fire code enforcement.
- Understand how codes and standards are adopted within states.
- Describe the differences between prescriptive and performance-based building and fire codes.
- Understand fire inspectors' ethical values and the professional development necessary for the inspection task.
- Review the political, business, and other interests that influence the code enforcement process.
- Understand the basic requirements for the preparation of a line item budget.

Professional Levels of Job Performance for Fire Inspectors as Cited in NFPA 1031 and NFPA 1037

- NFPA 1031 Fire Inspector I *Obj. 4.2.5 Application of codes and standards*
- NFPA 1031 Fire Inspector II *Obj. 5.23 Investigate complex complaints*
- NFPA 1031 Plan Reviewer I *Obj. 7.2.4 Determine the applicable codes and standards*

- NFPA 1037 Fire Marshal *Obj. 5.2.4 Establish a budget*
- NFPA 1037 Fire Marshal *Obj. 5.2.5 Monitor the condition of an approved budget during the budget period*
- NFPA 1037 Fire Marshal *Obj. 5.5.3 Prescribe professional development programs*
- NFPA 1037 Fire Marshal *Obj. 5.6.1 Codes, standards, and jurisdictional requirements*
- NFPA 1037 Fire Marshal *Obj. 5.6.2 Manage a process for the adoption, modification, and maintenance of codes*
- NFPA 1037 Fire Marshal *Obj. 5.6.8 Manage the compliance interpretation process*
- NFPA 1037 Fire Marshal *Obj. 5.6.11 Generate jurisdictional requirements for administrating the regulatory management programs*

Resource**Central** For additional review and practice tests, visit **www.bradybooks.com** and click on Resource Central to access text-specific resources for this text. To access Resource Central, follow the directions on the Student Access Card provided with this text. If there is no card, go to **www.bradybooks.com** and follow the Resource Central link to buy access from there.

Introduction

As we have discovered, fire prevention regulations are usually born out of great tragedy. Disasters often create a public sentiment in which there is a political reaction to create, change, or alter fire prevention and building codes. The impetus for change may come from citizens, media, industry, politics, or the fire services itself. It is important for fire inspectors to understand that all codes are political documents that will change based on the environment and public sentiment within the community. Fire officials and fire inspectors must always survey the political landscape to understand where and when change may come and how to influence or direct it positively for public safety. Fire inspectors are a vital part of a community's fire defense plan; therefore, they need to understand how building and fire code changes occur to both the benefit and the detriment of fire safety. Fire inspectors in small communities must be creative in their approach to the political realities of fire code enforcement. Business owners often do not understand the reasons behind the fire code or the need for correcting code violations. The fire inspector must build allegiance with customers through a professional, yet understanding approach to the inspection process and help educate them on the importance of correcting code violations to avoid fire risk. Small-town politics can be very tricky to negotiate on these matters, requiring fire inspectors to be skilled in effective communications and attitude to convince the business owner that they are on the same side.

Code Development

Building and fire codes have existed all the way back to the Roman Empire. In colonial America, the founders quickly realized the need for fire codes to prevent major conflagrations. Most large cities like Philadelphia enacted local rules about fire safety and fire prevention. It was not until the 1800s and the Industrial Revolution, however, that formal organizations began to develop to address the rising concerns of fire and building safety in America. The majority of code development at the time came from the private rather than the public sector. The insurance industry led the way, with the National Board of Fire Underwriters assisting in the development of the National Fire Protection Association (NFPA) and Underwriters Laboratories in electrical and fire safety and insurance companies like Factory Mutual building the foundations for community fire protection rating and automatic fire suppression system standards. Model building and fire code associations began to develop, and uniform building and fire safety regulations began to be adopted by local and state governments. Moving into the twenty-first century, builders, government, and industry began to push for more uniform building and fire codes through the consensus code-making process. This push for uniformity takes us to today, where there are essentially only two nationally recognized building and fire codes under the ICC and NFPA. Let's examine the four major private sector organizations that shaped the development of the United States building and fire code industry.

FACTORY MUTUAL

Zechariah Allen, a textile mill owner, was the founder of **Factory Mutual** in 1835. He founded Factory Mutual because he could not acquire a reduced fire insurance rate from his insurance company. Allen worked with other factory mill owners to form Factory Mutual insurance company, which required owners to have regular fire inspections by the company. Factory Mutual (also called *FM*) was the first insurance company to hire engineers to conduct these fire inspections for the company. The first two engineers hired by FM were C. J. Woodbury and John R. Freeman, who played a pivotal role in developing municipal water supply systems and calculating fire flow requirements used today in modern fire protection hydraulics. Factory Mutual today is called *FM Global* and is still the only fire insurance system to conduct fire safety research on products, listing or labeling those products with the FM label. Factory Mutual Research is a world leader today in identifying building product safety.[1]

NATIONAL BOARD OF FIRE UNDERWRITERS

The **National Board of Fire Underwriters** (NBFU) was established in response to the Great Portland, Maine fire of 1866. The fire began when a firecracker was thrown in a boat yard. The fire quickly spread to buildings and consumed the entire eastern part of the city of Portland. The original mission of the NBFU was to control and equalize the cost of fire insurance rates, but as time progressed, the NBFU's role became mainly fire loss prevention. The NBFU was responsible for the development of the first National Electrical Code in 1897 and the first National Building Code in 1905 in the United States.. The National Board of Fire Underwriters existed until 1965 when it became the American Insurance Association (AIA). The AIA later merged into the Insurance Services Office, or ISO. The ISO now conducts community fire defense surveys for the fire insurance industry under the **Fire Suppression Rating Schedule** and the **Commercial Rating Schedule** to determine a community's fire risk.[2]

UNDERWRITERS LABORATORIES

William Henry Merrill founded **Underwriters Laboratories (UL)** in 1894. Merrill was an electrical engineer from Boston who was hired by the NBFU to inspect the electrical wiring of 100,000 Edison incandescent light bulbs at the "Palace of Electricity" during the 1883 World's Fair in Chicago. Merrill's relationship with the NBFU provided seed money to develop the Underwriters Electrical Bureau, which later was named Underwriters Laboratories. UL's role in electrical testing expanded by the early 1900s, and they were testing other types of fire protection equipment. In 1903, UL printed the first "Tin Clad" fire door standard. UL continues today to be the ensurer of safety for fire protection and electrical equipment. The UL approval labels are recognized worldwide.[3]

Factory Mutual ■ Started in 1836 by Zechariah Allen, Factory Mutual represented a group of insured that were protected risks. Today, FM Global insures risk worldwide and also provides independent laboratory analysis through FM Global Research.

National Board of Fire Underwriters ■ Created in 1866 by a national board of members of the insurance industry, the National Board of Fire Underwriters (NBFU) began an evaluation of communities' abilities to prevent and suppress fires.

Fire Suppression Rating Schedule ■ An evaluation method employed by the Insurance Services Office (ISO) to evaluate community fire protection classifications. There are ten public protection classes for insurance rating purposes.

Commercial Rating Schedule ■ Commercial fire insurance rating for fire properties protected by automatic fire suppression systems.

Underwriters Laboratories ■ A laboratory nationally recognized by model codes that lists or labels fire protection and electrical equipment based on safety performance standards.

NATIONAL FIRE PROTECTION ASSOCIATION

The National Fire Protection Association, or NFPA, was founded at the turn of the twentieth century, in 1896, based on the need for safety standards in the fields of electricity and fire protection. Two gentlemen whose paths crossed at that time helped the NFPA to become the current leader in standard-making organizations. William Henry Merrill, who established Underwriters Laboratories, provided testing and research to the Joint Conference of Electrical and Allied Interests in 1896, which led to the NFPA's publishing the first edition of the National Electric Code. During the same year, in New York City, John Freeman and C. J. Woodbury presented a report on criteria for automatic sprinkler systems leading to the development of NFPA 13, *Standard for the Installation of Sprinkler Systems*. This meeting of these fire safety leaders established the foundation of the National Fire Protection Association. The NFPA held its first annual meeting in May of 1897, and Umberto Crosby was named the first president of the NFPA. Crosby outlined the organization's key principles, which are still followed by technical committees in the development of NFPA standards today:

> To bring together the experience of different sections and different bodies of underwriters, to come to a mutual understanding, and, if possible, an agreement on general principles governing fire protection, to harmonize and adjust our differences so that we may go before the public with uniform rules and conditions which may appeal to their judgment is the object of this Association.[4]

National Fire Protection Association standards are the very foundation of all modern building and fire codes today.

The Inspection Agency Environment

While fire code and standard development was driven by private sector industries, the enforcement authority of codes has always been considered a function of government, as it serves the public interest. Fire inspection authorities exist at many different levels of local, state, and federal government. These inspection agencies are not always fire departments. Many federal and state fire authorities employ fire inspectors or fulfill other important aspects of fire safety.

FEDERAL FIRE AGENCIES

The federal government has numerous fire agencies spread throughout several departments including Homeland Security, Agriculture, Labor, Health and Human Services, and Defense. Let's examine some of the main federal fire agencies.

Department of Homeland Security

The Department of Homeland Security is in charge of the U.S. Coast Guard, Federal Emergency Management Agency (FEMA), and the U.S. Fire Administration (USFA). Both FEMA and the USFA provide leadership, direction, and training to the emergency response community in the areas of fire prevention, code enforcement, and disaster mitigation. The USFA maintains the national fire statistics from the National Fire Incident Reporting System (NFIRS) and the National Fire Academy and Learning Resource Center at the National Emergency Training Center in Emmetsburg, Maryland. The USFA is the federal central fire focus for state and local emergency services. The U.S. Coast Guard provides fire inspection and fire safety on all water vessels in U.S. waters as well as hazardous materials entering any U.S. ports.

Department of Agriculture

The Department of Agriculture (USDA) is the home of the U.S. Forestry Service and fulfills the mission of ensuring the management of federally owned forests and grasslands with regard to fire safety. The USDA works hand in hand with the states through their departments of agriculture and local fire agencies to prevent and mitigate wildfires. The U.S. Forestry Service plays an instrumental role in the continued development of the *National Incident Management System*, which was born out of the original Forestry **FIRESCOPE** program of 1972. FIRESCOPE was created after a devastating California forest fire had destroyed over 200 structures and killed 16 people. The USDA also oversees the National Wildfire Coordination Group, which is instrumental in the development of the National Incident Management System.

Department of Defense

The Department of Defense has fire inspection responsibilities for all military installations for all branches of the armed services including the Navy, Army, Air Force, and Marines.

Department of the Interior

The Department of the Interior has several divisions with interest in the prevention of fires. The National Park Service provides aviation and other fire assets, including fire inspectors and fire investigators. The U.S. Fish and Wildlife Service provides fire management and education, and the Bureau of Indian Affairs maintains fire plans and the library of the National Wildfire Coordination Group. All fire activities are coordinated through the National Interagency Fire Center.

Department of Health and Human Services

The Department of Health and Human Services has connections to fire prevention efforts through public education on burn injuries and firefighter-related deaths. The National Institute of Occupational Safety and Health (NIOSH) is charged with investigating all firefighter line-of-duty deaths.

STATE FIRE AGENCIES

State government also may have fire authorities that have code enforcement or fire inspection responsibility. Often, these responsibilities may be shared at both the state and municipal levels of government. These agencies may include state fire marshals offices, state fire academies, state environmental agencies, or state police. As an example, many states have adopted statewide building and fire prevention codes that are enforced through a system of interagency responsibilities among local, county, and state enforcing agencies. The states usually retain concurrent jurisdiction on all enforcement activities and control the licensure and certification process required to perform any duties as a fire inspector or plans reviewer within the state. While each state is different in its approach to fire inspections, most states adopt some educational requirement as described in NFPA professional qualification standards.

The Fire Inspector's Role

The fire inspector has the most important and critical role in the enforcement of fire prevention or building codes. Without a knowledgeable and skilled workforce, fire and building codes will not be properly applied or may not be applied at all in a community. Fire inspectors are expected to meet the professional standards established under their state or local regulations. Most states base fire inspector job performance requirements on the NFPA Professional Qualifications Standards. NFPA 1031, *Standard for Professional Qualifications for Fire Inspector and Plan Examiner* establishes the knowledge, skills, and abilities necessary to perform as a fire inspector or plan reviewer. Certification exams may be required by the jurisdiction or may be achieved directly through the NFPA's certification process. The NFPA 1031 standard contains job performance requirements, or JPRs, that are evaluated based on specific fire inspection tasks that must be performed by the fire inspector (see Figure 2.1). Each task has a performance requirement that can be

FIRESCOPE ■ Started in 1972 in response to California forest fires, FIRESCOPE began the development of incident command systems and was the forerunner of the National Incident Management System, or NIMS.

4.3.2 Building Construction and Occupancies Classification   Skill Sheet #8

Candidate:_____   Date_____

SS#_____

STANDARD 4.3.2 NFPA 1031, 2003 Edition	**Task:** Compute the allowable occupant load of a single-use occupancy or portion thereof, given a description of the occupancy so that the calculated allowable occupant load is established in accordance with applicable codes and standards.

Performance outcome: The candidate, given a detailed description of a single-use occupancy, shall compute the allowable occupant load for this occupancy or portion thereof, using the description and applicable codes and standards to establish occupant load.

Condition: The candidate will complete all elements of the assigned task. Include candidate's narrative on task completion forms, photos/drawings etc., department policy or procedure.

Equipment Required: Scenario, applicable codes and standards, pen/pencil, paper. Computer, if applicable.

No.	Task Steps	First Test		Retest	
		Pass	Fail	Pass	Fail
1.	Calculate allowable occupant load for single-use occupancy.				
2.	Use given description of occupancy to establish occupant load.				
3.	Use applicable codes and standards to establish occupant load.				
4.	Utilize proper occupant load formula for calculation.				
5.	Accurately calculates occupant load.				
6.					
7.					

Supervisor/Proctor/Comments:_____

_____

_____

_____   _____   _____   _____
Supervisor/Proctor   Date   Candidate   Date

_____   _____   _____   _____
Re-Test Supervisor/Proctor   Date   Candidate   Date

FIGURE 2.1 Example of Job Performance Requirement skill evaluation from NFPA 1031-2003. Reprinted with permission from NFPA 1031-2003, 2009, Standard for Professional Qualifications for Fire Inspector and Plan Examiner

5

evaluated based on the proper tool selection, observed conditions, and a series of task steps necessary to be effective. The key element is that these job performance requirements are fair and measurable in evaluating the knowledge, skill, and ability of the fire inspector. Fire inspectors may be certified by the NFPA through their Pro-Board process, which consists of three steps. First, the fire inspector must complete two case studies and take a competency exam within four months of filing the application. Second, the fire inspector must complete a practicum of inspecting specific types of structures to prove competency with fire code application in specific building uses and fire system evaluation. All of this information is forwarded to the NFPA for certification. The final step, once the inspector is certified, is to complete continued education over the three-year renewal period. The certification is valid for three years and then must be renewed. Renewal of professional certification requires the submission of a minimum of 15 points to a maximum of 60 points of continuing education credits. CEUs are college credits issues based on contact hours of training. The NFPA certification program awards 1 point for each contact hour. Certification renewal requirements are divided into five separate areas of fire inspector performance: training, teaching, professional practice, writing for publications, and association memberships. The candidate is only permitted to submit a specific minimum and maximum number of points in each specific area. Over 26 states currently use the NFPA 1031 program in certification of Fire Inspectors I and II and Plans Reviewers. Life safety educators have to meet the qualifications under NFPA 1035, fire investigators must meet the qualifications under NFPA 1033, and fire marshals are evaluated under NFPA 1037, *Standard for Professional Qualifications for Fire Marshal.*

Fire inspectors also require specialized professional training in a number of NFPA standards in order to perform efficiently. The fire inspector must have a comprehensive understanding of NFPA 13, *Standard for the Installation of Sprinkler Systems,* and NFPA 25, *Standard for the Inspection, Testing, and Maintenance of Water-Based Fire Protection Systems,* for the installation of automatic sprinkler systems and maintenance aspects of water-based fire protection systems. Fire inspectors must also receive training in NFPA 72, *National Fire Alarm and Signaling Code,* for installation and testing of automatic fire alarm systems. Inspectors need to be competent in citing and interpreting their fire prevention code, whether the jurisdiction uses NFPA 1, *Fire Code* or the ICC *International Fire Code.* Fire inspectors must have good organizational skills and must be effective communicators both verbally and in writing in order to be efficient with the public in code compliance. Fire inspectors must be self-motivated, energetic, and ethically and morally strong. These traits are necessary because fire inspectors generally carry out their work by themselves and may be confronted with situations such as attempted bribery or people offering privileges that could constitute official misconduct. Fire inspectors must also be empathetic to businesses and work with property owners to gain code compliance in the field.

The Fire Code Development Process and the Legacy Codes

Fire codes exist to manage the safety of the community. Fire inspectors must understand that fire prevention codes are political documents and may change based on the public and political will. As we have seen, most of the early fire codes were based on the requirements established by the insurance industry. The model code was produced by the American Insurance Association and was adopted and used in many parts of the country up until the 1970s. The AIA stopped publishing the code in 1981, as other model building codes became more widely accepted and adopted. Let us examine how the current fire codes developed from the insurance industry to the public safety sector as we see them today.

BUILDING OFFICIALS CODE ADMINISTRATION

The Building Officials Code Administration (BOCA), located in Chicago, Illinois, produced model codes starting in the 1950s. The BOCA *National Building and National Fire Prevention Codes* were primarily used in the midwestern and eastern parts of the United States. The membership of the organization was primarily building code officials and associate members such as architects, engineers, and industry members. Initially, the codes were revised every five years, and later every three years, to stay current with industry changes. Code changes could be submitted by anyone; however, only member code officials would have the final vote on change acceptance.[5]

INTERNATIONAL COUNCIL OF BUILDING OFFICIALS

The International Council of Building Officials (ICBO) produced the *Uniform Building Code* and the *Uniform Fire Code*. The first editions were printed in 1927, and the code was used primarily in the Pacific Coast and Rocky Mountain states. This code had a development process very similar to the BOCA codes and placed a heavy emphasis on automatic sprinkler protection as an alternative to passive fire protection features.

SOUTHERN BASIC BUILDING CODE COUNCIL

The Southern Basic Building Code Council (SBBCC) started publishing the *Standard Building Code* in 1945. The SBC was used extensively in the south and gulf region and had a heavy emphasis on high wind and hurricane protection requirements. It was similar to the other model codes in that a code change process was used to update the code periodically.

THE INTERNATIONAL CODE COUNCIL

In 1990, the three model code legacy organizations—BOCA, ICBO, and SBCCI—merged and formed the International Code Council (ICC) to produce a single series of international building and fire codes. The ICC was officially formed in 1994 and produced the first international code series known as the I-Codes in 2000. Since their development, all fifty states have adopted the I-Codes. New York City also has adopted the international code, which is the first time in the city's history that a model code has been adopted. Puerto Rico also has adopted and uses the international code (see Figure 2.2). The International Fire Code, or IFC, is currently adopted in 43 states, the District of Columbia, and the city of New York (see Figure 2.3).

The International Code Council development process is predicated on the following principles that came from the legacy model code groups:

1. *Openness of Process:* Anyone can submit a code change proposal to the ICC. The change is directed to a committee, who must consider all aspects and views on the proposed change before voting. The committee may vote to implement, modify, or

FIGURE 2.2 International Building Code Adoption— International Code Council. *Reproduced with permission. All rights reserved. www. ICCSAFE.org, Washington, DC, International Code Council*

INTERNATIONAL BUILDING CODE ADOPTION MAP
The IBC is in use or adopted in 50 states, the District of Columbia, the U.S. Virgin Islands, NYC, Guam and the Northern Marianas Islands.

IBC administered statewide

IBC administered at the state and/or local level

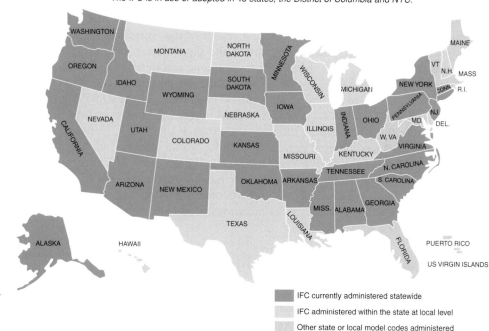

INTERNATIONAL FIRE CODE ADOPTION MAP
The IFC is in use or adopted in 43 states, the District of Columbia and NYC.

- IFC currently administered statewide
- IFC administered within the state at local level
- Other state or local model codes administered

FIGURE 2.3 International Fire Code Adoption— International Code Council. *Reproduced with permission. All rights reserved. www.ICCSAFE.org, Washington, DC, International Code Council*

reject the change but must give reasons why, and the vote may be overridden at the final hearing by the general membership.

2. *Transparency:* The process must be transparent in that all votes are posted and documented. All final decisions occur at an open public annual meeting, when final adoption votes are cast.

3. *Balanced and Fair:* The committees are to be balanced between members who have a general interest, user interest, producer interest, or multiple interests. Public safety officials must comprise one-third of each committee. Members who have a conflict of interest on any issues abstain from voting in the committee on those issues.

4. *Due Process:* Disagreement with the results of the voting by the committee may be disputed, and decisions will be rendered by the ICC.

5. *Consensus:* The process of approval, modification, or rejection of code changes is made by simple majority on the committee. All final votes are conducted at the annual meetings, and only public safety officials may vote, which eliminates vested interests and ensures that the public interest in safety will be served.

The ICC I-codes are revised on a three-year cycle based on this consensus process.

Code development is divided into two groups, and the consensus process is divided into a code development process and a final action process. The hearings on each process occur in the spring and fall of each year during the first two publication years of the I-codes. The unveiling of the new code editions takes place at the annual meeting on the third year. This process is designed to have the public hearings on proposed code changes at the same time each year and should allow better participation in the code development process. The ICC hopes this will increase the attendance at the annual code hearings and allow for better participation.

NATIONAL FIRE PROTECTION ASSOCIATION

The National Fire Protection Association (NFPA) has been producing fire and building code standards since 1896. The NFPA standards were referenced in all of the model codes, and NFPA originally was a participant in the International Code Council merger process to establish a single uniform building and fire code for the United States. The NFPA removed itself from the process due to disagreements with the ICC, and several lawsuits ensued between the ICC and the NFPA. At the core of the issues was the direction being taken by the ICC for writing fire code content and the code approval process, which ran counter to the way the NFPA developed its fire code standards. The NFPA decided to establish its own building code, NFPA 5000, and the fire prevention code, NFPA 1. Many national organizations attempted to stop the NFPA, especially the American Institute of Architects (AIA) and Building Owners and Managers Association International, or BOMA, because the NFPA used a different code format than the ICC. BOMA, who represents building owners, pointed to many items that would increase building construction costs if states adopted the NFPA 5000 code. Initially, the state of California did adopt NFPA 5000, but when the governor's office changed, it reverted to the ICC I-code series. While many of the disagreements and issues between the ICC and the NFPA have been resolved, there are still two separate organizations producing two building and fire codes in the United States.

Performance vs. Prescriptive Codes

Prescriptive building codes existed primarily in the early days of building and fire code enforcement and still exist in the model building codes used today, but to a lesser extent. **Prescriptive codes** are specification driven and allow little flexibility in design. As an example: In NFPA 13, *Standard for the Installation of Sprinkler Systems*, pipe schedule tables may still be used. These tables are specification driven. The standard also allows the hydraulic calculation of automatic sprinklers on a pressure loss basis but established specific performance objectives that still must be attained. While codes and standards are moving in the direction of **performance codes**–based design, they are still very specification driven. Many of these specifications come from fire testing procedures for assemblies based on specification-oriented fire tests. The building code may require a one-hour firewall performance rating, and the architect may specify any one-hour assembly he chooses to meet the performance requirement. The final one-hour assembly selected, however, will have to meet the prescriptive specification from the testing laboratory that tested the final design.

Great advances are being made in advanced computer technology, and many dynamic fire models are currently being used in the evaluation of fire safety performance measures for the determination of building system performance. The National Institute of Science and Technology (NIST) and the Fire Research Division have developed many useful tools to assist architects and engineers in the performance evaluation of building design. These tools include the following:

- FDS (Fire Dynamics Simulator)
- CFAST (Consolidated Model of Fire and Smoke Transport)
- ASCOS (Analysis of Smoke Control Systems)
- ASET-B (Available Safe Egress Time – BASIC)
- CCFM (Consolidated Compartment Fire Model version VENTS)
- ELVAC (Elevator Evacuation)
- FPETool (Software and Documentation)

All of the above fire models are available free of charge from the federal government. As these fire analysis computer systems become more advanced they may eventually replace traditional building and fire codes in evaluating building system fire performance. Computer modeling may at some point replace traditional building and fire codes. There are currently computer-modeling programs such as EXODUS, developed by the fire safety engineering group at the University of Greenwich, which can be used to evaluate the

prescriptive code ■
Code oriented by specification requirements to achieve a specific fire protection objective.

performance code ■
Code based on fire and life safety evaluation rather than specifications on design configuration.

Traditional construction has been based on prescriptive building code requirements for many years. Prescriptive codes specify the particular building materials and methods of construction to be used in construction of buildings. As an example, on the east coast most residential construction in cities were row homes or attached town houses. The prescriptive codes at the time required masonry firewalls and party walls between dwelling units. These walls typically extended above the roofline to form a parapet and prevented fire in one dwelling from extending to the roof next door. These buildings were traditional ordinary construction with exterior masonry walls and fire cut wooden floor and roof joists. Firefighting experience in these structures demonstrated that rarely would the fire extend beyond the origin dwelling unit, as the firewall and masonry nonbearing walls were noncombustible and the fire could not spread unit to unit very easily.

In today's performance building codes, residential townhouse or row homes are required to meet performance standards such as a one-hour firewall between dwelling units. The firewall may be constructed of combustible wood frame and fire-resistant drywall or any other method that satisfies the one-hour fire resistance requirement. The walls may also terminate at the underside of the roof level and are not required to have a parapet. The parapet is eliminated provided that fire-resistant treated plywood is applied four feet on each side of the firewall. What the performance code requirement has allowed is the use of combustible materials in the exterior walls and roofs as long as the fire resistance rating is maintained. In March 2003, a fire occurred in Atlantic City in townhouses constructed under these performance code standards. The outcome of the fire was the death of one resident and the total destruction of three adjoining units. Interestingly, the performance measures worked well in that the firewalls between units remained standing and intact; however, the fire was able to spread past several fire walls via combustible exterior vinyl siding and across the roof once the fire-resistant plywood had burned through. While these townhouses complied 100 percent with state building code standards, the firewall performance in an actual fire was significantly ineffective compared to the older specification-driven prescriptive code. Innovation in building materials and acceptance in performance code is a good thing, but fire inspectors must be aware that sometimes new methods are not necessarily better and should raise concerns when fire experience raises issues of questionable performance.

evacuation time of building occupants. Building designers must be sure, however, that the proper fire scenarios are being examined with proper evaluation of fuel packages for the long-term life of the building. The World Trade Center is a great example of this, considering that when the building was designed, office furnishings were metal and the largest airplane flying was a Boeing 707. In 2001, office furnishings were mostly synthetics and plastics, and the Boeing 767 used against the towers was about 60,000 pounds heavier than a Boeing 707. These factors were not considered when the building was designed in the late 1960s because they didn't exist at that time. The problem with performance criteria is that we cannot predict technology and building materials changes in the future.

Code Interpretations

Building and fire codes are not always clearly understandable by both the public and enforcing agencies, often leading to misinterpretations. Building and fire code formal interpretations may be necessary to clarify the meaning and intent of a code requirement and how and when it should be applied. The interpretation process may occur at different levels of government. Generally, when a state adopts a building or fire code, a process is placed within the regulation for formal interpretation of any disputes in the code. The process of interpretation may include formal code interpretations, formal technical opinions, or commentaries from the code writers. The authority having jurisdiction may also provide code interpretation. It is beneficial for fire inspectors to review code interpretations to bring clarity to their enforcement efforts. The ICC publishes code commentaries and code interpretations for use by code enforcement officials in the model code process, as does the NFPA. Older code commentary reference books are very useful in understanding the intent and purpose of the code application. States also may have local code changes

formal technical opinion ■ A binding interpretation of a code section and its proper application, usually issued by a state agency or code authority.

or requirements that will require some formal interpretations. Some states have rules within the administrative code for interpretation that apply to all enforcing agencies for both building and fire prevention. The state may issue interpretations as **formal technical opinions** to local officials on particular code requirements. These formal technical opinions are binding on all the enforcing agencies within the state. Fire officials may interpret the code and ensure that all of the members of their organization use that interpretation in the enforcement of a code provision. This uniformity of opinion helps to promote consistency in enforcement and eliminates improper application of code sections. In all cases, the interpretations of the code writers should be complied with to ensure fairness to the public in applying code regulations. Lastly, the board of appeals or the state or federal courts may establish precedents in a particular code enforcement matter. Boards and judges may make rulings on the proper interpretation of regulations, and these interpretations may alter the way the code can be enforced or may alter the code itself. Judges evaluate the code enforcement issue by what the adopted statute permits. You can think of the statute as a picture frame. The building and fire codes are prepared as administrative regulations, or the painting of the picture inside the frame. The courts allow the government to alter the painting of the picture by altering the regulation; however, you cannot paint outside the statutory frame. Most building and fire codes have a **severability clause** for such occasions where the courts may rule the regulations as outside the boundary of the statute.

severability clause ■ A legal term that deems that if any part of a code regulation is deemed unenforceable or unconstitutional, it does not affect the remainder of the regulation and in effect is severed from the same.

FIGURE 2.4 Blue lights installed under local code amendment to assist firefighters and improve efficiency. *Courtesy of J. Foley*

Developing Local Code Strategies

Often, fire prevention codes may permit the addition of local amendments to the code regulation. For example, the fire official in a state may adopt local code provisions based on unique fire problems or conditions in their jurisdiction. These local amendments must strengthen the fire code and can never circumvent any code requirements. The local code provisions must be duly adopted by municipal ordinance and must not conflict with any provision of the State Uniform Construction Code. Many times, these local amendments enhance firefighters' safety in responding to emergencies or help the local enforcing agency to identify changes in occupancy conditions that might affect the performance of fire protection systems. As an example, casinos are very large, undivided areas, and when firefighters respond, they may have to connect hoses to standpipes located throughout the casino area. The challenge is finding these connections dispersed 130 to 200 feet apart on the casino floors. Standpipe connections are usually obscured by slot machines or table games. As an example, a fire official can prepare a local code amendment requiring the installation of 4-inch blue lights 24–48 inches above the hose cabinet and on all four sides of columns if the standpipe is located there (see Figure 2.4). This requirement helps firefighters identify the standpipe locations and expedites fire suppression.

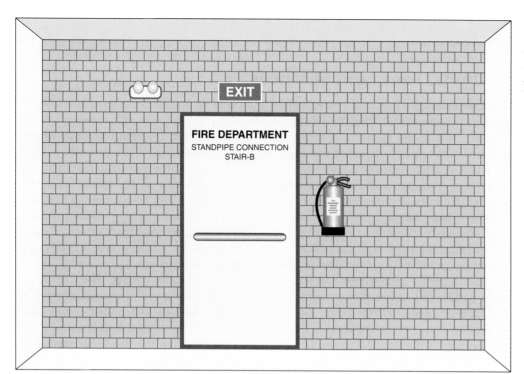

FIGURE 2.5 Standpipe door location markings enforced by local code amendment. *Courtesy of J. Foley*

Another example of improving efficiency and safety is a requirement that all stand-pipe and sprinkler system fire department connections have a posted sign displaying the minimum operating pressure required at the point of connection to support the system. Design professionals, such as architects or engineers, can calculate the required pressure to be placed upon each sign. This example of a local requirement enables fire department pump operators to eliminate the guesswork for operating pressures when connecting to support these systems. Additionally, requirements can be instituted for new construction that the fire department standpipe connections are to be placed within 50 feet of the clos-est fire hydrant. The idea is to enable a single pump operator to make up the connections without the assistance of the remaining fire crew. Other changes that could potentially be proposed are low-light marking systems in egress stairs and mechanical rooms where exit signs may be obstructed by overhead equipment, the marking of stair doorways that contain fire department standpipes if they are not immediately obvious to firefighters (see Figure 2.5), and the installation of elevator lobby Knox boxes for firefighter over-ride keys.

Another area of concern that can be addressed by local code amendment is fire alarm system audibility. Often, buildings are expanded or modified to the extent that the fire alarm may not have sufficient audibility levels to be effectively heard above ambient noise. Discretionary regulations may be adopted to provide authority to the enforcing agency to request a fire alarm audibility test. The test is conducted by first taking a 24-hour time analysis of ambient noise levels at specific locations between fire alarm speakers. Sound pressure measurements are taken every hour and recorded. The high, low, and average decibels are recorded, and then the fire alarm sound pressure is recorded at the same location. The fire alarm must be at least 15 decibels above the average ambient sound level. Through this alarm system investigation, many deficien-cies of the fire alarm systems can be identified due to sound competition and sound attenuation.

Fire officials should examine local safety issues that fire inspectors encounter or that may directly affect firefighter safety and address them, if possible, through local fire code amendments. Local amendments strengthen areas not identified in the model codes.

Budgets, Fees, and Finances

We have examined how code enforcement and building and fire codes are developed as well as the important role the fire inspector plays in the code enforcement system. None of this would exist, however, without proper finance through budgets and fees. Finance is an important part of the code enforcement equation. While it is nice to think that fire and building code enforcement serves a more noble purpose, managing the costs of operation of a fire prevention bureau will always be a critical factor for the fire official. Fire officials must recognize that fire prevention is a business, and like all businesses, it will change based on current economic conditions. The product produced and delivered is fire safety, and it is not free. The fire code enforcement efforts are paid for by either municipal tax dollars or fire inspection and permit fees. Fire officials must operate with business savvy because government administrators will always be looking for additional revenue streams or areas to cut costs in the city budget. The fire official must be able to justify expenses within the budget for number of personnel, equipment replacement, and overtime costs. Fire prevention bureaus may have their own budget or may be a line item in the total fire department budget. Some state regulations require the budget for the fire prevention bureau to be separate from the fire department budget and may also require the establishment of local fire inspection fees in addition to state-mandated fees. Collecting and accounting for fees requires the fire official to maintain an accounting system for revenue generated and an annual budget to monitor bureau expenditures.

Preparing a budget can seem like a daunting task to the fire official, but it is a necessary task nonetheless. The budget process usually involves an internal fire department review, a city administrator review, and approval by the mayor. After the city administration has approved the budget, it is submitted to a city council or alderman for final approval. The council usually will hold budget hearings and will allow public comment on the budget. Generally, city councils will either approve or modify the budget for implementation in the next fiscal year. Funds may be spent only after the budget has been adopted.

There are many types of budgeting methods that can be employed by fire officials:

- *Program Budgets*—These budgets encompass the entire program, reflecting the program's performance objectives and tying expenditure to those objectives. The value of program budgeting is that it enables you to contrast programs based on meeting established objectives.
- *Performance Budgets*—These budgets identify each performance measure accomplished annually, such as the number of fire inspections conducted, and tie the unit cost to the expenditure cost. This type of budget allows you to see the value versus performance of each individual measure.
- *Line Item Budgets*—The line item budget is the most common method used in government. The line item budget identifies expenditure to specific items over the course of the fiscal year. A line item budget allows you to see exactly where and what money is being spent on. Line items also make it easier for governments to analyze and reduce spending in specific categories. Monies attached to each line item may be moved only with approval of budget administrators, and unspent monies usually indicate that the line should be decreased in the next budget cycle.

Fire officials may use program or performance budgets internally to determine program and performance efficiencies, but for most government agencies, the line item budget most likely is what is to be provided to the local government. State government usually requires municipalities to use specific budgeting documents in the preparation of the budget. The budget generally consists of four parts:

1. *A Budget Narrative and Summary:* Describes the organization and justification for changes in the budget as compared to the previous year.
2. *Salary and Wage Accounts:* Include all employee costs including stipends, health care, longevity, educational increments, proposed raises, overtime, and any other costs related directly to the employee's services.
3. *Operating Expenses:* The annual operational costs, including publications, licenses, vehicles, office equipment, clothing or uniforms, postage, copiers, telephones, electricity, and any other hard costs that must be paid to maintain an effective operation.
4. *Capitol Improvement Budget:* Generally long-range expenditures that include larger items that have useful lives over 5 years such as vehicles. Capitol improvements usually involve bonding of expenditures and may be dealt with at the same time as the annual budget, or as a separate budget. Capitol improvement is usually part of a long-range planning process.

LINE ITEM BUDGETS

The **line item budget** should reflect the inspection program and give the budget reviewer a good idea of the program's accomplishments and values. Each line usually has a justification space for the particular expenditure. Line item budget forms usually show the previous year's expenditures so that increases or decreases in lines become very evident to the budget reviewer (see Figure 2.6) The problem with line item budgets is that often, padding is added to the line in anticipation of the cuts that will follow in the budgeting process. This tendency causes budget numbers to increase artificially rather than based on actual program needs. The reason for this is political expediency in that politicians always want to hold the budget in line or cut it before a reelection. Budget cuts and increases often are driven by election cycles. Generally, politicians do not approve budget increases in election years, as this gives their competitors election issues to raise against them. After all, no one wants to elect an official who raises taxes or increases budgets. This tends to cause a cat-and-mouse game of padding budgets to reduce the impact of anticipated election-year cuts. While this may not be true all of the time and in every municipality, it is prevalent, and you must be aware of the political side and consequences of the budget process.

line item budget ■ A budget by department and category of expenditures. Line item budgets include analysis of previous spending, current appropriation, and future spending.

State budgets usually run on fiscal years rather than annual years. The majority of state budgets run from July 1 to June 30 every year. Municipal budgets, however, may run on calendar years. A typical state budget process might be as follows: The municipal budget must be submitted by all mayors in January and must be approved by all city councils by March. The state usually provides waivers from these schedules, and often budgets are not finally approved until late April or early May. Budget extensions are a common practice, and even the federal government has failed to produce timely budgets. When budgets are not approved, the government exists on temporary budgets based on the quarter of the year it is operating in. Sometimes, the temporary budget may still be in effect to the end of the second quarter of the year. After the budget is finally approved, purchasing must begin immediately. This can be done in June, July, and August. Fire officials must be prepared to purchase needed items quickly because by late September or early November, budget purchasing often is closed except for emergency purchases. The budget closure allows the government to prepare for the next budget cycle. It is critical to your organization that you understand your municipal budget cycle so that you can plan each year's needs properly. If you don't plan ahead, your organization may suffer by not receiving necessary equipment, training, or operational needs.

Account		1998 Amended Budget	1999 Amended Budget	1997 Actual Budget	1998 Actual Expenses	1999 Actual Expenses	1999 Projected 10/1-12/31	Total 1/1 to 12/31	2000 Division Head increases	Justification
5101	S&W	$ 647,151.00	$ 759,439.00	$ 558,224.00	$ 647,150.00	$ 543,108.00	$ 45,259.00	$ 588,367.00		
5103	Overtime	$ 7,760.00	$ 10,000.00	$ 9,338.00	$ 7,759.00	$ 4,464.00	$ 3,700.00	$ 8,164.00		
5105	Out of Title	$ 5,589.00	$ 6,300.00	$ 4,141.00	$ 5,589.00	$ 2,328.00	$ 900.00	$ 3,228.00		
5107	Longevity	$ 51,628.00	$ 54,604.00	$ 39,911.00	$ 51,627.00	$ 37,280.00	$ -	$ 37,280.00		
5108	Holiday	$ 52,069.00	$ 62,078.00	$ 49,973.00	$ 52,069.00	$ -	$ -	$ -		
5109	Other	$ -	$ -	$ 6,204.00	$ -	$ -	$ -	$ -		
5114	Education	$ 37,596.00	$ 39,804.00	$ 31,066.00	$ 37,596.00	$ 29,620.00	$ -	$ 29,620.00		
5199	Allotment	$ -	$ -	$ -	$ -	$ -	$ -	$ -		
S&W Totals:		**$801,793.00**	**$932,225.00**	**$698,857.00**	**$801,790.00**	**$616,800.00**	**$ 49,859.00**	**$666,659.00**	$ -	
5201	Advt & pub	$ 671.00	$ 700.00	$ 635.00	$ 671.00	$ 671.00	$ 671.00	$ 671.00		Fire Prevention Week Ad
5202	Travel Exp	$ 300.00	$ 200.00	$ 130.00	$ 275.00	$ 200.00	$ 200.00	$ 200.00	$ 800.00	October 6,2000
5230	Printing	$	$	$	$	$				
5250	Dues & Membership	$ 400.00	$ 400.00	$ 41.00	$ 356.00	$ 300.00	$ 100.00	$ 400.00		
5251	Education	$ 600.00	$ 600.00	$ 600.00	$ 600.00	$ 600.00	$ -			
5252	Confer/Meetings	$ 600.00	$ 500.00	$ 300.00	$ 350.00	$ 250.00	$ 250.00	$ 500.00		
5265	Maint MV	$ 225.00	$ 225.00	$ 225.00	$ 125.00	$ 125.00	$ -	$ 125.00		
5270	Maint Office equip	$ 128.00	$ 200.00	$ 150.00	$ 130.00	$ 75.00	$ -	$ 75.00		
5285	Rent	$	$	$	$ -	$				
5290	Prof Contracts	$	$	$	$ -	$				
5292	Medical Se	$	$	$	$ -	$				
5301	Agric	$	$	$	$ -	$				
5303	Purchase	$	$	$	$ -	$				
5304	Dry goods	$ 7,650.00	$ 10,200.00	$ 7,650.00	$ 7,650.00	$ -	$ 11,900.00	$ 11,900.00	$ 13,600.00	Clothing Allowance (16 members)
5316	Office supplies	$ 2,000.00	$ 2,000.00	$ -	$ -	$ 1,750.00	$ 350.00	$ 2,000.00		(Anticipate 2 retirements)
5317	Data Processing	$	$	$	$ -	$				
5319	Educ & training	$ 1,000.00	$ 900.00	$ 900.00	$ 900.00	$ 900.00	$ -	$ 900.00		
5320	Electrical	$	$	$	$ -	$				
5322	Gen hardware	$	$	$	$ -	$				
5328	Food	$	$	$	$ -	$				
5333	Fire & Safety	$ 2,500.00	$ 2,000.00	$ 2,000.00	$ 2,000.00	$ 2,000.00	$ -	$ 2,000.00		
5336	Photo & PR	$	$	$	$ -	$				
5410	Vehicles	$ 15,000.00	$ 15,000.00	$ 15,000.00	$ -	$ 8,700.00	$ -	$ 8,700.00	$ 54,000.00	2 Cars and 1 Pick up truck w/cup
5416	Office Equip	$	$	$	$ -	$				(2-prev & 1 Arson)
5420	Elec Lights	$	$	$	$ -	$				
5433	Fire & other	$ 400.00	$ 800.00	$ 600.00	$ 500.00	$ -			$ 1,400.00	Flashlite & Safety Equipment $ 100.00 per men
5436	Photo & other	$	$	$	$ -	$				
Other Exp Totals:		**$ 31,474.00**	**$ 33,725.00**	**$ 28,231.00**	**$ 13,557.00**	**$ 15,571.00**	**$ 13,471.00**	**$ 27,471.00**	**$ 69,800.00**	
Fire Inspection Totals:		**$833,267.00**	**$965,950.00**	**$727,088.00**	**$815,347.00**	**$632,371.00**	**$ 63,330.00**	**$694,130.00**	**$ 69,800.00**	

FIGURE 2.6 Example of line item budget format.

Budget Planning

The budget planning cycle begins with an understanding of the revenue that your organization generates both tangibly and intangibly. Fire officials must analyze their organization and anticipate revenues and expenditures as well as the value of the product they deliver. They need to examine fee structures to ensure that these cover the cost of the required inspections or at least offset the majority of operational costs (see Figure 2.7). In city government, when revenue is generated by an agency, the expectation is that costs will be covered or additional revenue will be generated that can be used in other areas of city government. The reality is that rarely will inspection fees completely offset operational, salary, and benefit costs in career fire departments. There will always be a pressure to downsize staff to meet revenues and make fire inspections revenue neutral for city government. Fire officials must be prepared to demonstrate the value of career staff and the workload or product that is produced. This can be accomplished only by analysis of the organization and the benefits to the community.

Fire officials need to understand the adage that "time is money." A complete analysis of the inspection program is required to determine how many inspections, re-inspections, complaints, and other emergencies are dealt with by the organization. There is need also to account for the lost time that is often overlooked during fire inspections. Lost time includes inspection preparation, travel time, report writing, and paperwork processing as well as code research. Keep in mind that every hour spent represents budget expenditures. Better accounting of time demonstrates professionalism, competence, and instills confidence in budget reviewers that the numbers projected represent money well spent by the organization (see Figure 2.8).

FIGURE 2.7 Example of cost analysis from an inspection permit.

Building Type	Number	Int Insp	Reinsp	Total Insp	Prep/hr	Travel/hr	Insp hr	Reinsp	Hr Resch/hr	Report/hr	Total/hr
Assembly < 300 occ.	100	100	50	150	0.5	0.5	1	0.5	0.5	1	4

Fee Analysis	Total/Manhours	Rate/hr	Total Cost	Cost per Insp	Annual Insp Fee	Profit	Loss
Assembly < 300 occ.	600	$42.00	$25,200.00	$168.00	$170.00	$2.00	$(2.00)

Budget Cycle

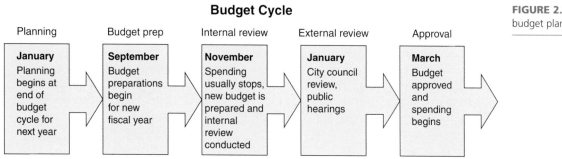

FIGURE 2.8 Example of budget planning cycle.

Summary

This chapter has reviewed how building and fire codes developed in the United States and how they were influenced by the insurance industry. We have examined the beginnings of the code enforcement industry and how it has developed into an international set of building and fire code regulations. We have reviewed the basic role of government and how the local enforcing agency and the fire inspector fit into the overall regulatory process, as well as how performance must be justified through the budgeting process. The systems of code regulation will continue to evolve and reshape in the future. As engineers, architects, and building construction professionals continue to push the envelope of advanced building design, there will be significant changes in the way fire inspections and code enforcement activities are conducted.

Fire has a significant cost to the country and every community. In 2008, the NFPA reported the direct and indirect losses from fire at 362 billion dollars or 2.5% of the gross national product. Fire departments and fire prevention inspection programs also are part of that cost and must justify costs every year when budgets are submitted. As agency competition for tax dollars increases, every cent placed in a municipal budget becomes a tug of war over who will get the funding. Fire marshals and fire officials must understand their budget process and maintain accurate financial records on the fees they generate, plan their budget effectively, and manage the operational costs of the bureau. Budgets must be justified by the programs and services provided to the community. Fire marshals and fire officials must measure their results and must constantly evaluate and improve their efforts in all facets of their daily work. A key element is customer service and satisfaction, as this helps to build customer support for programs. Fire marshals and fire officials must have the business acumen to translate work results through statistical analysis into budget support. This is especially necessary in difficult economies and political election cycles. Budget support must be both internal in the fire department and external in the community. Building relationships and support is necessary in both areas. Respect for fire prevention efforts must be built within fire departments by tying inspection results to firefighter safety, and support must be gathered in the community by reducing violations, educating customers, and reducing fire incidents and fire deaths and injuries.

Review Questions

1. Review the regulatory authority of your fire prevention code. At what level of government does the authority to conduct fire inspections come from?
2. Review the basic requirements of a Fire Inspector level 1 under the NFPA 1031 standard. Devise an evaluation system or method to evaluate the job performance knowledge, skills, and abilities, based on the NFPA 1031 requirements.
3. Contrast and compare the four model code groups that eventually became the International Code Council.
4. Prepare a line item budget for your fire prevention bureau. What items should be included?
5. Develop a method of analysis for determining inspection fees in your jurisdiction.

Suggested Readings

Crawford, James. 2010. *Fire Prevention Organization and Management.* Upper Saddle River, NJ: Prentice Hall.

NFPA 1031, *Standard for Professional Qualifications for Fire Inspector and Plan Examiner.*

NFPA 1037, *Standard for Professional Qualifications for Fire Marshal.* 2007 edition.

Endnotes

1. FM Global History, "The Rich History of FM Global" http://www.fmglobal.com/page.aspx?id=01070000.
2. Arthur E. Cote, "History of Fire Protection Engineering," *Fire Protection Engineering Magazine*, http://www.fpemag.com/articles/article.asp?i=375.
3. Ibid.
4. Percy Bugbee, *Men Against Fire. The Story of the National Fire Protection Association. 1896–1971* (Quincy, MA: NFPA, 1971), 3.
5. http://www.iccsafe.org/AboutICC/Documents/GovtConsensusProcess.pdf

3

Understanding Statistics and Community Risk Reduction

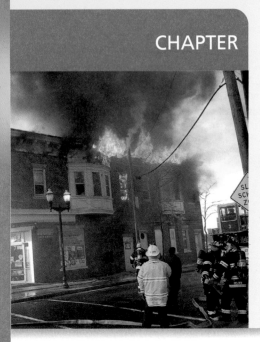

Courtesy of J. Foley

OBJECTIVES

After reading this chapter, the reader should be able to:

- Understand where to locate information necessary to define and analyze community fire risk.
- Understand basic statistical analysis and how to present information visually and effectively.
- Understand the eleven core measurements used to evaluate a fire prevention program's effectiveness.
- Be aware of the need to analyze the social and political landscape as it applies to community risk evaluation and reduction.

Professional Levels of Job Performance for Fire Inspectors as Cited in NFPA 1031 and NFPA 1037

- NFPA 1031 Fire Inspector I *Obj. 4.3.10 Emergency planning and preparation*
- NFPA 1031 Fire Inspector II *Obj. 5.3.4 Recommend modifications to codes and standards of the jurisdiction*

- NFPA 1037 Fire Marshal *Obj. 5.3.1 Community planning processes, emergency planning processes*
- NFPA 1037 Fire Marshal *Obj. 5.3.3 Manage a data and information management program*
- NFPA 1037 Fire Marshal *Obj. 5.3.4 Interpret data and information*
- NFPA 1037 Fire Marshal *Obj. 5.3.5 Conduct risk analysis*
- NFPA 1037 Fire Marshal *Obj. 5.3.6 Evaluate risk management solutions*
- NFPA 1037 Fire Marshal *Obj. 5.3.7 Integrate the risk management solution*
- NFPA 1037 Fire Marshal *Obj. 5.3.10 Develop a plan given an identified fire safety problem*
- NFPA 1037 Fire Marshal *Obj. 5.5.6 Forecast organizational professional development needs*

Resource**Central** For additional review and practice tests, visit **www.bradybooks.com** and click on Resource Central to access text-specific resources for this text. To access Resource Central, follow the directions on the Student Access Card provided with this text. If there is no card, go to **www.bradybooks.com** and follow the Resource Central link to buy access from there.

Analysis of Local Fire Risk

To be successful in fire prevention, fire inspectors must have a firm understanding of the fire problems in the communities they serve. Communities are complex systems composed of different social, educational, religious, and racial backgrounds. The fire inspector must have a clear understanding of the customer and the particular types of fires or fire risks that may occur. Risk analysis will help the fire inspector target audiences for fire education programs more effectively than the traditional "one size fits all" approach. This will give the fire inspector a clearer understanding of the community's needs, which will lead to a more successful and relevant approach to fire inspections and public fire safety education program delivery.

The first step in performing any analysis is to gather information from identified sources of data within the community. A good place to start gathering data is the local tax office. The tax office records will show all of the real estate within the community and the types of land parcels that are present. Tax records identify properties by categories such as residential, farms, apartments, commercial, industrial, nontaxable, government, public land, public schools, and churches as well as vacant parcels of land. Tax records usually are identified by block and lot numbers as well as assigned street address. A tax map will show the block and lot numbers that tie to the deed and property record of the land owner. Tax records are an excellent starting point for determining the types and numbers of properties within the community (see Figure 3.1). Tax records provide owner

OWNER INFORMATION	
Owner Name:	
	CENTRAL PIER
Owner Address:	601, North Drive, Malibu California
PROPERTY INFORMATION	
Property Location	1400 BOARDWALK
County	01 – Atlantic
District	02 – Atlantic City
Block Number	1
Lot Number	125
Qualifier	
Property Class	4A – Commercial
Land Description	137.5XTO PR HD LN
Building Description	S2S/CENTRL PIER
Acreage	
Land Value	$3,000,000
Building Value	$1,000,000
Net Value	$4,000,000
Prior Year's Taxes	$100,430.63

FIGURE 3.1 General property records will show location, owner, and property values.

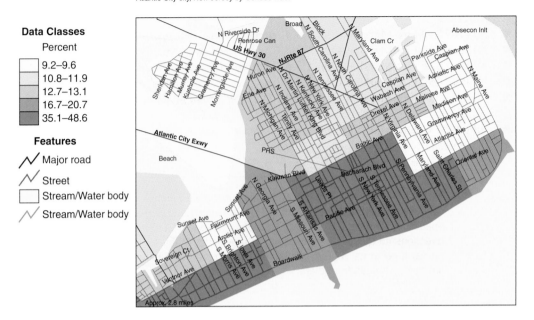

TM-H002. Percent of Housing Units Vacant: 2000
Universe: Housing units
Data Set: Census 2000 Summary File 1 (SF 1) 100-Percent Data
Atlantic City city, New Jersey by Census Tract

Data Classes

Percent

9.2–9.6
10.8–11.9
12.7–13.1
16.7–20.7
35.1–48.6

Features

Major road

Street

Stream/Water body

Stream/Water body

FIGURE 3.2 U.S. Census map of Atlantic City—number of vacant housing units. *Courtesy of the United States Census Bureau*

information, location of the property, block and lot, and approximate value of the property including the land value and building value. The tax records also will show the amount of real estate taxes paid. This information can be helpful in determining fire losses as well as the value of property protected by the fire department or inspected by fire prevention when analyzing programs.

Another source of valuable community information comes from the U.S. Census Bureau. The **census data** for your community can be accessed online at the United States Department of Commerce, Bureau of U.S. Census website (http://www.census.gov/). The census data will provide valuable information about the community's makeup including population profiles, housing and building stock, economics, education, and much more. The census databases allow you to build thematic maps from the data files that graphically plot information about areas of the community based on many different factors such as socio-economic levels, education, or property values. As an example, Figure 3.2 shows the percentage of vacant housing units in Atlantic City, New Jersey. This information is useful for developing programs for securing abandoned properties or for targeting potential areas in an arson prevention program. Vacant properties usually present community complaints for fire risk, health code violations, and criminal activities, including arson. These maps can help visualize and direct resources to the areas with the highest vacant building rates.

The census data is also useful in targeting public fire safety education efforts in that the age, racial, and educational makeup of the community can be determined. This information can assist fire educators to target appropriate audience information as well as fire risk exposures. This data again can be manipulated graphically to identify trends within the community for inspection, public education, or arson awareness efforts.

The most common graphical representation of data is referred to as a **histogram**. Histograms are helpful in looking at a set of data as a picture rather than values or numbers. The census data can be used to identify different aspects of a community, such as the

census data ■ Information collected by the Department of Commerce describing population, housing, and economics of the United States.

histogram ■ A column graph where the height of the column indicates the relative number, frequency, or value of a variable.

number of the housing units within the community or how many housing units are vacant versus occupied. A histogram can be used to visually represent the data. As an example, Figure 3.3 shows the number of residential housing types by number of units per building in Atlantic City, New Jersey. By examining the data on the years in which housing units were constructed, it is possible determine the approximate number or the percentage of units that may have been constructed under old building codes rather than more modern building codes (see Figure 3.4). One can see that 63 percent of the housing units were built before 1970 under older local building codes, and 47 percent were built under the more modern BOCA and IBC codes. This information can be helpful in determining the type of construction within the jurisdiction, from balloon frame, to platform frame, to lightweight truss. This can provide a more global perspective of the percentage of buildings constructed under different versions of building codes.

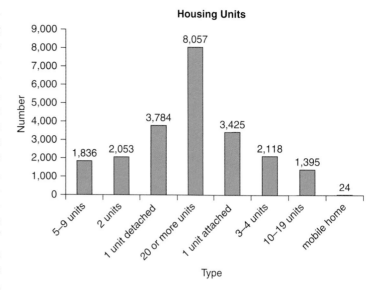

FIGURE 3.3 Example of a housing unit histogram.

Another useful tool that is available to fire inspectors for free is global positioning software such as Google Earth or Microsoft Live. These programs provide overhead aerial photographs of buildings and streets within a jurisdiction. In some cases, street-level photographs are attached to the aerial building file, and you can examine the building from four different aspects—north, south, east, and west. These programs are very similar to the old Sanborn insurance maps. Sanborn mapping was originally created to determine fire insurance rates in congested urban areas. Modern Sanborn maps use the same GPS technology as the free software tools described earlier. Typically, Sanborn maps show building height, building footprint, number of stories, type of construction materials, and general building use. Google Earth (see Figure 3.5) can provide similar information from both the overhead aerial photographs and street-view photographs. Additionally, the program has measuring tools available that allow the user to measure the building's footprint to determine square footage or the distance to exposed structures for firefighting purposes. It is a helpful free tool that can assist fire inspectors

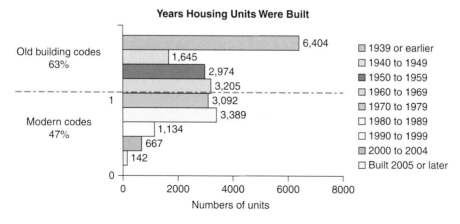

FIGURE 3.4 Bar graph of year in which housing units were built.

FIGURE 3.5 Google Earth images can be used to determine rough square footage using measuring tools. Here, Central Pier at 1400 Boardwalk is 146' × 474' = 69,204 square feet. ©2011 Google

in gathering useful data. Google Earth also has been found to be useful in fire investigations, as you can get before and after overhead views of the structure. This is a very useful tool, especially if the structure has collapsed during the fire.

The more data that a fire marshal or fire official collects on the community, the more effective that individual can be in identifying fire trends and potential fire problems. Fire marshals and fire officials must keep abreast of what types of buildings or use groups are having fires in their jurisdiction, and how those fires were caused. Most fire departments employ the National Fire Incident Reporting System, or **NFIRS**, to collect and analyze fire data. NFIRS reports have coded fire data identifiers that provide useful information to fire officials. This information includes fire protection system performance, materials

NFIRS ■ National Fire Incident Reporting System, used by local, state, and federal governments to collect and analyze fire data.

realized service demand ■ These are the historical values of demands for service that have occurred in the past. Realized service demands tell us where we have been but do not necessarily reflect future occurrences.

first ignited, ignition causes, and room of ignition, to list a few examples. NFIRS reports help to identify the hazardous act and the ignition sources that start the fire. The data collected by NFIRS is also historical data and quantifies the **realized service demands** of the fire department. Good data collection and completion of NFIRS reports by fire officers assist the fire marshal or fire official in evaluating the community's fire risk and targeting fire prevention efforts.

Community fire risk assessment is based on local fire data analysis, and that data should be compared to the state and national fire data banks to establish baselines. Reviewing fire statistics identifies trends that may involve building and fire code deficiencies, fire safety education issues, fire protection system maintenance problems, or juvenile and arson-related fire trends. The NFIRS system provides the fire official with comparative data on state and national fire incident reporting and enables the fire department to make local comparisons. For example, in New Jersey almost 100 percent of all career and volunteer fire departments complete and submit NFIRS fire incidents reports to the state. The State Division of Fire Safety was able to gain reporting compliance through requiring data submission in order to apply for state grants for such equipment as thermal imaging cameras and computers. Connecting grant approval to the submission of fire reports greatly enhanced NFIRS reporting in New Jersey.

As we examine community fire risk, fire inspectors need to analyze two important parts of the fire risk equation. The first part is the NFIRS fire data that will tell us what our present and past fire experience has been. These statistics help build a baseline for comparison and help to identify changing trends in fires by occupancy groups. Keep in mind that effective fire prevention efforts require identifying three key components: a fuel source, an ignition source, and the hazardous acts that bring them together. If the fire inspector can eliminate or control any one of these three factors, the fire will not occur. Reviewing fire incident reports helps to identify these causes of fires and the lessons to be learned from these incidents. It is not a good idea, however, to depend on fire data exclusively to identify all of the potential fire risks. Doing so would be like driving a car forward while looking only in the rear view mirror. The same is true with community fire risk. The fire official and inspector need to examine the **latent service demands** or risk potentials that have not been realized yet. The latent service demand is

latent service demand ■ Latent service demand is potential risk that has not been realized through occurrence. This includes potential exposures to perils.

a measurement of the potential for emergency incidents to occur and the likelihood of that occurrence within the community. Latent service demand will reflect the gaps between controlled risk and unprotected risk. A good example is the terrorist attack on the World Trade Center in New York City on September 11, 2001. While the potential always has existed for the collapse of a high-rise building, this had never happened in a fire until that time. In fact, World Trade Center building 7, which was not struck by an airplane, collapsed after an uncontrolled fire had raged in the building for several hours. The New York City Fire Department certainly cannot build its resources to avoid this sort of incident in the future; however, it can plan and manage its available resources in a more productive manner to reduce the potential risk and harm in the future. Latent service demand includes the "what-if" questions that need to be asked. These are the fire scenarios that have not yet been experienced in the community but have the potential to occur. These potential problems need to be identified as part of community risk assessment. Often, benefit can be derived from another community's problem caused by a tragic event. Reviewing fire journals and trade publications can provide insight into risk problems before they occur in jurisdictions. The lessons learned need to be examined and applied to foster better fire prevention efforts. Latent service demand requires identifying target hazards and looking for the potential risks that may exist or may occur and have a negative consequence. A complete risk analysis will help develop reasonable risk avoidance strategies that can be put in place before a fire incident. As an example, the state of New Jersey legislated a smoking ban in all places of public assembly. When people could no longer smoke inside a building, they would go outside and smoke, often discarding lit cigarettes into landscaped areas. Fire departments began to see an increase in exterior mulch fires, and in some cases, these fires extended into buildings and caused extensive damage. A risk avoidance strategy was developed to request that owners avoid applying mulch in areas prone to smoking and that smoking areas be specifically designated as such with cigarette receptacles to control ignitions. This strategy began to reduce the number of exterior fires to which fire departments were responding.

Fire officials must also consider other types of emergencies besides fires that will require fire prevention strategies for risk avoidance. These may include strategies for emergencies such as the loss of water supply to fire suppression systems or the failure of audible fire alarm systems in high-rise buildings or hospitals. Risk factors have to be identified before any risk reduction strategies can be developed to reduce the potential risk. Keep in mind that evacuation of buildings may not always be a viable risk avoidance option; alternate methods of providing temporary fire protection to keep a building occupied may have to be considered because evacuation is simply not possible. The risk management graph in Figure 3.6 may help in establishing priorities for planning purposes. Low-frequency occurrences of high-risk consequences must be given top priority in action plans. Fire prevention policy and procedures for these non-fire emergencies should use both risk avoidance measures and risk control methods that may be employed during the preplanning for such events. Often, fire inspectors will be required to deal with situations of diminished fire protection in an occupied building during weather emergencies or other constraints. Plans for dealing with these risks should be developed and committed to written emergency action procedures. These procedures may include fire watch policies, fire department support of fire protection systems, ignition reduction policies, or other mitigation and fire control counter measures required to diminish fire risk potential.

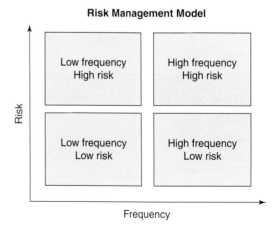

Risk Management Model

FIGURE 3.6 Risk management graph

Organizing Statistical Data

The organization of data can be a daunting task for most fire officials. The idea behind statistical analysis is to view data in a meaningful way and to be able to present it or employ it to the benefit of your organization. This will require a quick review of mathematics related to basic statistical analysis. The ultimate objective of analyzing statistics is to allow yourself and others to make better and more informed decisions on organizational matters and how they might impact public safety.

Analyzing data is performed in four steps: collecting, categorizing, analyzing, and decision making. In the first step, the data is collected from files, tax records, fire reports, and other information sources within the community. This data must be relevant to the problem to be examined. The second step is to categorize the data and turn it into information for further analysis. Items that will help analyze the specific problem area may be categorized by type, count, percentage, or rate. The data analysis will involve the mathematics of statistics and the development of graphical representations of the data. In the third step, the organized and categorized data are examined to understand what it indicates about the community fire risk or fire inspection and fire safety education program performance. In the fourth step, decisions must be made based on what has been learned from the data. Fire officials can make very sound decisions provided that they are based on accurate and understandable data analysis.

When speaking about statistics analysis, there are two different types of data to be considered. The first is **descriptive statistics**, which are numbers that describe something in mathematical terms. Examples would be averages, percentages, proportions, or counts. Usually, descriptive statistics are employed to represent data. Another type of statistic used is **inferential statistics**, which infer conclusions about a larger population from a smaller sampling. The use of inferential statistics allows projection of a small sample to forecast potential future trends. An example of this is political polling of data in a national election where the general mood of the country can be inferred from a smaller sampling of likely voters.

When examining data about fire prevention inspections or fire education programs, the first piece of information to be collected is the count of these activities. This could be the number of public assembly buildings, the number of three-unit apartment buildings that we inspect, the number of people attending a fire prevention program, or the number of false alarms that occur in buildings with fire detection systems. All statistical analysis begins with some form of a count.

Table 3.1 shows all the residential dwelling units by type in Atlantic City, New Jersey, according to the 2010 U.S. Census. You will notice that some of the data in the table is a direct count such as one-unit attached dwellings, while others are banded within a fixed range like 10–19 units. Banding data sometimes helps to make smaller samples more visible in a graphical representation like a pie chart (see Figure 3.7). Pie charts give us some idea of how large the category is compared to the whole sample. The data from the table can also be represented graphically in a pie chart as a percentage of the whole, which again can provide useful information for applying resources or looking at potential risk exposure (see Figure 3.8). Data can be grouped, as it is in this example to simplify 22,692 individual pieces of data into eight broader categories. Grouping allows the data to be simplified while still representing graphic scale to the information. Displaying the percentage assists the fire inspector in visualizing how many units of the total number are represented within each category. Note that the

TABLE 3.1	Residential Dwelling Units
1 unit detached	3,784
1 unit attached	3,425
2 units	2,053
3–4 units	2,118
5–9 units	1,836
10–19 units	1,395
20 or more units	8,057
Mobile home	24

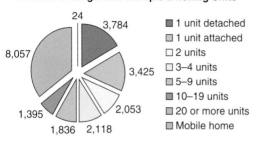

Number of Single and Multiple Dwelling Units

- 3,784
- 24
- 8,057
- 3,425
- 1,395
- 2,053
- 1,836 2,118

Legend:
- ■ 1 unit detached
- ▨ 1 unit attached
- ☐ 2 units
- ☐ 3–4 units
- ▨ 5–9 units
- ■ 10–19 units
- ▨ 20 or more units
- ☐ Mobile home

FIGURE 3.7 Pie chart of single-family and multifamily housing units, displaying the distribution of housing in relation to the whole.

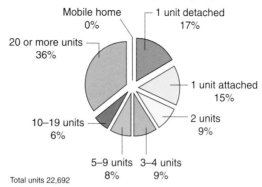

Percentage of Housing Units by Type

- Mobile home 0%
- 1 unit detached 17%
- 20 or more units 36%
- 1 unit attached 15%
- 10–19 units 6%
- 2 units 9%
- 5–9 units 8%
- 3–4 units 9%

Total units 22,692

FIGURE 3.8 Pie chart showing the percentage of types of housing units. Note that mobile homes show as "0%" because the number is less than <1%.

mobile homes category equates to 0 percent even though there are 24 units. This is because the 24 units are only a fraction of a percent. This will occur when large and small samples are mixed.

Helpful information can be determined from the analysis of the statistical mean, median, and mode of a data set. The **mean** is the average or the center of a set of numbers; it is the balancing point between the high numbers and low numbers. Adding all of the data values and dividing the sum by the number of observations determines the mean or simple average:

mean ■ Simple average that is calculated by summing all of the data values and dividing by the number of observations.

EXAMPLE

$Students\ grades = 93, 86, 86, 75, 94$

$$X = (93 + 86 + 86 + 75 + 94)/5 \quad 434/5\ X = 86.8$$

The mean can only be calculated with **continuous variables**. A continuous variable is one that takes on some type of measurement, like the time or the number of occurrences. **Categorical variables**, or qualitative variables, are those that define by data type or group, such as the residential dwelling units listed in the legend in Figures 3.7.

The **median** is the center of a set of numbers in which there is the same amount of numbers to the left as there are to the right, or above and below. The median is the mid-size number or 50 percent mark in the set. In the previous example, 86 percent would be the median. The median can only be calculated with continuous variables The **mode** is the number that appears most frequently. The mode is usually the peak of the histogram. The mode is also the only one of the three averages that can be applied to categorical variables and continuous variables. Once the mean within a set of data is established, it is also useful to know how data are spread around that mean. It is necessary to determine whether the data are gathered tightly or spread out far from the mean. This question can be answered by a mathematical calculation called **standard deviation**, which will tell us how much of that data is clustered around the mean. In a typical normal data distribution, 68 percent of all data will fall within one standard deviation of the mean, and 95 percent of all data will be within two standard deviations of the mean.

The formula for standard deviation is as follows:

$$S_d = \frac{\sqrt{sum - (X - mean)^2}}{N - 1}$$

continuous variables ■ Numerical values that reflect some form of measurement.

categorical variables ■ Qualitative variables that are not measurable but classify data into groups or categories.

median ■ The middle value in a group of numbers where the same number of observations exists above and below.

mode ■ The most frequent value that appears in a data set, usually represented as the peak in a histogram.

standard deviation ■ Identifies how data is distributed around the mean in a distribution. The standard deviation is the square root of the variance of data.

Let's do the calculation step-by-step on a sample set of data: 1, 2, 3, 6, 7.

Step 1. Calculate the mean $X = (1 + 2 + 3 + 6 + 7)/5 = 19/5$, $X + 3.8$; this is the mean.

Step 2. Find the sum of each value minus the mean:

$$1 = (1 - 3.8)^2 = (2.8)^2 = 7.84$$
$$2 = (2 - 3.8)^2 = (1.8)^2 = 3.24$$
$$3 = (3 - 3.8)^2 = (.8)^2 = .64$$
$$6 = (6 - 3.8)^2 = (2.2)^2 = 4.84$$
$$7 = (7 - 3.8)^2 = (3.2)^2 = 10.24$$
$$\text{SUM} = 26.80$$

Step 3. Determine the standard deviation, which is the square root of the sum divided by the number of values minus 1.

$$S_d = \sqrt{26.80/(5 - 1)} = \sqrt{26.80/4} = \sqrt{6.7} = 2.59$$

Of all the data, 68 percent falls between 1.21 and 6.39, or 1 standard deviation.

The standard deviation describes the dispersion of the data set around the central point of mean.

Another analysis that may be needed is the comparison of one group of data to another group of data. In this case, we may use proportions or percentages to illustrate the differences. The mathematical calculation for proportions is as follows:

$$\frac{N_1}{(N_1 + N_2 \dots)}$$

As an example, suppose we had a building inventory of 7,780 buildings and we knew that 180 of those were public assemblies. We can determine the percent of buildings that are public assemblies from the following proportion:

$$\frac{180}{(180 + 7,600)} = \frac{180}{(7,780)} = 0.023 \times 100 = 2.3\%$$

The answer is that 2.3 percent of the buildings are places of public assembly. Percentages are very useful in examining data such as completed versus noncompleted inspections, or abated versus nonabated violation notices.

As an example, in the data set above, suppose that we complete inspections of only 98 of the 180 places of public assembly:

$$98/180 = 0.54 \times 100 = 54\%$$

The inspection completion rate for public assembly is 54 percent completed, or, put differently, 46 percent of public assemblies have not been inspected.

rate ▪ Rates are measurements per 1,000 or 100,000 population that can be used to compare experiences in other jurisdictions.

The last mathematical term we will examine is **rate**. The rate is important as it allows for comparisons to other organizations' data to see how well your fire department compares. Rates can be used to determine the code compliance rate, the rate of inspections compared to population, the rate of fires compared to particular use, groups, and much more. Rates are determined by dividing the number of occurrences by the total population. As an example, if 18 fires are experienced in the 180 places of public assembly discussed above, the inspector can calculate the rate of public assembly fires per 1,000 places of public assembly in the community as

$$18/180 = 0.10 \times 100 = 10\%$$

of the public assemblies experienced a fire in this community.

To express this as the number of fires per 1,000 population, the resident population of 42,000 is divided by 1,000, which gives the answer as 42. This means that there are 42 groups of 1,000 people in the community. We can now divide the number of incidents by the population groups:

$$18/42 = 0.43 \text{ public assembly fires per 1,000 population}$$

The rate is helpful because it can indicate the level of potential exposure of the population to each type of fire in the community. The fire marshal or fire official often must defend budgets and fire prevention programs to municipal officials and community business groups. Understanding how to present fire statistics in a useful and illustrative manner provides a better understanding of the fire problem and better community support. Using comparative statistics helps fire marshals and fire officials examine the fire problem in a broader spectrum by allowing comparisons to other cities of equal size or to the state or national fire data. The differences and similarities in those statistics can help support new programs or maintain existing ones.

Core Measurements of Fire Prevention

The general goals of fire prevention are twofold: first and foremost, to prevent fires from occurring; and second, to ensure that the destructive forces of fire are reduced through the applications of fire prevention technologies. When examining data for fire prevention activities, useful information must be provided so that fire officials or fire marshals can make good decisions. When considering what to measure, keep in mind that the metrics must reflect effectiveness, efficiency, and the equity of the work performed. Effectiveness is how well the data defines accomplishing program goals, efficiency is the measurement of how well resources have been matched to meet those goals, and equity measures the cost of those resources compared to the benefits provided to the community. It must be verified that taxpayers' dollars are being well spent. The question becomes what core measurements need to be examined in fire prevention to make these determinations? How do we measure how many fires have been prevented if they never occur? According to the National Fire Protection Association and the Fire Protection Research Foundation's final report "Measuring Code Compliance Effectiveness for Fire Related Portions of the Codes,"[1] the methods of determining measurement of fire prevention activities have been around for about 30 years. In the 1970s, the NFPA and the Urban Institute, a Washington, DC, think tank on social and economic issues confronting the nation, published the following three reports on fire prevention inspections and the measurements of effectiveness:

- 1974: "Measuring Fire Protection Productivity in Local Government"
- 1976: "Improving the Measurement of Local Fire Protection Effectiveness"
- 1978: "Fire Code Inspection and Fire Prevention: What Methods Lead to Success?"[2]

The final report placed fire inspection metrics into three categories and identified eleven core measurements to determine the effectiveness of fire inspection programs. The three categories of fire inspections are as follows:

1. New construction
2. Existing buildings undergoing rehabilitation
3. Existing structures

The eleven core measurements were based on three areas of fire inspection, which are process, impact, and outcome. Process measurements examine the quantity and quality of the fire inspections being performed. Process evaluations respond the quickest to change as they are measured against fire loss. Process changes made in one year should begin to be reflected in fire data the following year. Outcome evaluation provides information on the achievement of the stated goals of a program. Impact evaluations are a reflection of both process and outcome. Impact evaluations describe any change in the target audience behavior due to either a process change or an outcome change, or both. Keep in mind that fire prevention program success is a combination of removing hazards by fire inspection and changing people's behavior and attitudes through educational improvement. Measurements of people's motivation and education are very difficult and complex to obtain because of the many uncontrollable variables in the process. The data are difficult to collect, which makes any meaningful analysis very difficult to quantify. Simply put, there are just too many variables to effectively target in order to determine what caused a specific change in behavior.

Table 3.2 contains the eleven core fire prevention measurements divided into three categories by fire inspection types: existing buildings, new and existing buildings, and new construction.

Data Presentation

The format of data in a presentation is important to communicating the results of analysis effectively. Often, fire officials or fire marshals have to make presentations for the fire chief or city council to justify expenditures on fire prevention programs. The rule you should follow is, keep it simple! Often, presenters get carried away making bar graphs, pie charts, and PowerPoint slides with numbers, and the presentation becomes complicated—losing people's interest. Data presentation is most effective when it is simple, straightforward, and uncomplicated. If too much information is placed on a graph, the information can become lost in the graphics and notes. Presentations should be short and to the point—no longer than 10–15 minutes. Too many presentation slides become overwhelming to the audience, and people lose interest in the message. As you develop your information, sort it into the most pertinent facts related to the questions asked or anticipated during the presentation. Simple design and uncomplicated graphics will sell the point for you. If you are using PowerPoint, do not place too many bullets on a slide, make sure the graphics can be read easily, and be concise on the information presented. In today's data-overloaded society, "less is more."

When preparing statistical data information, you should present it in one of three ways:

1. *Raw number or count of events:* This tells how many of a certain occurrence or data element happens. Raw number may be the number of fires, false alarms, or inspections completed.
2. *Percentage:* This tells you the relationship of the raw number to a total sample. Examples include the percentage of structure fires versus incidents, the percentage of false alarms versus incidents, and the percentage of completed inspections versus all inspections. The use of percentage allows you to gauge the number of occurrences or performance in relation to the total workload. The percentage represents mathematically where work is being performed or not performed.
3. *Rate:* This tells you the rate per 1,000 or per 100,000 population and allows the measurement of performance based on a fixed sample. The use of rate allows a comparison of local, state, and national statistics to determine whether performance is consistent or inconsistent with the rest of the state, nation, or cities of comparable size. The comparison of rates can demonstrate either exceptional performance or areas lacking in performance. Using rates per 1,000 or 100,000 population allows for equalization of the statistic to a smaller sampling group.

DISPLAYING DATA

When displaying data, you need to think about what information you are trying to convey and what method presents it in the simplest terms. The display of the information

TABLE 3.2		**Core Fire Prevention Measurements by Fire Inspection Type**		
1	Existing Structures	Structure fire rate per 1,000 inspectable properties	Use five-year average on small samples, may exclude intentional fires	Fatal fires should have a separate matrix using #4, 5, 6,
2	Existing Structures	Estimate monetary value per additional inspection by major property group	Insurance payouts, direct and indirect loss estimates *Link to # 4*	Value of annual inspection (Fire loss) × (% preventable)/ # Occupancies
3	Existing Structures	% of fires that were preventable by inspection, educational program, or element of inspection	*Link to # 2* on estimating the value of an additional inspection	
4	Existing Structures	% of fires that were preventable, pending uncorrected violations at the time of the fire	Focus on post inspection correction of violations	
5	Existing Structures	% of fire in properties subject to inspection but not in inspection files	Code by reason not listed	
6	Existing Structures	% of inspections beyond departments target inspection cycle	Analyze by major occupancy groups	
7	New & Existing Structures	List inspectable properties structure fires over $25,000 loss matrixes with fire severity and major hazard	Insurance loss estimates both direct and indirect *Link to # 8*	Work with insurance companies to get best estimate
8	New & Existing Structures	Number of violation found per inspection separately for a) Sprinkler violations b) Evacuation violations	Major reasons for multiple deaths in inspectable properties *Link to # 7*	
9	New & Existing Structures	% of inspections conducted by properly certified inspectors	Analyze separately for major occupancy groups, initial and follow-up inspections	Vary based on type of inspection of company inspection program conducted
10	New & Existing Structures	% of inspections conducted by full-time inspectors	Separate by major occupancy groups	
11	New Construction	Number of building system features for which inspections and approvals were not completed	Building systems or features for which no timely inspection has occurred.	

Courtesy Fire Protection Research Foundation. Copyright © 2008, National Fire Protection Association

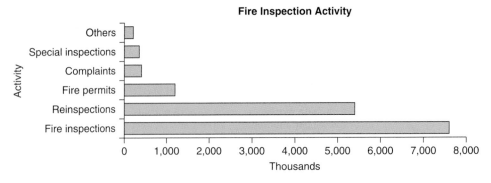

FIGURE 3.9 Bar graph of fire inspection activities.

should lead the viewer to the conclusion you are trying to portray. Different types of graphs are useful in presenting different types of data.

BAR CHARTS

Bar charts are often used to display data in a simplistic fashion that visually represents the numbers by the length of the bar. The bar graph is best used with categorical data and shows the categories by rank or bar length (see Figure 3.9.) The bar chart illustrates activities or problem areas and can assist in decisions or determination of problem priority.

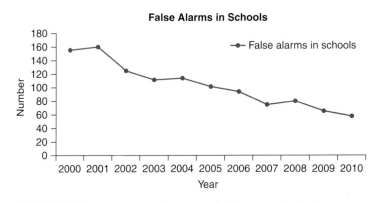

FIGURE 3.10 Line chart expressing changes in false alarms in schools.

LINE CHARTS

Column and line charts are often used to show changes in data over time. These types of charts help illustrate trends or the impact of a program (see Figure 3.10.) In the example, we have examined the effects of an implemented false alarm education program. The line graph illustrates the reduction of false alarms over time.

COLUMN CHARTS

Column charts are used when we want to compare data from different groups such as Figure 3.11. The charts help to identify how inspection work is progressing and may also indicate the need for additional motivation or training by examining inspector performance.

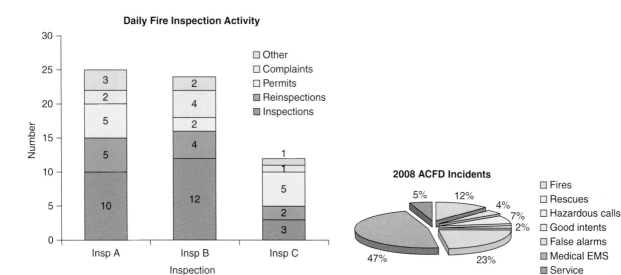

FIGURE 3.11 Column charts are most useful in the comparison of data such as inspections performed by different inspectors.

FIGURE 3.12 Pie chart showing percentage of fire incidents.

PIE CHARTS

Pie charts are useful in describing elements of a problem as part of a whole. Pie charts often are used to represent percentages of each data element (see Figure 3.12.) Pie charts, however, are not always the best way to display data, as pie charts are not effective when comparing several groups of figures.

PICTOGRAPHS

Pictographs are also useful in that they can represent data over a geographic area, as we described earlier in Figure 3.2. Pictographs can be of a local area, a county, a state, or the entire country. It is useful sometimes to show comparisons such as fire death rate per 1,000 population over a map to get a broader understanding of how your data fits the rest of the country. Many useful maps already exist in the census database for communities such as Figure 3.13, which shows the percentage of population in Atlantic City below the poverty

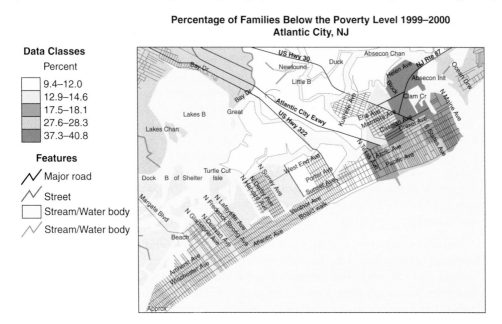

FIGURE 3.13 U.S. Census map showing poverty level by census district. *Courtesy of the United States Census Bureau*

level. This information is useful because, statistically, these groups are more affected by fire, and education programs can be more easily targeted once a clear understanding is achieved on where to start in the community.

Community Risk Avoidance

We have discussed the analysis of data to determine and identify fire risk and how to analyze the effectiveness of fire prevention efforts. Let us take a moment to examine community risk and risk avoidance. The NFPA reported in 2010 that U.S. fire departments responded to 1.3 million fires resulting in 3,020 civilian fire fatalities, 17,720 civilian fire injuries, and 72 firefighter deaths. While the 2010 fire statistics were 1.3 percent lower than those for 2009, fire death rates increased by 3.7 percent and fire-related injuries by 3.9 percent. The NFPA Fire Clock illustrated the following:

24 seconds	A fire department responds to an emergency.
50 seconds	An outside fire occurs.
65 seconds	A structure fire occurs.
85 seconds	A home fire occurs.
30 minutes	One civilian is injured.
2 hours 49 minutes	One fire death occurs.

These fires resulted in 12.5 billion dollars in direct property loss nationwide.[3]

In 1973, "American Burning" identified risk reduction strategies as a recommendation to reduce the nation's loss from fire. In 2002, the "America Burning Recommissioned, America at Risk" report stated that in many ways, the fire problem identified in 1973 still existed 30 years later. The Commission drew the following two conclusions:

1. The frequency and severity of fires in America do not result from a lack of knowledge of the causes, means of prevention or methods of suppression. We have a fire "problem" because our nation has failed to adequately apply and fund known loss reduction strategies. Had past recommendations of America Burning and subsequent reports been implemented, there would have been no need for this Commission. Unless those recommendations and the ones that follow are funded and implemented, the Commission's efforts will have been an exercise in futility. The primary responsibility for fire prevention and suppression and action with respect to other hazards dealt with by the fire services properly rests with the states and local governments. Nevertheless, a substantial role exists for the federal government in funding and technical support.
2. The responsibilities of today's fire departments extend well beyond the traditional fire hazard. The fire service is the primary responder to almost all local hazards, protecting a community's commercial as well as human assets and firehouses are the closest connection government has to disaster-threatened neighborhoods. Firefighters, who too frequently expose themselves to unnecessary risk, and the communities they serve, would all benefit if there was the same dedication to the avoidance of loss from fires and other hazards that exists in the conduct of fire suppression and rescue operations.[4]

The Commission reported that fire deaths are preventable and should receive the same public outcry as drunk driving or children being killed by hand guns. In the report, fire prevention and risk reduction strategies were identified as key elements for reducing fire incidents. The commission expressed the need for better commitment from the federal government and better coordination of federal agencies to further reduce the fire problem.

FIGURE 3.14 Risk analysis matrix.

HAZARD	HAZARD FREQUENCY			PROBABILITY			POPULATION AFFECTED	CONSEQUENCE					VULNERABILITY
	LOW	MODERATE	HIGH	LIKELY	POSSIBLE	UNLIKELY		DESTRUCTION (high = 3, mod = 2, low = 1)	ENVIRONMENTAL (high = 3, mod = 2, low = 1)	ECONOMIC (high = 3, mod = 2, low = 1)	SOCIAL (high = 3, mod = 2, low = 1)	PLANNING (high = 3, mod = 2, low = 1)	
FIRES													
RESIDENTIAL													
R-1	X				X		100–7,000	X 2		X 2	X 1	X 1	6
R-2		X		X			25–300	X 2		X 2	X 2	X 1	7
R-3			X	X			15	X 3		X 1	X 3		7
													0
ASSEMBLY													0
A-1	X				X		100–1,500	X 1		X 1	X 1	X 1	4
A-2		X			X		300–1,000	X 1		X 1	X 1	X 1	4
A-3			X	X			25–300	X 2		X 2	X 2	X 1	7
A-4	X						100–300	X 2		X 2	X 1		5
													0
BUSINESS		X			X		25–50	X 3		X 3	X 3	X 1	10
MERCANTILE		X			X		300–1,000	X 3		X 3	X 3	X 1	10
FACTORY													0
F-1	X				X		100	X 1	X 1	X 3	X 2	X 1	8
F-2	X				X		50	X 1	X 1	X 3	X 2	X 1	8

SCALE 5 TO 8 = LOW
9 TO 11 = MEDIUM
12 TO 15 = HIGH

The report identifies that while effective mitigation strategies have been developed since the "America Burning" report was published in 1973, neither Federal Emergency Management Agency nor the United States Fire Administration has effectively implemented them. This has mostly been due to ineffective funding by Congress and failure of advocacy on the part of these federal agencies.

Risk reduction recommendations that were suggested included long-term strategies on the implementation of fire sprinkler and fire alarm technology. Recommendations were made to implement all hazard mitigation programs and training curriculums. The commission recommended better funding for computer systems to provide more efficient data collection and analysis. It was recommended that FEMA and the USFA take a stronger role in setting the fire research agenda with NIST and that they assume more active involvement with the development of building and fire codes. Fire prevention training and fire prevention accreditation were recommended as well as a better delivery of programs through the USFA and the National Fire Academy.

While community fire risk hits hardest at the local level, the solutions must be resolved on broader bases through state and federal government interaction. Without adequate, broad-based support, it is difficult for local agencies to mount sustained efforts to maintain long-range prevention strategies. Local fire officials or fire marshals must identify their fire problems because if the problems are not identified, funding is not provided to fix them.

Risk is described as facing a hazard, danger, or peril that may result in death, injury, or property destruction. When we examine community risk, we need to consider two important aspects of the risk: the *probability of occurrence* and the *risk consequence*. Risk reduction planning must consider both risk controls and consequence management. A simple matrix can be constructed to identify a hazard, the expected frequency of occurrence, and the anticipated severity or consequence (see Figure 3.14). Based on that matrix, we can determine the proper protective measures or controls to reduce the risk exposure. Matrixes can be established based on building uses or general hazard categories. The idea is to develop a useful way of identifying risk and vulnerability. While fires are often deemed as accidents, there is really nothing accidental about them. All fires are caused by a series of consequences that allow fuel, oxygen, and heat to come together. The process of risk analysis involves breaking the potential occurrence down into understandable segments so that controls may be improvised to address or reduce the risk. The elements may fall into any of these categories:

- Social and physical environment
- Human factors
- Unsafe acts
- Unsafe conditions

As we identify these factors, we can identify controls to reduce the risk. Controls must be targeted to individual behaviors, items, or elements and systems. The control actions may involve changing potential ignition sources, separating fuel and ignition sources, changing the fuel types, or modifying the ignition and fuel propagation potentials. The means to implement these actions against the target can be technological, legal, economic, or social and political or a combination of all (see Figure 3.15). In this matrix, actions are identified in rows, the means to accomplish those actions are in columns, and the target of those actions and means are found in the depth of the cube. There is a minimum of 48 alternative solutions,[5] although not all may be possible, for different reasons. The

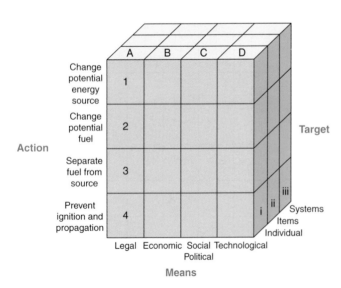

FIGURE 3.15 Alternative policy matrix. *Courtesy of RAND*

matrix helps examine solutions to problems by looking at the appropriate actions, means, and to achieve the desired outcome. As an example, using the matrix in Figure 3.15, if there is a problem of people smoking in a public building and causing fires, a no-smoking policy can be implemented in the building. The action is found in block 3, "separate ignition source from fuel." The means to accomplish this is instituting a no-smoking regulation directed at the target (column A, "Legal"). The target in this case is the individual (small 'i'). There are also other possibilities such as manufacturing a fire-safe cigarette, where the action would be "change potential energy source" (block 1), the means would be "technological" by changing the cigarette's burning potential (column D), and the target would be the "item" or cigarette (small ii). The purpose of the matrix is to help identify different alternatives to solving problems.

In many cases, technological changes from the private sector have significantly reduced fire in the United States. As an example, changes such as clothes irons that shut off automatically when unattended or knocked over and portable space heaters equipped with tipping switches or ceramic elements that lessen ignition potential have come about through industry innovation. Bedding and mattresses are now fire ignition resistant, and furniture is being manufactured with fire retardants making it less capable of being ignited by cigarettes. The installation of smoke detectors and automatic sprinklers has reduced fire fatalities and property damage significantly. In the 1970s, civilian fire fatalities exceeded 12,000 per year.[6] In 2010, the NFPA reported the civilian fire fatality rate at 3,020 deaths. That is almost a 75 percent reduction in the loss of life from fires, and much of it is due to advances in making appliances and the living environment safer. The advances made in the last three decades are significant, but there is still a long way to go. Fire prevention officials need to evaluate risk in their communities by analyzing the fire data and clarifying potential strategies to address and reduce risk exposure.

Community Tolerance and Social Influences

As we have explored in this chapter, knowledge is power, and good data analysis provides good knowledge for decision making. Fire officials or fire marshals, however, must still understand the political and social implications of instituting policy changes. All communities have **unprotected risk**, which is the imbalance between risk control and risk exposure. The community decides the level of **acceptable risk** it is willing to take. A contractor once said to the author, "You fire guys are all the same; you want a concrete building underwater to make it fireproof." While the concept is interesting, it certainly is not a practical solution. The statement does, however, say something about the perception of code-enforcing agencies as overreaching in prescriptive measures. This perception is fueled by economics versus the perceived value of what is being corrected. What makes this perception more difficult for the public is that they see new buildings having to meet stringent standards, while old buildings are allowed to stay at *status quo*. This dichotomy makes individuals believe that the inspection agency is picking on them, and this perception often will gain political momentum. This is why education of the public in fire prevention concepts becomes so important.

As a case in point, there was a situation in which a permit fee for special events was viewed by several small contractors as excessive. The fee was established by city ordinance as part of the fire prevention code. The contractors attended city council meetings and complained about the cost and the inflexibility of the fire marshal in reducing or waiving the fee. Several council members were business owners and were subject to the enforcement of the fire prevention code. One council member, who had a record of performing work without proper building permits, seized this opportunity to attack the fire marshal's office in a public forum. A council member even commented that fire marshals behave like "Gestapos." He further stated that city council could easily replace fire personnel with more business-friendly civilian inspectors. This created an unfavorable view of the fire marshal's office by the public in attendance, and the point of the fee discussion

unprotected risk ■ The imbalance between risk and risk control measures.

acceptable risk ■ The amount of unprotected risk a community will tolerate as acceptable loss.

Making a Case for Fire Department Staffing

Presenting statistical information can always be a challenge for fire chiefs and fire officials. This is especially true when presenting information to decision makers who may already be leaning one way or another on a decision affecting the fire department. Often these decisions are based on economics and staffing of personnel. In Atlantic City the fire chief came to the author one day and requested me to help the administrative staff prepare a short presentation to city council on fire department staffing and why each shift required 47 firefighters on duty. The council was in an election-year budget cycle, and they were looking for cost reduction areas. The council member could not understand why so many firefighters were necessary in the six city fire stations. The challenge of the presentation was that it had to make the staffing issue clear in a short period of time to officials who understood very little about firefighting other than its significant cost in tax dollars. The committee knew that the case had to be made in a simple fashion in no more than the allotted ten-minute period.

The key element was to define the fire problem clearly and to identify why 47 firefighters were necessary to protect the city. The approach was to first enlighten them on the fire problem and how the city compared to other cities of similar size. The issue in resort communities is that one must also take into account transient populations that require fire services but are not necessarily year-round residents. We began the presentation with some easy-to-understand bar graphs on the fire problem and a comparison to state and other comparable jurisdictions. We reinforced those facts with National Fire Protection Standards on staffing practices and emphasized both resident and firefighter safety. The major question to which the council wanted an answer was "What do the 47 firefighters do, and why do we need them?" We addressed this by preparing several slides describing several fire scenarios of a single-family home, a commercial business, and a high-rise residential building. Each slide was presented with the company, apparatus, and the four firefighters assigned so the council could visually understand where the firefighters were and how companies were organized. These slides were further supported by a graphic of each fire scenario demonstrating each company and firefighter assignment on that particular type of fire. This made understanding the body count very simple, and council members could visually see with each specific alarm level the remaining companies in the city. The result of the time spent in planning and presenting the information in a logical and simple fashion convinced the council to look for savings in other areas and not the fire department. The moral of this story is that when presenting information, the program must stay on point, simplify complicated data, and be kept short.

was totally missed. The city council certainly had the authority to alter local ordinances and fees as they see appropriate, and the fire marshal's office could only impose fees that the ordinance requires and could not change fees at their discretion. The fact was that two council members, based on a dislike of fire codes that applied to them, saw an opportunity to attack an effective public safety agency. The point of this story is that political will and social support in the community can turn on a dime in each election cycle. The way this unwarranted attack was addressed was to appear at city council, politely clarify the issue, and show a willingness to work with small contractors to recommend appropriate fees to city council to amend the ordinance.

When looking at solutions to reduce fire risk, one must be aware of the community's acceptance of these solutions. Policy changes that are viewed as too expensive, burdensome, or business restrictive will bring political pressure and ultimately fail when implemented. As another example, one of the council members mentioned earlier proposed an ordinance change eliminating all permit fees from the city ordinance. This was done when the chief fire marshal was on vacation. The ordinance was stopped by several telephone conversations with the council president explaining that first, under state law, state-mandated inspection and permit fees could not be eliminated; second, the fees being eliminated would reduce the city's revenue by $275,000 per year, which offset the budgeted costs of the fire prevention bureau. The ordinance was quickly rescinded by the city council president. While most political bodies look out for the public interest, be aware that once in a while, you may encounter self-serving individuals in positions of authority. The most effective way of dealing with them is having a good grasp of fact, figures related to the issue, and the effect on increased risk.

Summary

The examination of data provides fire officials, fire marshals, and fire inspectors with useful information and knowledge about their communities. The strategic goals of a fire prevention program depend on congruent objectives that must be measurable to define the effectiveness of the program. While it is never possible to know exactly how many fires are prevented, there are core measurements that will give at least some measurable information on inspection program performance. The use of statistics assists in defining data into useful information for decision making. Fire officials and fire marshals must learn how to effectively present data in clear and concise terms to different audiences to address questions about the fire problem and their organization.

Fire officials and fire marshals must constantly be "testing the water" to see where political and societal will is within a community. The one sure thing is that it changes over time and swings like a pendulum. There will be opportunities to make positive changes to fire codes, and there will also be time periods where the *status quo* is essential. Fire officials and fire marshals need to understand that data analysis provides power and knowledge in presenting or dispelling issues that may arise in fire code enforcement. Remember that being right does not make you bulletproof; therefore, it is important to use professionalism, tact, and discretion in public presentations and forums. Keep in mind that it is hard to refute good information that is clearly presented. Professional competence brings credit to your organization.

Review Questions

1. Name and identify several sources of information that may be useful to a fire prevention inspection program.
2. Go to the U.S. Census website (http://www.census.gov/) and search your community's census information. Create a thematic map of your community showing the statistics, population, property value, education, and income.
3. Gather the realized service demands from your fire department's NFIRS reports. Construct a short presentation using the data that demonstrates the fire problem in your community.
4. Examine and identify the latent service demands in your community, list and describe the unprotected risks, and identify possible controls that may reduce the risks.
5. Describe, giving examples, the differences between descriptive statistics and inferential statistics.
6. In the following table, calculate the false alarm rates.

CITY	POPULATION	# OF FALSE ALARMS	RATE/1,000 POPULATION
A	27,409	206	
B	32,177	306	
C	14,381	122	
D	41,916	486	

7. Construct a line graph illustrating the effect of a kitchen fire suppression system inspection program that was instituted in 2006. Use the following data points:
   2006, 21 fires; 2007, 20 fires; 2008, 26 fires; 2009, 24 fires; 2010, 20 fires; 2011, 21 fires.
8. Calculate the mean number of sprinkler system activations from the following data: 2011, 3; 2010, 5; 2009, 5; 2008, 7; 2007, 4; 2006, 5.
9. In question #8 above, what are the median and the mode of the data set?
10. Determine the distribution of data around the mean in question #8 by calculating the standard deviation.

Suggested Readings

Federal Emergency Management Agency. 2002. *America at Risk; America Burning Recommissioned*. Washington, DC: FEMA.

Federal Emergency Management Agency. 2004. *Fire Data Analysis Handbook,* 2nd ed. Washington, DC: FEMA.

Hall, John Raymond, and The Urban Institute. 1979. *Fire Code Inspections and Fire Prevention: What Methods Lead to Success?* Boston, MA: NFPA.

National Commission on Fire Prevention and Control. 1973. *America Burning*. Washington, DC: U.S. Government Printing Office.

National Fire Protection Association and Fire Protection Research Foundation. 2008. *Measuring Code Compliance Effectiveness for Fire-Related Portions of Codes*. Quincy, MA: The Fire Protection Research Foundation.

Endnotes

1. John R. Hall, Jennifer Flynn, and Casey Grant, *Measuring Code Compliance Effectiveness for Fire-Related Portions of Codes* (Quincy, MA: NFPA and The Fire Protection Research Foundation, 2007).
2. Ibid., 10.
3. Michael J. Karter, *Fire Loss in the United States–2010* (Quincy, MA: NFPA, 2010).
4. *America at Risk, America Burning Recommissioned*, FA-223/June 2002 (Washington, DC: FEMA, 2002)
5. Warren E. Walker and Jan M. Chaiken, *Fire Department Deployment Analysis: A Public Analysis Case Study* (New York: The Rand Fire Project, North Holland Pub., 1979).
6. *America Burning- Report of the National Commission on Fire Prevention & Control* (Washington, DC: U.S. Government Printing Office, 1973).

Courtesy of Michael Ruley

KEY TERMS

achieved power, *p. 60*

ascribed power, *p. 60*

force field analysis, *p. 75*

link pin theory, *p. 61*

organizational culture, *p. 59*

OBJECTIVES

Upon completing this chapter, the reader should be able to:

- Explain the code enforcement system and the fire inspector's role.
- Recognize ethical practices for code enforcement officers.
- Describe the political, business, and other interests that influence the code enforcement process.

Professional Levels of Job Performance for Fire Inspectors as Cited in NFPA 1031 and NFPA 1037

- NFPA 1031 Fire Inspector I *Obj. 4.3.10 Emergency planning and preparedness*
- NFPA 1031 Fire Inspector II *Obj. 5.2.3 Investigate complex complaints*
- NFPA 1031 Fire Inspector II *Obj. 5.2.4 Recommend modifications to codes and standards of the jurisdiction*
- NFPA 1031 Fire Inspector II *Obj. 5.2.5 Recommend policies and procedures for the delivery of inspection services*
- NFPA 1037 Fire Marshal *Obj. 5.2.2 Establish personnel assignments to maximize efficiency*
- NFPA 1037 Fire Marshal *Obj. 5.2.3 Establish a strategic and operational plan*
- NFPA 1037 Fire Marshal *Obj. 5.3.7 Integrate a risk management solution*

- NFPA 1037 Fire Marshal *Obj. 5.7.6 Evaluate fire and life safety programs*
- NFPA 1037 Fire Marshal *Obj. 5.4.2 Present safety proposals to community groups*
- NFPA 1037 Fire Marshal *Obj. 5.5.4 Implement professional development programs*
- NFPA 1037 Fire Marshal *Obj. 5.5.5 Evaluate organizational professional development programs*

Introduction

The role of the fire inspector is unlike any other position in the fire department. Fire inspectors must be self-starters, be good communicators, have excellent reading and comprehension skills, and must also have good leadership skills in order to deal with the public, government, and business owners. There is an old adage that leaders are born and not made; however, leaders can be made if they are provided with the right guidance and knowledge in the initial stages of their careers as a fire inspector. Building and fire code enforcement in the protection of life and property from fire is critical in every community. Leadership in fire prevention does not come from the fire suppression side of the fire service; rather, it must be cultivated and developed within the fire prevention bureau. Every new fire inspector that is assigned to fire prevention duties will quickly realize that this job requires great individual knowledge, skills, and abilities in managing daily situations with the public that they were never exposed to in the firehouse. The fire official's and fire marshal's duty is to be a mentor and develop the next level of fire prevention professionals within their organization. To be effective, fire inspectors must start by broadening the skills necessary at the entry level to attain a level of knowledge, skill, and ability to step into their supervisor's shoes in the future. Every fire officer's goal should be to bring their personnel up to their highest potential. This journey begins by understanding the mission, vision, and values of the organization as well as the ethical principles of sound management. At each level of career advancement, the fire inspector will broaden his or her understanding and implementation of sound management and leadership skills that are being cultivated at the entry level.

The Organization

One of the most important lessons any fire official, fire marshal, or fire inspector can learn is the culture of his or her organization. Every entity, including fire departments, has an organizational culture that must be negotiated to maintain operational effectiveness. In the public sector, there is not only the culture within the fire department to deal with but also the culture of the governing body itself. The governing body may be the mayor and city council, a mayor and alderman, a township council, a board of supervisors, county executives and freeholders, or any other elected body that governs your community. The governing body's culture may shift with each election, and this can have a positive or negative effect on the organization. Fire officials, fire marshals, and chief fire officers must constantly be aware of these shifts and the potential effects they may have on the fire department's organization.

When we speak of **organizational culture**, we are talking about the core values of the fire department. These values may differ based on the geographic location of the department, the traditions of the department, and most of all, the fire department's leadership. In 1974, the "America Burning" report pointed to the need for fire departments to be proactive in fire inspections, building code enforcement, and the reduction of arson incidents in the United States. How each fire department responded to this call varied from community to community.

In examining the history of fire prevention, it is clear why fire departments were given this mission: They are the largest stakeholders next to the building occupants as far as negative outcomes that can occur once a fire starts. Although great strides have been made in improving firefighter safety, there is still no better protection than stopping the fire before it begins or ensuring that it will be of a smaller magnitude. Early fire departments across the nation established fire prevention bureaus to conduct fire inspections in specific occupancies based on the many tragic historic fires that occurred; however, they often were staffed by injured or sick firefighters who could not perform the fire suppression function any longer. This practice helped build a culture within the fire service that the fire prevention bureau was not an important function of the organization but a place for the sick, lame, and lazy. This belief also kept fire prevention bureaus from being large enough to achieve their mission effectively; it is difficult to recruit firefighters into an organization that has a negative image within its own culture. Staffing resources traditionally have been directed to the suppression of fires rather than the prevention of fires. "America Burning" indicated that for every public fire protection dollar spent in 1974, only 5 cents was spent preventing fires while 95 cents of was spent suppressing fires.[1]

The field of fire prevention has made great progress since 1974 in improving the image of fire prevention bureaus and staff; however, the mission of fire prevention must be recognized as a critical core value of the fire department as a whole. Rarely do men and women join a fire department to enter the field of fire prevention, so a key element is how effectively fire prevention services are integrated into recruit firefighter curriculums. The overwhelming culture within the fire department is that of firefighting—emergency responders—and that is the primary reason that the fire department exists. This cultural bias may lead to the activities of the fire prevention bureau being viewed as less important in the organization or as an area for members who are not really firefighters. While this bias does not exist in every fire department, it does exist. Some may even view fire prevention as a deterrent to the firefighting functions because fire prevention actually reduces the number of fires within a community.

Fire prevention administrators must ensure that the fire prevention mission is institutionalized into the larger organization culture. All levels of the organization from the newest firefighter to the chief of the fire department must recognize the importance of fire prevention's core values. If a fire prevention administrator fails to change the culture within his or her fire department, then the fire prevention bureau will always be subject to less than positive change, such as reduced staffing and reduced program financing. Without strong leadership, the fire prevention bureaus will struggle to stay on mission.

Another aspect that fire officials must consider is the fire department's role in the realm of local government. Every fire department operates under the direction of an elected or appointed government official whether it is a mayor, city administrator, public safety director, alderman, councilperson, fire commissioner, or a board of supervisors. The fire department and the fire prevention official may have to redefine their mission with each change of administrative or elected officials. It is not unusual for fire marshals or fire officials to serve under many different elected administrations, each having a different concept of what services the fire department should or should not

1

organizational culture ■ The core mission, values, and accepted norms of an organization.

provide in the community. At the beginning of each elected official's term, the fire prevention official must be prepared to give an overview of the importance of the mission of the fire prevention bureau and what services the office provides. It is important to lay this groundwork to demonstrate a professionally run organization before any issues arise, as they may with newly elected government officials. What is meant by "issues" is fire prevention inspections that place the fire department in the public eye in a high-profile manner. Business owners often make complaints about fire inspection violations, fire inspector attitudes, or permit and penalty fees they are required to pay. The mayor, administrator, or councilperson generally contacts the fire chief for an explanation to gain some resolution for his or her constituents. The fire chief usually will then ask the fire official to investigate and file a report on the particular complaint. Sometimes, complaints are legitimate, but often they are not. People naturally try to manipulate the system by going to the new elected official for assistance on fire code issues. This is especially true when the code compliance issue costs significant capital to make the improvement. The fire official must be prepared for these challenges with proper facts and a professional approach to problem solving. This effort begins with a clear understanding of mission and vision by all members of the fire prevention bureau and the fire department as well as communicating that understanding in a professional manner to the citizenry and the elected officials.

The Team Leader

Chief fire marshals or fire officials and senior fire inspectors are the team leaders of fire prevention bureaus. Leadership is the act of influencing others to perform work or to conform to a specific standard of care, such as influencing an business owner to correct fire code violations. As the leader, the fire inspector may have a position of power over the public or sometimes other members of their department. How the fire inspector uses that leadership power determines his or her success or failure in any given situation. Power in the workplace can be either **ascribed power** or **achieved power**. Ascribed power is the power and authority you have gained by position as an inspector, that is, by rank or promotion. Ascribed power in the fire service usually comes in the form of promotional examinations or assignments and appointments. This is the power you receive legally from the authority granted by your employer or the statutory authority expressed in the fire prevention code. Achieved power is power you have derived on your own through advanced education or achievements in your field, as well as through the respect of your peers. Leadership effectiveness requires both ascribed and achieved power.

achieved power ■ Power derived from knowledge, experience, or reverence by others. Achieved power is self-derived.

ascribed power ■ Power that is achieved by position or promotion from one's employer; this is power provided by outside sources.

Leaders, additionally, may have five other powers as supervisors. They are the following:

1. *Legitimate Power:* Legitimate power is ascribed. This is the power by position in the organization, or rank, and also the statutory authority granted to an individual under the fire prevention code.
2. *Reward Power:* The power to reward is also ascribed. This is the ability to reward employees for good work and behavior or to apply an acceptable variation to a fire code requirement.
3. *Coercive Power:* The power to punish bad behavior also is ascribed. In fire codes the inspector may have the ability to institute monetary penalties
4. *Referent Power:* This is power by charisma that comes from being a likable person that people will follow. Referent power is achieved.
5. *Expert Power:* This also is achieved power from your education or recognized knowledge in the fire prevention field.

In general terms, managers have at least three ascribed powers, but true leaders have all five.[2]

The fire department is considered a paramilitary organization, which means that its management model is similar to those of the armed forces. It is expected that management will be top down, orders are not to be questioned but followed, and rank automatically commands respect. Although unity of command is a very important concept in emergency work, in the field of fire prevention, which is mostly nonemergency, it does not translate as effectively. Leaders must be aware that the business office environment is markedly different than that of the firehouse. In offices, work groups usually contain both uniform and civilian personnel. This will require a more business-oriented management philosophy to be successful.

In 1961, Rensis Likert, an organizational psychologist who studied management styles, introduced the **link pin theory** of management. Likert proposed that employee-centered supervision was far more effective than job-centered supervision. Likert asserted that there are four types of management models:

1. *Exploitive authoritative system (I):* In this model, management is strictly top down with no input from subordinates. The sole effort is completion of the job with either threats or fear being used to complete the tasks.
2. *Benevolent authoritative system (II):* In this model, management is still top down with no subordinate input, but subordinates may be rewarded for efforts if management desires to do so. This is the theory of the benevolent dictator.
3. *Consultative system (III):* This management model includes consideration of subordinates' ideas but is still top management oriented.
4. *Participative (group) system (IV):* This model of management includes subordinates in the actual decision-making process and rewards them for their efforts, although major decisions are still made by senior management staff. Management has free-flowing communications with the subordinates to accomplish the tasks.

Likert viewed participative management as groups of teams of workers with the supervisor being the "linking pin" between each layer of the organization. This provides two-way flow of information freely within the organization.[3]

Douglas McGregor from the Massachusetts Institute of Technology, Sloan Business School, developed the management theory on how managers view their employees in 1960. This was known as "Theory X and Theory Y" (see Figure 4.1). Theory X concludes that managers view employees as lazy and not willing to perform work unless closely supervised. Theory Y concludes that managers think that employees are self-motivated and, given the right environment and atmosphere, would perform work with very little supervision This theory was very popular in management circles in the 1970s and 1980s. Basically, it identified two distinct management models and philosophies on how managers and employees co-exist. Robert Blake and Jane Mouton, two psychologists at the University of Texas in 1964, developed a managerial grid that expanded Theory Y (see Figure 4.2). This was referred to as the *Blake Mouton grid*. The Blake Mouton grid identified five major management styles or philosophies through the completion of questionnaires regarding supervisory work. The grid examines the manager's style, looking at the concern for people and the concern for production. The grid identifies the person's general management style based on a scale of 1 to 9 in the categories of concern for production

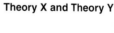

Theory X and Theory Y

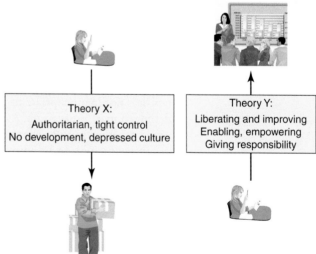

Theory X:
Authoritarian, tight control
No development, depressed culture

Theory Y:
Liberating and improving
Enabling, empowering
Giving responsibility

FIGURE 4.1 Douglas McGregor Theory X /Theory Y. *© 2002 Alan Chapman. Based on Douglas McGregor's XY-Theory. www.businessballs.com. Reprinted by permission*

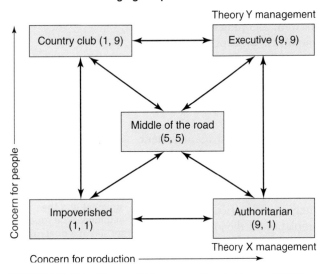

Managing People and Production

FIGURE 4.2 Blake Mouton Grid. *Based on: Blake, R.; Mouton, J. (1964). The Managerial Grid: The Key to Leadership Excellence. Houston: Gulf Publishing Company*

and concern for people, with 1 being a low score and 9 being a high score.[4]

The five Blake Mouton Management styles are as follows:

- *Authoritarian (9,1)* This manager places a high emphasis on getting the job done. The human element is not allowed to interfere with productivity. Emphasis is on what is to be done and how it is to be done.
- *Country Club (1,9)* This manager has a high emphasis on creating comfort and a friendly work environment. The emphasis of work is on the "why" and "what" related to a particular needed result. This manager lets the subordinates handle the "how's."
- *Executive (9,9)* A high morale and a high commitment to production are generally achieved by these managers. A participatory approach allows the leader to create a situation in which subordinates can take part. This develops transformational leaders who build cohesive teams where everyone owns the result.
- *Impoverished (1,1)* These managers place low emphasis on both people and work production. The leader may exert no leadership at all and plays a passive role.
- *Middle of the Road (5,5)* These managers balance the need for good morale with the need for good production. They place emphasis on the "what," "why," and "how" depending on how each individual feels. This is a compromising style of management.

In 1999, Blake and Mouton added two additional management styles to the grid:

- *The opportunistic style:* This manager exploits and manipulates individuals. Managers using this style do not have a fixed location on the grid. They adopt whichever behavior offers the greatest personal benefit at the time.
- *The paternalistic style:* This manager prescribes and guides individuals. This style was added to the grid theory before 1999. It alternates between the (1,9) and (9,1) locations on the grid. Managers using this style praise and support workers but discourage any challenges to the manager's way of thinking.

While the executive manager (9,9) is the most overall effective in organizational efficiency, it is not inappropriate to adopt other management styles in certain situations, and sometimes it may be necessary. As an example, the authoritarian method is necessary on fire grounds or other rapidly changing emergencies because of the highly dynamic changes that can occur minute by minute. Emphasis has to be on controlling the emergency. Another situation might require the country-club method; as an example, if the manager needed a fire inspector to work a late shift every week. The leader could establish the rule that the shift must be rotated to all members fairly, but also may allow the inspectors to decide who will cover each shift monthly. The leader has no strong feelings on the final outcome of who is assigned, as long as the shift is covered. The inspectors make the decision every month on who will cover the shift, which allows participation and flexibility in their schedules. In these cases, the employees can make the decision without the leader's involvement. Even the impoverished method has a time and a place when it is necessary to remain at the *status quo*. This method is desirable after major changes in the organization to allow the changes to institutionalize. Learning leadership skills begins at the fire inspector level. In their work, fire inspectors must deal with complex code enforcement issues involving people, processes, and structures. A fire inspector must be capable of communicating effectively with plant managers, supervisors, and other professionals as a leader on an equally competent level. Leadership skills are necessary job performance requirements under NFPA 1031 at fire inspector levels II and III, and are absolutely necessary for fire marshals under NFPA 1037 who administer fire code enforcement programs. Good leadership skills begin with an understanding of self-evaluation and identifying one's dominant management style so as to be able to adjust one's style to fit different situations in the workplace or in the field.

Mission and Vision

Mission and vision are two critical components for effective management in the fire department and fire prevention bureau. It is critical to any organization to have a clear understanding of what business it is in and where it is going. Mission and vision must be communicated from the chief of the organization to the newest member. In the fire service, it often is believed that the mission is clearly understood by every member because of the paramilitary structure of the fire department itself; often, however, the mission and the vision have not been effectively communicated to firefighters. Frequently, newer firefighters complain about doing fire inspections or other activities related to fire prevention education. These complaints are largely based on a lack of understanding of the fire department's mission. Every fire department has a mission statement, but it is how well the goals and objectives of that statement are institutionalized into the organization that breeds success and brings credit to the organization. Mission statements are more than just words—they need to be the actions that every member of the fire department carries out every day. The mission statement should not be a complicated document, but should express the organization's core beliefs and values in a clear and concise format. The statement should reflect the goals of the organization and the level of customer service to be provided. The mission statement should be prominent in the training of firefighters. As an example, the Los Angeles Fire Department ties its mission statement to three areas; it focuses on the core values of the organization, both externally to the community and internally to its members, and relates that to its mission goals, including the prevention of fires. Additionally, the Philadelphia Fire Department displays a series of flags in its training academy to instill the fire department's core values. The flags display words such as "dedication," "courage," "service," and "knowledge," and all recruit firefighters go through a training session on the meaning behind each flag and must discuss what that term means to them. It is a great way of instilling core organizational values in new recruits. People want to belong to organizations that have strong core values and also value their individual contributions to the organization. Mission statements need to be about "doing the right things all the time," even when the public or the supervisor is not watching. Those unnoticed and unplanned deeds often bring credit to both the individual and the organization over time.

It is important for organizational leaders to communicate the mission statement through their vision. Vision is looking past the horizon of today and determining where the organization is going in the future and how it is going to get there. Vision is instilling core values in your members by rewarding the right actions and correcting the wrong actions. In fire prevention, the landscape is continually changing, whether it is political, economic, or technological. The leader's role is to try to anticipate what is coming in the future and determine how to adapt the organization to accomplish the organizational goals. Often, changes are forced by outside influences beyond one's control. Fire service leaders must constantly reexamine fire programs to stay ahead of the change curve. One thing is certain in your fire service careers, and that is that change will happen! As a leader you must both embrace and direct change or suffer the consequences when it negatively affects your organization. Fire officials and fire marshals need to stay cutting edge on all issues that can confront their organizations. They must look for and anticipate the win-win solutions when confronted with conflicts, both within the fire department and externally. Good organizational skills don't just happen—they are the culmination of individual efforts and commitment to a set of core values and beliefs. According to Michael McCann, there are seven core values[5] necessary to be effective and successful as a leader. Let's explore each one.

MAINTAIN YOUR INTEGRITY

Integrity is your most important asset. Good leaders are truthful and honest, and they admit when they make mistakes. Members of your fire department watch a leader's actions to see if they reflect his or her words. Fire prevention work often presents situations that can impair one's integrity. You must always be on guard to prevent this. You must be aware of situations that may question your integrity and avoid those situations. Sometimes, you may know the person who is in violation of the fire code personally, in which case you need to remember that "business is business" and everyone must be treated the same under the rule of law. You need to be fair in your decisions and withdraw your involvement when there is a conflict of interest in any matter. Ethical management is a priority to effective leadership, so you must protect your integrity at all times.

DECLARE YOUR VISION

You must communicate your vision to upper management and every member within your organization. A great opportunity to do this is when the annual reports are produced and reviewed each fiscal year. The annual report should not just be a look in the rearview mirror to see what was accomplished but a challenge to direct change in the upcoming year. Goals and objectives should be redefined to keep the mission on task. Problem areas should be identified and positive solutions created to address those concerns. A good leader needs to view the big picture and adjust the details as change occurs, and not get bogged down in the nuts-and-bolts issues that are better solved by others. Declare your vision and be the spark plug that inspires others to help you achieve it.

SHOW UNCOMMON COMMITMENT

Commitment means taking on the hard tasks first. The leader of an organization should be the first one into work in the morning and the last one to leave at night. If you lead by example, people will naturally follow you. Do not avoid the hard jobs during the day, and go the extra mile when you really have to. When your subordinates see that the boss is committed to a project, they will follow more enthusiastically.

EXPECT POSITIVE RESULTS

Positive attitudes yield positive results. As a leader, you must be the drumbeat of your organization. It is always easy to be negative toward a particular organizational change, but you have to be able to adapt and make a positive result out of what may be a negative situation. Positive expectations are contagious; they make the workers around you feel that everything is going to be ok, even in difficult situations. When you think about the people in your life you would most like to emulate, most will have the trait of always being positive and expecting that attitude from others.

TAKE CARE OF YOUR PEOPLE

In any organization, the most important asset is the worker. Leaders and managers must accomplish work or production through the efforts of other people. This can be effectively accomplished only if you take care of the people who work for you. This lesson ties back to your integrity; you need to defend your subordinates and reward them when they do good work, and you need to mentor and encourage them when they need improvement. Effective leaders should treat their co-workers and subordinates the way they would like to be treated, with respect and concern for their well-being.

DEDICATION TO DUTY

Being dedicated to your duty and position is an important lesson in leadership. Organizations depend on leaders when situations arise that need their expertise. It is not unusual for the chief fire marshal to be called at 2 or 3 a.m. by a battalion chief or deputy chief needing assistance with a fire code situation in the city. The fire marshal must realize that in a leadership position these calls will happen often when they are not on duty. His or her support and assistance to the caller helps build relationships, especially with new or acting officers who may be unsure of how to address a dangerous code problem. It is important for callers to know that no matter what time it is, they can contact the chief fire marshal and not be criticized and that they will receive assistance and support.

LEAD FROM THE FRONT

It is an old saying that you cannot lead from the rear. As a leader, you are expected to be the face of your fire prevention organization. When your organization is praised, the leader needs to share that praise with the inspectors and co-workers, as it is their efforts that bring the credit to the organization. When criticism comes your way, the leader needs to take it for the entire organization. While these principles seem relatively simple, they can be very difficult in given situations. People follow leaders who are not afraid to roll up their sleeves and get dirty. Never ask a member to do a job that you would not do. Leaders must always be willing to jump in and assist in performing difficult tasks, whether they are dealing with a building code issue or investigating a fire scene.

Fire inspectors, like firefighters, should strive to improve their knowledge and develop leadership skills. Understanding management theory and applying the core principles of leadership will help focus the fire inspector in developing the skills necessary to advance in their careers in the fire prevention field. As a famous president once said:

> "If your actions inspire others to dream more, learn more, do more and become more, you are a leader."
>
> —John Quincy Adams

Organizational Evaluation

As a leader of a fire prevention organization, you must evaluate your organization's effectiveness on a regular basis. Fire prevention is a multifaceted discipline that falls into four major categories: fire inspections, plan review, fire investigations, and fire safety education. Each element of the fire prevention bureau must be evaluated based on assigned personnel, resources, and the activity levels required to maintain operational efficiency. The fire inspector is by far the most important asset in the fire prevention organization, so let's focus on the inspector's needs first. Depending on the size of the fire prevention organization, each task may be assigned a supervisor and a specific number of personnel to carry out the mission. In smaller organizations, an individual fire inspector may be required to perform all of the program tasks in a more generalized approach. Large staffs may allow for specialization, and the fire marshal or fire official gains the benefit of fire inspectors becoming very knowledgeable and proficient in a specific aspect of the fire prevention bureau. Many large departments have specific units for inspection of high rises, healthcare facilities, hazardous materials, fire safety education, and fire investigation. This staffing method allows for more targeted training of fire

5

inspectors in specific areas. Smaller staffed fire prevention bureaus do not have the ability to specialize, and so the fire inspector in these offices becomes a "jack of all trades." While greater exposure helps the inspector to become well rounded, he or she generally lacks the in-depth knowledge of the specialist. Fire marshals or fire officials must try to maintain a balance between specialized and general staffing to keep the organization effective. One way to maintain this balance is to periodically rotate fire inspectors through the different assignments within the bureau to expose them to each area of fire prevention. Fire officials must understand, however, that not every member is capable of performing equally in each task area of assignment. Another consideration in the rotation of staff is the inspector's experience and skill level. Very experienced fire inspectors should be maintained as mentors and trainers in each of the bureau sections. The fire marshal or fire official must ensure that a high degree of efficiency is maintained and only move personnel when other fire inspectors are sufficiently skilled to replace them in the section. Knowledge and skill development should be predicated upon nationally recognized standards. The NFPA 1031 standard for professional qualifications for fire inspectors should be used as the benchmark to measure each fire inspector's skill level. The NFPA 1031 standard divides fire inspector skill levels by job performance requirements that are based on knowledge, skills, and abilities that the inspector must attain at each specific level. Many states also have training and certification requirements that must be followed, and these generally will be based on the NFPA 1031 standard or some other adopted professional qualification for code enforcement programs. Each state may adopt a different version or year of the 1031 standard; therefore, check your local jurisdiction for its particular requirement.

The NFPA standards that apply to fire prevention personnel are as follows:

NFPA 1031   *Standard for Professional Qualifications for Fire Inspector and Plan Examiner, 2009 Edition*
- Fire Inspector I
- Fire Inspector II
- Fire Inspector III
- Plan Examiner I
- Plan Examiner II

NFPA 1033   *Standard for Professional Qualifications for Fire Investigator*
- Fire Investigator

NFPA 1035   *Standard for Professional Qualifications for Fire and Life Safety Educator, Public Information Officer, and Juvenile Firesetter Intervention*
- Public Fire and Life Safety Educator I
- Public Fire and Life Safety Educator II
- Public Fire and Life Safety Educator III
- Public Information Officer
- Juvenile Fire Setter Intervention Specialist I
- Juvenile Fire Setter Intervention Specialist II

NFPA 1037   *Standard for Professional Qualifications for Fire Marshal, Chief Fire Marshals*
- Fire Officials
- Fire Marshals
- Fire Prevention Administrators

The NFPA professional qualification system builds each higher level of fire inspector I certification upon the previous level's job performance requirements and demonstrated skills and abilities. Therefore, a Fire Inspector level III must meet all of the requirements for levels I and II before applying for level III certification. As an example, a new fire inspector would meet the requirements for Fire Inspector level 1, which are very basic fire inspection techniques. Level I includes no plan review requirement and basically involves identifying fire hazards, preparing inspection reports, and general verbal and written communications skills. Fire Inspector level II requirements apply to intermediate fire inspectors and include skills such as code interpretation, occupant load verification, evaluation of life safety and emergency plans, occupancy classification, and some advanced fire protection system knowledge. The Fire Inspector level III requirements are geared to advanced or senior level fire inspectors and include such skills as permit inspections, initiating legal actions, recommending program budgets, evaluating evacuation plans, and ensuring compliance with construction documents. The next two levels of NFPA 1031 are plan reviewer and plan examiner. These skills require a more in-depth knowledge of the building codes and general building construction methods as well as the installation and testing of fire protection systems and equipment. Additional NFPA professional qualification standards apply to specialized areas of fire prevention such as fire investigators and life safety educators. The NFPA defines the skills required for fire investigators in NFPA 1033 and for life safety educators in NFPA 1035. These skills again are directed to specific specialized knowledge necessary for a fire inspector to perform these specialized functions.

Chief fire marshals' knowledge, skills, and abilities are defined in NFPA 1037, *Standard for Professional Qualifications for Fire Marshal*. The NFPA 1037 fire marshals standard is geared to chief officers and fire officials in charge of fire prevention bureaus and describes the job performance requirements and requisite knowledge, skills, and abilities needed for the bureau supervisor. These skills include establishing strategic organizational goals and objectives, risk identification, record systems, media strategies, managing inspection programs, code interpretations, handling citizen complaints, budgeting, managing education programs, and much more.

Along with regular fire departmental training and evaluation, the fire inspector may be required to meet other certification or accreditation measures established within a state or by the local jurisdiction. These may include completion of state certification and examination program, or they may involve meeting the requirements of an outside accreditation agency, such as the International Association of Fire Service Organizations Pro-Board. Many states will also require the completion of certification testing by code writing organizations such as the *International Code Conference* or *ICC*. The ICC certification tests are based upon the user's knowledge of code application and general fire inspection principles taught through both the International Fire Service Training Association, or IFSTA, and the NFPA. Here is a sample breakdown of questions required for certification under the ICC certification process:

Fire Inspector I

60 multiple-choice questions
Open book—2-hour time limit

Content Area % of Total References

General Inspection Administration 15%, *2006 International Fire Code®*
General Provisions for Fire Safety 48%, *2006 International Building Code®*,
    chapters 1–10
Occupancies 20% *Fire Inspection and Code Enforcement (IFSTA)*, 6th edition
Regulated Materials and Processes 17%
 Total 100%

Fire Inspector II

50 multiple-choice questions

Prerequisite Certification:

Fire Inspector I (must be current) Open book—2-hour time limit

Content Area % of Total References

General Inspection Administration 16%, *2006 International Fire Code*®, including
    Appendix B
General Provisions for Fire Safety 36%, *2006 International Building Code*®,
    chapters 1–10
Occupancies 24%, *Fire Inspection and Code Enforcement (IFSTA)*, 6th edition
Regulated Materials and Processes 24%
 Total 100%

Fire Plan Examiner (equivalent to NFPA 1031, Fire Plan Examiner I and II)

60 multiple-choice questions
Open book—3-1/2-hour time limit

Content Area % of Total References

Administration 5%, *2006 International Fire Code*®
Occupancies 15%, *2006 International Building Code*®
Hazardous Materials 20%, 2002 NFPA 13, *Standard for the Installation of Sprin-
    kler Systems*
Fire Protection 35% 2002, NFPA 72, *National Fire Alarm and Signaling Code*
Egress and Safety 25%
 Total 100%

FM Certified Fire Marshal

60 multiple-choice questions

Prerequisite Examinations—Current certifications for
Fire Inspector II
Fire Plans Examiner (equivalent to NFPA 1031 Fire Plans I and Fire Plans II)
Legal/Management Module
Technology Module

Content Area % of Total References

Public Information and Media Relations 5%, *2006 International Fire Code*®
Fire Origin Determinations 12%
Fire Cause Determination 7%
Fire Scene Documentation 6%
Evidence Retrieval and Chain of Custody 6%
Interview Techniques and Suspect Questioning 9%
Criminology 10%
Due Process 6%, Local Statutes, Fourth and Fourteenth Amendments
Fire Investigation 10%
Fire Play, Fire Setter, and Arson Prevention 7%, *Introduction to Fire Origin and
    Cause (IFSTA)*, 3rd edition
Public Relations 5%, *Fire and Life Safety Educator*, 2nd edition (IFSTA)
Development of Lesson Plans, Fire Material, and Education Programs 12%, *2006
    International Building Code*®
Data Analysis 5%, 2004 NFPA 921, *Guide for Fire and Explosion Investigations*
 Total 100%

Properly trained and certified fire officials and fire inspectors enhance the organization's credibility and professionalism, both within the fire department and in the community. Proper training and continuing education programs in the field of fire prevention are the foundation on which effective organizations are built.

The next area of organizational evaluation that requires examination is the types of programs and the consistency of delivery of those programs by the fire inspectors. Each community will have different fire safety needs that should be based on community risk assessment. The data analysis performed will help the fire official or fire marshal in applying resources to programs for specific community risk reduction. Generally, the majority of staff resources will be directed to the fire inspection program. Fire inspection is the area where significant impact can be made in both the elimination of fire hazards and reduction of fire starts. Fire inspection programs have the added benefit of keeping firefighters safer in emergency response. Fire inspection is also the most labor-intensive area of a fire prevention program. Conducting plans review of fire protection systems will also be a high priority as the approval of building permits prompts new construction and community growth. Sufficient staffing resources must be applied to ensure that plans are properly reviewed and inspections are timely to avoid delays in construction. Plans review demand tends to be cyclical depending on the growth and development of the community. Fire inspections on new construction also tend to be a high-profile activity, as political bodies want to see new growth to increase the tax base and revenue. Increased growth represents job creation in the community. The plan review process is conducted to ensure compliance with building and fire code standards, but reviews must be accomplished in a manner that doesn't generate complaint to the governing body that construction is being stifled by the permit, plan review, and inspection process. This may mean that the fire marshal or official must provide additional staff resources as necessary to complete these reviews and inspections in a timely fashion. It is equally important that all personnel assigned to the plan review function are knowledgeable and technically competent in the review and inspection process. Note that not all fire departments will have plans review responsibility; in many cases, inspections and plan reviews are assigned to municipal building departments or may go to other outside third-party agencies based on the complexity of the structure. In some states, offices in state government may reserve plan review responsibility for certain specific types of construction including casinos, schools, nuclear plants, refineries, and other complex buildings. The state further regulates the local agencies' plan review and inspection responsibilities based on licensing of the agency and the local inspectors' certifications. The licenses are broken down into residential commercial specialists, industrial commercial specialists, and high-rise, high-hazard specialists. Only agencies with the proper licensed staff can review and inspect those levels of construction. If the fire department is not involved in the plans review and inspection process, it should maintain close contact with the appropriate agency that is conducting the fire protection inspections. Fire officials should request copies of acceptance test reports and other appropriate documentation for their inspection files; these will be needed for future fire inspections and the maintenance aspects of the fire code.

The fire marshal or fire official must also provide trained staff to conduct fire investigations. Although many departments may not have the legal authority to conduct criminal investigations for arson, it is important that all fires be properly investigated to determine the cause and origin of the fire. The investigation of fires helps determine the adequacy of fire code requirements and identifies the need to establish public fire safety education programs targeted to particular issues like cooking or the use of electrical space heaters. Fire investigation normally is initiated at the fire company level by either the deputy chief, battalion chief, or company-level officer responding to the fire call. The assistance of a fire investigator may be requested in order to determine the origin and cause of the fire. Fire prevention bureaus charged with this duty need to ensure that the personnel assigned to this responsibility have adequate training or certifications complying with NFPA 921, *Guide for Fire and Explosion Investigations*. Fire investigators may be required to be certified through other local or state agencies in criminal justice to ensure the continuity of the investigation in case the fire scene turns out to be a crime scene. In criminal investigation, special care must be taken to preserve and collect evidence and document the fire scene. In any case, all fires will require investigation to ensure that fire code compliance requirements are working, fire public education gaps are being identified, and the crime of arson is being deterred.

Lastly, fire officials or fire marshals need to evaluate public fire safety education programs they deliver. Usually, the fire education safety programs receive the least amount of resources in fire prevention bureaus. This is generally because the other areas of fire inspections, fire investigation, and plan review generally are required by statute and have specific time constraints to meet statutory regulations. In fire prevention codes, public fire safety education is more implied in the regulation than required by statute. These implications usually come in the code sections dealing with changing human behaviors, such as fire drills, fire evacuation planning, smoking materials, housekeeping hazard identification, and so on. A fire inspector simply citing these types of violations will not ensure code compliance—additional education programs must follow to close the loop in changing people's attitudes and behaviors toward fire safety. Because most fire departments in the country become very active with the public during fire prevention week in October, that initiative rarely is maintained throughout the year. A well-trained fire safety educator can be invaluable to the fire prevention effort in delivery of targeted programs to change community and human behavior. The combination of a general fire safety presence in the community with targeted programs raises the value of fire prevention for its citizens. A public education program allows the community to give feedback to the fire department on safety concerns and provides a visible fire department to the public. For example, a public fire educator may be assigned to attend community group meetings of homeowner associations and other residential community groups. He or she can actively participate in fire extinguisher training and Community Emergency Response Teams (CERT) training programs with businesses. Fire inspectors are an ever-present entity in the local public and private schools and day-care centers and can become a valuable resource. Public fire safety education programs are a high-profile method of connection to your community and help provide support for the organization at the grassroots level. These programs should not be overlooked when managing fire prevention staff resources.

The last area of evaluation is the clerical staff and physical plant. In many cases, fire inspectors prepare their own fire inspection reports, or computer systems may be used to generate reports; however, a skilled and dedicated office staff is still necessary to ensure program efficiency. A record-keeping system of hard files must still be maintained, and documents must be recorded and filed to maintain an accurate public record system. Support staff must be capable of handling the public on the telephone and completing permit application forms, while recording citizen complaints and other daily requests that come into the fire prevention office. Staff must maintain accounting systems to accept and deposit permit and inspection fees and violation penalties that require payment. All financial documents must be properly processed and expedited effectively, as fire inspections are time sensitive. Clerical staff must have adequate workspace and a work environment conducive to good performance. The clerical staff has to be of adequate size to handle the workloads generated by the fire inspection workforce on a daily basis. The fire marshal or fire official must evaluate the record-management systems and streamline the system to provide work efficiency. Workspace efficiency also is important to expedite the workload. Offices need to have filing cabinets and sufficient file room space for record management. Equipment storage areas and adequate space for photocopiers, fax machines, and other resources must be provided. All workspaces must be of sufficient size to handle the assembly of paperwork for processing or other required activities. The office environment affects the employees' attitudes and ability to work efficiently and should not be overlooked.

It is the leader's job to see that others complete their work efficiently. To accomplish this task, it is necessary to have the proper tools for each job. Fire marshals and fire officials should evaluate clerical tasks and the necessary tools and equipment needed to make the job simple, safe, and efficient for the workers. Fire inspectors will need vehicles, lights, clipboards or portable computer tablets, reference libraries for codes and standards, efficient work-spaces with proper lighting, and specialized inspection equipment for difficult jobs. Tools and equipment are an investment in the total fire prevention system that cannot be over-looked. During budget preparation, equipment desires have to be identified not as luxuries but as necessities for the organization's efficient and effective operation. Fire marshals and fire officials should gather input from the fire inspectors, who are expected to use the equipment, to ensure they have employee buy-in. Often, the fire department purchases new equipment without getting input from firefighters, and the tools or technologies are ignored, discarded, or both because the users do not like them or they simply do not work very well. In fire prevention, this is especially true when purchasing computer technologies—if the users do not have sufficient input in the investment, they will not use the technology effectively, and it may never be implemented properly. Employees are more open to change when they are solicited for ideas and management implements some of those ideas. When members feel more valued, they generally work harder, and change is easier to accomplish.

Preparing for Change

Change is unavoidable; it can occur slowly over time or suddenly due to some outside pressure or force. Change occurs with technological advances and alters the way we think and perform work. Many examples can be given of where technology changed the workforce. In 1913, the Ford Motor Company employed more than 13,000 people in its Dearborn, Michigan, plant. Today, the same plant employs 3,000 workers and is far more efficient.[6] The same was true in the steel industry that once provided thousands of jobs but now has very few to do the same amount of production work. These workplace changes occurred because of improvement in technologies and machines that allow better performance of repetitive tasks with fewer product defects. We can foresee that at some point buildings will have the capability of self-inspection through computer technology. As an example, addressable smoke detectors can automatically adjust their sensitivity settings when they get dirty, and they notify the fire alarm control panel that service is required. Automated cameras can monitor building areas for violations and overcrowding. We already see cameras replacing police officers and issuing tickets at intersections for motor vehicle violations. Technology will eventually become cost effective enough to serve such purposes and change the industry forever. The question to be answered is what changes are coming to fire prevention organizations and how are we going to adapt to those changes?

Change can be a welcoming breath of fresh air, or can create apprehension and concern in people's minds, depending on the type of change that is anticipated. It is helpful for fire marshals and fire officials to understand the process of change. Peter DeJager, a change management speaker and consultant, wrote for Associate Xpertise Inc. a publication entitled "The Canadian Association" in January 2004. Mr. DeJager explained that there are three distinct categories of change:

Type I Change: *That which is done to us.*
Type II Change: *That which we do to ourselves.*
Type III Change: *That which we do to others.*

Mr. DeJager pointed out that Type I change is the most difficult to accomplish because it runs counter to human nature. People like to be in control of their own destinies and they also like to feel they are part of the change process. As the chief administrator in a fire prevention bureau, we often must make changes in work rules, staff placement, or program performance. Because we are making the changes, we view them as Type II or Type III changes because they are our ideas and we have already rationalized and accepted them. The staff, however, may view these changes as Type 1 and may resist the changes or make them more difficult to implement. This is an important concept to understand when the changes being made are significant and directly affect the subordinates such as a shift

A management challenge for fire marshals or fire officials always arises when new technologies such as handheld computers are introduced into the workplace. The challenge for the manager is how to get everyone on board with the change and how to get them to accept and work with these technologies. It is difficult because human nature is generally resistant to change, and personnel may be taken out of their comfort zone. In the future, implementing computer technology may get easier as younger fire inspectors may likely already be accustomed to technologies they have grown up with. The challenge is with the older senior fire inspectors who have used pencils and paper for years and are now confronted with a computer tablet and understanding a whole new language related to computers. Often these individuals will say that the equipment doesn't work properly or isn't fast enough and that they can do it quicker using the existing method. Try not to discount these types of complaints; try to get to the core issue with these individuals and figure out how to address it or make the technology work better for them. Keep in mind that sometimes their concerns are legitimate and may indicate some software or hardware performance problems or adjustments.

The Atlantic City Fire Department began using handheld computer technology in 1994 and has gone through two technology upgrades in both software programs and hardware. These changes are never easy, especially for a mixed cadre of talent from those with good computing skills to those who ask, "What is software?" To bridge these gaps it is important for the fire marshal or fire official to include all of the personnel, especially secretarial and support staff, in the analysis and implementation of the technology product. When individuals having difficulties are included in the discussion and possible solution they can see that their recommendations are not being dismissed and they are more likely to work harder to implement the technology. In implementing any type of technology change the manager will have a better experience if he or she includes the users from the onset of the project, listens to and addresses their concerns, and makes them feel they are an important part of the process and not a victim of it.

reorganization that moves staff members, changing paperwork procedures, or implementing other fire department directives from the fire chief. In order to effectuate the change and have it accepted, the rest of the staff must be assisted to view the change as Type II. To accomplish this, discussion and input must be allowed into the particular issue or implementation plan. Fire officials must understand that changes that occur are entirely out of one's control and those changes will have to be made regardless of disagreement. Any changes cause some level of upheaval in the organization, but eventually it subsides after emotions settle and the change becomes institutionalized and accepted policy. Leaders also must ensure that the change is necessary. All too often, procedures are changed rapidly without regard to the impact on every component of the organization. This happens frequently in fire departments because the larger component is fire suppression as opposed to fire prevention. A fire chief may implement a staffing rotation that works for the firefighters and company officers, but has a negative impact on fire prevention staff. The fire marshal or fire official must actively discuss these situations with the fire chief to avoid the negative impact. Unnecessary changes need to be guarded against. Finally, in internal organization changes, if you have been successful with a process or procedure currently in place, do not change it for the sake of making change. Keep in mind that fire inspectors may change things because they have defined a problem and are trying to solve it. If you wish to implement such changes, you need to define the problem clearly among the staff so they can clearly see the solution.

Organizational Development

Organizational development (OD) is the improvement of organizations through planned actions that make the organization healthier in fulfilling its mission. Organizational development may be performed by outside OD consultants to improve profitability, efficiency, morale, team building, and conflict resolution or to solve other organizational problems. It is also helpful in understanding how an organization operates.

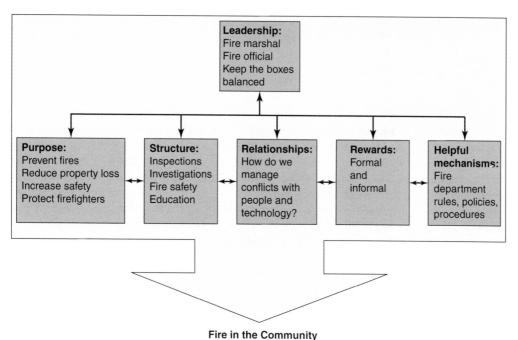

FIGURE 4.3 Weisbord, Six-Box Organizational Model. *Based on: Marvin R. Weisbord, "Organizational Diagnosis: Six Places to Look for Trouble with or without a Theory," Group & Organization Studies 1,4 (December 1976): 430–447*

In 1970, an organizational design consultant, Marvin Weisbord, developed a six-step model to diagnose organizational effectiveness (see Figure 4.3). The model describes six areas where things must go right for the organization to be successful.[7] These six areas are as follows:

1. *Purpose:* The organization's members are clear about the organization's mission, purpose, and goals.

   Goal fit: How appropriate is the organization's goal to its environment?
   Goal clarity: Are goals clear enough to provide guidance to organization members?
   Goal agreement: To what extent do organization members share the organization's goals?

2. *Structure:* How is work divided up in the organization? Is there an adequate fit between the organization's purpose and the internal organizational structure?

   Draw the organization chart.
   Determine the dominant type of structure (i.e., functional, project, program, or hybrid).
   Assess the rate of change of the environment, technology, and departments.
   List issues of the formal and informal systems.

3. *Relationship:* Describe the relationships between individuals, units, or departments that perform different tasks, and between the people and requirements of their jobs.

   Organization members (peers and manager/staff):
   Units executing different tasks:
   People and technology systems, equipment, and methods.

4. *Rewards:* The consultant should diagnose the similarities between what the organization formally rewards or punishes employees for doing.

   What the organization needs to reward.
   What the organization pays, both in real terms and psychologically.
   What circumstances make the organization member feel rewarded or punished.

5. *Leadership:* Should balance the issues among the other boxes.

6. *Helpful mechanism:* Helpful mechanisms are policies and procedures necessary to survive. These include planning, control, budgeting, and information systems that help organizations accomplish work.

Assist in the coordination or integration of work.
Assist in monitoring the organization's work.
Help deal with issues from scanning and diagnostics activities.

The six-box model shows the interrelationship of the five work components with leadership at the center. The idea of the model is to evaluate the organization and the interrelationships within each work component to help identify changes or improvements to the organization. The examination of the formal and informal structures in each work component box identifies congruency within the organization. Organizational problems arise when the formal structures are far removed from the informal structures. As an example, if a policy is in place that requires a certain fire inspection procedure to be followed, but the fire inspectors state that it is not enforced or enforced only in certain inspections, that signals incongruence in the inspection procedure. It is the difference between the way a fire inspection is supposed to be done (*formal*) and the way it is really done by the fire inspector (*informal*). Analysis by organizational development consultants using special questionnaires and reviewing policy and procedures can help narrow any inconsistencies between formal and informal policies and procedures. Once the analysis is performed, problems that were identified can be corrected with the implementation of an action plan to improve the organization's effectiveness.

Dealing with Problems and Issues in the Workplace

Dealing with problems or issues in the workplace is one of the most difficult tasks that the fire marshal or fire official must address. Problem solving takes some analysis and understanding of human behavior as well as knowledge of technical issues. Sometimes, technical issues are far easier to resolve than employee issues. To this point, we have discussed the qualities of leadership as vision, integrity, ethics, and providing a comfortable and efficient work environment. Now we must consider issues that arise in a workplace that are difficult and will require problem or conflict resolution. Issues that may arise in fire prevention usually occur in two general areas: internal to the organization, and external from the surrounding environment. Internal issues usually consist of personnel, process, and communications problems. External issues usually are personnel driven, technical, communications, or political in nature. In any case, the fire marshal or fire official will have to resolve the issue and provide guidance and a solution to maintain organizational efficiency.

A key objective for fire marshals and fire officials to pursue is not to make decisions in haste, and to remember there are two sides to every story—and sometimes the truth lies in between. The need to gather all the facts and to analyze them before formulating a solution is extremely important. Decisions made in haste or under pressure usually turn out to be bad decisions. You should avoid making decisions in moments of anger or heightened emotional states. There is nothing worse than making a decision and then immediately changing it because it was not vetted thoroughly. The other side of decision making is that you should not procrastinate in making a decision for long periods of time. Procrastination shows indecisiveness and avoidance of dealing with difficult issues.

To help with decision making, you can use the six "D's" of good decision making:

1. *Decisiveness:* Make good decisions, analyze your options, and be decisive.
2. *Deliberation:* Perform a careful analysis, and deliberate actions for 24 hours. It is a good idea to sleep on decisions if you can; it often clarifies them.
3. *Details:* Employees can be upset by small, insignificant details. Keep this in mind and ask for input when appropriate. Be flexible in the small details if it resolves the larger problem.

4. **Determination:** Be determined to do things right the first time. Changing your mind immediately after making a decision shows you have not analyzed the problem sufficiently.
5. **Diligence:** Be careful and purposeful in your decision making. Consistency in decisions builds trust and support from your staff.
6. **Discipline:** Training that develops self-control, character, orderliness, fairness, and efficiency can help you make better decisions.

Decision making is a learned art; accept that some decisions, no matter how hard you try, can still be wrong. Just remember, if you own it and it is proven wrong, fix it. Learn from the mistakes you make, and do not repeat them.

Developing Active Solutions

Problems are those bumps in the road that keep people from attaining their goals. The ultimate goal of any decision is to move over or around the bump and continue on a path to success. Active problem solving should look at the positive, negative, and interesting aspects of each problem and try to attain a win-win solution. The use of simple problem processing methods can help:

1. Place problems in a priority of order and tackle the hardest ones first.
2. Look for alternate solutions. Place them in an action plan and see if they work. Keep in mind that as you gain more information, you can alter the solution.
3. Perform a **force field analysis** of the problem. Force field analysis was developed by Kurt Lewin[8] to examine the balance between driving forces and restraining forces. This helps to identify the restraining forces and to identify possible solutions in the tug of war around an issue (see Figure 4.4). If we increase the driving forces and decrease the resistance of the restraining forces, change will occur. If we do not, the change stays in equilibrium.

   force field analysis ■ (Karl Lewin) Analysis of driving and restraining forces that keep change from occurring within an organization.

4. Negotiate agreement between the enabling and restraining forces identified in the force field analysis; the best solutions are to weaken the restraining forces to move the process forward. Lewin noted that increasing the enabling forces generally will produce more resolution in the restraining forces to keep the *status quo.* He equated this phenomenon to Newton's law of physics—for every action there is an equal and opposite reaction.

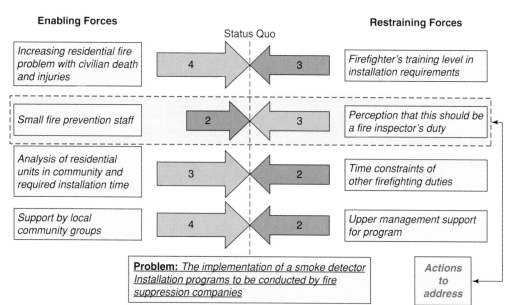

FIGURE 4.4 Lewin Force Field Analyses. *Based on: Lewin K. (1943). Defining the "Field at a Given Time." Psychological Review. 50: 292-310. Republished in Resolving Social Conflicts & Field Theory in Social Science, Washington, D.C.: American Psychological Association, 1997*

5. Establish a task force to produce the action plan; getting team involvement reduces the friction of change. Test the solution to see if it works. Often, implementing the policy as a trial enables employees to adjust to it or identify flaws that need to be addressed further.
6. Analyze the results of the solution to determine if the problem was solved.
7. Categorize the remaining problems to identify where they are best solved:
    a. Problems you can address.
    b. Problems that require others to address them.
    c. Problems that you must live with that are beyond the organization's ability to control.

Fire marshals and fire officials should not try to solve problems in a vacuum. Use other chief officers as a sounding board for ideas and possible solutions, and don't forget the team approach. Consensus building can be very effective when applied through team building. As an example, in the Battle of Gettysburg in the Civil War, General George Meade was placed in charge of the Union Army of the Potomac three days before the battle. The first two days at Gettysburg, the Union army was pushed back to Cemetery Ridge by the Confederate army of Robert E. Lee.[9]

General Lee had never lost a battle. On the night of July 2, 1863 General Meade called a meeting of all his generals to vote on whether to stay and fight Lee or to retreat to better ground. Many historians viewed Meade's decision to vote as a weakness in his leadership; however, that vote allowed the new general to build consensus with his team. They all agreed to stay and fight. The result was the bloodiest battle of the Civil War and Lee's first defeat. That vote by Meade was a decisive step in changing the course of the war and ultimately saving the Union. Do not be afraid to team build; your power and authority are not diminished but strengthened in the eyes of your team.

Effect on Firefighter Safety

#3

One might ask, why is firefighter safety included in a discussion on organizational effectiveness? The answer is simple: because firefighter safety, just as the general public's safety, needs to be considered in the activities performed in fire prevention. The fire prevention mission is twofold:

1. To prevent the fire from ever occurring by conducting complete and thorough fire inspection and code enforcement practices, and by changing the public's fire safety attitudes and behaviors. Lower fire rates translate into reduced firefighter deaths and injuries.
2. Ensuring that if a fire does occur, all of the building and fire code requirements are in place to ensure that the fire will be contained or controlled to the point that firefighters can accomplish their tasks with reduced exposure to uncontrollable hazards.

The organizational message for fire inspectors has to ensure that the building intelligence issues and hazards to firefighter safety are being recorded and relayed through proper communications. Fire code definitions usually require that situations that interfere with fire department operations be identified as fire hazards and corrected. The fire marshal, fire official, and incident commanders must work cohesively within their assignments to support each other during emergencies. Fire inspection personnel are familiar with buildings and should be dispatched to fire scenes to assist the incident commander with building intelligence. This not only improves firefighter safety but also improves property conservation with proper utilization of in-place fire protection systems. Fire department officers must make sure that fire prevention and fire suppression units work together cohesively toward the common goal of reducing firefighter deaths and injuries.

Summary

Fire prevention organizations must be based on values and goals that are congruent to the mission of preventing fires and saving lives. The mission must include all affected parties including firefighters who are placed at risk. Effective organizations have strong ethical values that are translated from the top to the bottom of the organization. Fire inspectors should be provided with the requisite knowledge and skills to begin their career advancement in management of inspection situations and workplace functions. Fire inspectors usually work alone on the streets and must have the leadership skills to manage and negotiate the complex fire code issues that will

arise. Organizational leaders must understand the roles and methods of effective leadership, must lead from the front, and must be mentors to those coming up in the organization. The fire prevention and the building code environment landscape is always changing, and fire marshals or fire officials must maintain the organization's training, education, and professional competence in order to be successful. Leaders must understand the mechanisms of change and be prepared to manage change in effective ways. Organizations must be constantly analyzed and examined to ensure that the formal and informal procedures are congruent.

Review Questions

1. Identify the mission statement and organizational values of your fire department.
2. List and describe, using the Weisbord six-box model, the formal and informal policies and procedures and how they can be more congruent.
3. You are the leader of your fire prevention team, and the fire chief asks you to prepare a budget justification plan for a new copying machine.

What items and analysis should be included to justify the purchase to the fire chief?
4. Pick a problem within your fire department and conduct a force field analysis of the driving forces and the restraining forces that keep the change from being implemented.
5. In the previous question, identify how the balance of forces could be changed to implement the organizational change.

Suggested Readings

Barr, Robert C., and John M. Eversole. 2003. *Fire Chief Handbook*, 6th edition. Tulsa, Oklahoma: Pennwell Corp.

Blanchard, Kenneth, and Norman Vincent Peale. 1988. *The Power of Ethical Management*. New York: William Monroe & Company.

Dyer, William G. 1977. *Team Building: Issues and Alternatives*. Reading, MA: Addison-Wesley.

Edwards, Steven T. 2010. *Fire Service Personnel Management*, 3rd edition. Upper Saddle River, New Jersey: Pearson.

Nelson, Bob. 1994. *1001 Ways to Reward Employees*. New York: Workman Publishing.

Solomon, Muriel. (1990). *Working with Difficult People*. Englewood, New Jersey: Prentice Hall.

Waterman, Robert H., and Thomas J. Peters. 1982. *In Search of Excellence, Lessons from America's Best Run Companies*. New York: Warner Books.

Endnotes

1. "America Burning The Report of the National Commission on Fire Prevention and Control" (Washington DC: U.S. Government Printing Office, 1974).
2. Warner W. Burke, *Organizational Development Principles and Practices* (Boston MA: Little Brown & Co, 1982), 132.
3. Ibid., 54.
4. Ibid., 141.
5. Michael McCann, "Eight Universal Laws for Becoming a Great Leader," http://ezinearticles.com/?Eight-Universal-Laws-For-Becoming-a-Great-Leader&id=2531877, June 26, 2009.
6. Daniel Gross, "Greatest Business Stories of All Times," *Forbes*, 1996, John Wiley & Sons.
7. Warner W. Burke, *Organizational Development Principles and Practices* (Boston MA: Little Brown & Co, 1982), 170.
8. http://www.change-management-consultant.com/force-field-analysis.html
9. http://www.historynet.com/battle-of-gettysburg

5

The Fire Inspection Process

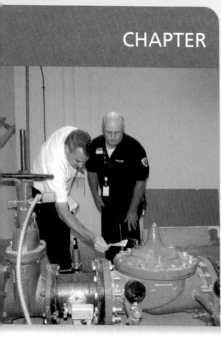

Courtesy of J. Foley

KEY TERMS

administrative warrant, *p. 91*

attractive nuisance, *p. 88*

certificate of inspection, *p. 104*

fire inspection, *p. 80*

lot and block, *p. 89*

right of entry, *p. 90*

systematic inspection pattern, *p. 95*

special inspection, *p. 86*

task force, *p. 86*

OBJECTIVES

After reading this chapter, the reader should be able to:

- Describe the basic goals and objectives of conducting a fire inspection.
- Understand how to achieve legal right of entry to conduct a fire inspection.
- Describe the importance of information gathered in pre-inspection file review.
- Conduct a systematic inspection of a property.
- Understand the importance of organization and notes to make a fire inspection report.

Professional Levels of Job Performance for Fire Inspectors as Cited in NFPA 1031 and NFPA 1037

- NFPA 1031 Fire Inspector I *Obj. 4.2.2 Recognize the need for a permit*
- NFPA 1031 Fire Inspector I *Obj.4.3.1 Occupancy classifications of a single use occupancy*
- NFPA 1031 Fire Inspector I *Obj.4.3.15 Determine code compliance*
- NFPA 1031 Fire Inspector II *Obj.5.2.1 Process a permit application*
- NFPA 1031 Fire Inspector II *Obj. 5.2.3 Investigate complex complaints*
- NFPA 1037 Fire Marshal *Obj 5.6.3 Manage a process of conducting compliant inspections*

- NFPA 1037 Fire Marshal *Obj.5.6.6. Manage a process of record keeping*
- NFPA 1037 Fire Marshal *Obj. 5.6.7 Manage a process of administering, evaluating and processing permits*
- NFPA 1037 Fire Marshal *Obj. 5.8.1 Codes, standards, and jurisdictional requirements*

Introduction

fire inspection ■ An inspection of property to determine compliance with a fire safety code or standard.

A **fire inspection** is the systematic examination of a building, the hazardous processes within the building, and the occupant's behavior in order to determine if they conform to an adopted fire safety standard. The fire safety standard that is applied is the fire prevention code of the jurisdiction. Fire prevention codes are maintenance oriented and are intended to maintain the requirements of the building code under which the structure was originally constructed. Fire inspections are the very backbone of an effective fire prevention program. The relationship between fire and building code enforcement and educating occupants to change safety attitudes and behaviors was identified at the 1947 Fire Prevention Conference in Philadelphia and developed into the three "E's" of fire prevention: engineering, enforcement, and education.

The building and fire prevention codes dictate the engineering concepts of safety building design and occupancy. The components of education and enforcement, however, will come from the fire inspector. The fire inspector must not only enforce the engineering regulations of the code but must educate the building's owner and occupants to avoid reoccurring violations.

The enforcement of the fire code has three distinct objectives:

- To separate potential ignition sources from potential fuel sources and correct any hazardous acts that bring them together;
- To ensure the safety of the occupants should a fire occur by making sure that the egress is properly designed and protected, that fire alarm systems operate, and that interior finish materials will not accelerate the spread of the fire; and
- To ensure that both passive and active fire protection features such as fire doors, fire dampers, automatic sprinkler systems, and fire pumps, to name a few, operate and are properly tested.

These objectives can be achieved only through a combination of enforcement of code regulations, basic knowledge of the building code requirements, and educating the building occupants in proper safety behaviors. Fire code enforcement centers on preventive measures that provide for the safe conduct and operation of hazardous processes, storage of combustible and flammable materials, maintenance of fire protection detection and extinguishing systems, safe exiting, and good housekeeping. Every fire inspection is also an opportunity to change the safety culture through education of the occupants.

Fire codes generally address these fire prevention issues by regulations that control either potential ignition sources or fuel sources by quantity and arrangement. The occupants within the building may be protected by both passive building construction features and active fire protection systems. The inspection of these features should include examining fire-resistance ratings of walls, fire doors, and windows; inspecting all components of the means of egress; and ensuring that the fire alarm, emergency voice communications, automatic sprinkler systems, and other fire suppression systems are properly maintained and in good working order. Lastly, the fire inspection should ensure the reduction of property damage and threat to life safety should a fire begin. This objective requires the fire inspector to employ a combination of education and enforcement of regulations such as keeping fire doors closed or providing clearance to portable or fixed heating appliances. Changing occupant behavior will require the fire inspectors to conduct follow-up inspections, including participating in fire safety and evacuation planning, conducting fire drills, training employees in the use of portable fire extinguishers and explaining why fire hazards need to be addressed. Identifying fire code violations is only one aspect of a fire inspection; the correction of the violations and changing the behavior and attitudes of the occupants is the bigger challenge. The fire inspector must incorporate all three "E's" in the inspection process to be effective.

The Fire Inspection Program

Each state, county, and municipality has a different approach to its fire inspection program regarding how and when fire inspections will be conducted. The scheduling frequency and types of buildings to be inspected usually are established by the fire codes adopting statute or ordinance. For example, in New Jersey, the state identifies buildings they deem *life hazard uses*. These buildings pay annual state inspection fees and are required to be inspected by local enforcing agencies annually or quarterly based on the specific building use. Additionally, all local enforcing agencies must establish inspection schedules and fees for all other types of occupancies they are going to inspect in their jurisdiction with the exception of owner-occupied one- and two-family dwellings. These additional fire inspections and fees are established by municipal ordinance and may be conducted on an annual or biannual basis. Most fire prevention bureaus divide up inspections by districts or zones and assign personnel to manage each geographical area. Larger fire departments may establish both inspection district and specialty inspection areas such as high rises, hospitals, schools, warehouses, industrial complexes, and public assemblies. Specialized inspection units usually consist of senior fire inspectors with specialized training and knowledge about these specific large and complex fire inspections (see Figure 5.1). This training may be provided internally by the fire department or externally through college programs or topic-related seminars.

FIGURE 5.1 Senior inspectors may be assigned inspection responsibility for specialized inspections such as hospitals. *Courtesy of J. Foley*

USE AND OCCUPANCY

All buildings are categorized by use and occupancy in the building codes. These categorizations are referred to as use groups and are determined by the building designer and approved by the building construction official when he or she issues a certificate of occupancy. All building uses are identified in one of the ten following categories:

- **Assembly:** Places where people gather for political, religious, entertainment, food, or drink purposes. (These places usually have large occupant loads.) These are further divided into Assemblies 1–5 based on the level of hazard and occupant loads.
    - Examples: A-1, movies; A-2, night clubs; A-3, restaurants; A-4, churches; A-5, stadiums
- **Business:** Commercial offices that provide services with limited amounts of storage. Examples: Doctor's or lawyer's offices, city halls, business offices.
- **Mercantile:** These are commercial facilities open to the public, selling goods and products. Examples: Hardware stores, supermarkets, department stores.
- **Institution:** These are facilities where occupants are restrained or need assistance to evacuate. There are three subcategories of institutional use: I-1, nursing homes; I-2, hospitals; I-3, prisons. This group also includes daycare centers, day nurseries, and senior daycare facilities.
- **Storage:** Storage facilities are generally not open to the public and are divided into two subcategories based on the flammability of the materials being stored. Examples: S-1, combustible and limited flammable liquids storage; S-2, noncombustible storage such as metal pipes.
- **Educational:** Educational use buildings go from kindergarten to the twelfth grade.
- **Factory:** Factories are places where raw materials are re-formed into some type of product. They are divided into two classes based on the combustibility and flammability of the materials used. Examples: F-1, combustible, limited flammable liquid use; F-2, noncombustible material use.
- **High hazard:** These are facilities using hazardous materials over the exempt amounts permitted by the building code. There are five subcategories by hazard: H-1, detonation hazard (explosives); H-2, deflagration hazard (flammable liquids); H-3, accelerated burning (combustible liquids); H-4, health hazards; H-5, hazardous production manufacturing facilities such as computer chips.
- **Residential:** These are buildings in which people reside and sleep and are divided into hazard classes based upon occupant familiarity with the building: R-1, hotels and motels; R-2, apartments and multifamily dwellings; R-3, one- and two-family homes.
- **Utility:** The utility category is for structures that generally are not occupied but require construction permits, Examples: water towers, fences, sheds, garages.

Lastly, the fire inspector will encounter mixed-use occupancies. Generally building codes allow mixed uses to be unseparated (with the exception of use H) if they are less than 10 percent of the total building floor area. Once the requirement is exceeded, fire separation walls or fire barriers must separate the uses. Fire inspectors must be capable of identifying the proper use of a building and must ensure that is the use identified by the building code official in the certificate of occupancy. Any change in the building use must go back to the construction official for evaluation and conformance with change of use building code requirements. Inspectors must pay particular attention to F-1, S-1, and M facilities as these can become H use by exceeding permitted quantities of hazardous materials.

Fire Inspection System Models

There are generally two different models that are used in establishing fire inspection programs. These are referred to as the *permit model* and the *inspection model*. In the permit model, occupancies with specific life safety hazards such as places of public assembly, hotels and motels, or warehouses with large quantities of flammable and combustible liquids of hazardous materials require permits to operate. The permit fees and thresholds are determined by ordinance or state regulation. It is the building operator's responsibility to complete the permit application and obtain the permit before operating the business. Permits differ from inspections in that it is unlawful to operate the business without the permit, and a violation of any condition of the permit can lead to immediate revocation of the permit by the fire marshal or fire official and effectively shut the business or process down. All permits must be inspected before allowing operation; therefore, owners usually will not prohibit fire inspectors from conducting the inspection even though they retain their constitutional rights to do so.

The fire inspection model differs in the context that the ordinance will identify the types of structures to be inspected and the inspection cycles for each building use group. This model may also generate fees annually for conducting the fire inspection. The inspection model allows greater flexibility in inspection scheduling, as it doesn't prevent the owner from operating. Violations under these models also do not shut the building operation down the way a permit violation would. Most fire prevention codes require some reasonable time to correct violations and have penalties or fines to encourage code compliance.

Many state and local jurisdictions use a combination of the two models, with permits required for certain processes or structures and others being identified by building use and inspection schedule. In either case, fire inspectors must understand and adhere to the model established under their jurisdictional codes.

ESTABLISHING INSPECTION PRIORITY

The consideration of inspection priority should be established by conducting a risk assessment of the community. Buildings should be sorted into priority categories as target hazards based on the number of exposed occupants in the building, the hazards of the processes being conducted or materials being stored, and the economic impact to the community if the building catches fire. These can be further categorized as life safety targets, economic targets, and special hazard targets.

Life safety targets should be given the highest priority in fire inspection performance. Life safety targets include places of assembly, large apartment and hotel complexes, high-rise residential and business facilities, outdoor sports complexes, schools, universities, and so on. These occupancies have large occupant load exposures should a fire occur; therefore, frequent fire inspections should be conducted to reduce fire hazards and ensure emergency preparation in the event of a fire.

Economic targets are facilities that have significant economic effect on the community should a fire occur. Economic facilities are those that may be significant to the tax base or may provide a large number of jobs in the community. As an example, Atlantic City has 13 casinos, but they represent about 78,000 direct jobs and 90 percent of the municipal tax base; therefore, rigorous inspections must be conducted to eliminate fire hazards. The same would hold true in California's Silicon Valley, or Seattle, and Florida's aerospace industry, or New York City's financial district on Wall Street. These facilities must be inspected on a regular basis to protect jobs and the health of the community and even the nation.

Lastly, there are special hazard targets, which are facilities that may have appreciable hazardous materials or that could cause an environmental or economic disaster should they be involved in a fire. As an example, a municipal waste treatment plant may have significant amounts of chlorine stored on the site. Special hazard targets may also include buildings vital to the quality of life in the community, such as city halls, police stations, firehouses, sewer treatment facilities, and electric plants. Special hazards should include all critical infrastructures, such as railroads and elevated highways. There have been many interruptions to interstate commerce due to fire hazards in and around these facilities, causing significant economic damage. Fire inspection schedules may be different based on the target group.

In March 1996, an automobile tire disposal company in Philadelphia located next to Interstate 95 was set on fire by an arsonist. The facility had collected and stored over 10,000 used tires and had placed much of this storage under the I-95 overpass next to the facility. The Philadelphia Bureau of Licensing and Inspections did not identify the potential problem under the interstate. On the day of the fire, the Philadelphia Fire Department saved the tire recycling building and the company, which employed about 18 workers; however, the burning tires did significant structural damage to the I-95 overpass and a piece of the highway. A section 63 feet by 128 feet had to be removed and replaced. This section of I-95 is traveled by between 130,000 to 150,000 vehicles per day. The highway was closed for weeks and cost 6 million dollars to repair, according to the Pennsylvania DOT. This fire also caused significant inconvenience to commuters, who had to find alternate routes to go to work, and interstate commerce had to be routed away from I-95 to more congested, smaller roadways. Fire inspectors must take care to pay attention to hazards that may impact infrastructure and cause the community economic harm.

Fire marshals and fire officials need to prioritize inspection work to maximize effectiveness of the enforcement program. By sorting the categories of buildings to be inspected, you can ensure that the high-profile occupancies, high life-hazard risks, and high-hazard structures get inspected on the proper cycles. The reality in fire inspection work is that rarely can any city fire department complete all of the required inspections within a 12-month period; therefore, establishing priorities helps manage the largest problems first.

ROUTINE INSPECTIONS

Routine fire inspections generally are established in two ways. In some cases, certain types of structures will be inspected at specific times of the calendar year. As an example, a school (see Figure 5.2) may get a complete inspection in late June so that corrections can be made before the school's opening in September. School reinspections will occur before the school is reoccupied. Another example is seasonal facilities that are common to seashore areas, as many of these facilities open on Memorial Day and close on Labor Day. Fire inspections may also be driven by specific holidays during the year. In September, as we all know, Halloween decorations begin to be displayed, and in November fire inspectors must prepare for the holiday season, which involves cut Christmas trees and

FIGURE 5.2 Schools and other seasonal occupancies must be inspected before opening for the season. *Courtesy of J. Foley*

decorations. This requires additional fire inspections or permit inspections to be scheduled during these months. Fire inspectors must be aware of the seasonal inspection cycles required within their jurisdiction.

Routine fire inspections may also be conducted by block inspections, where the fire inspector assigned to a specific street, district, or inspection zone goes building to building until all inspections are completed. These inspections are usually cold calls and generally are not scheduled ahead. Many jurisdictions prefer this method because it allows the fire inspector to see the property under normal operating conditions. The problem with cold calls is that the owner may not be available or the fire inspection may not be convenient on that day due to other daily business demands. Fire inspectors must realize this and be prepared to establish a firm inspection date within the next few days to complete the block inspection. Another consideration that fire inspectors need to be aware of is the date of the last inspections in that block. Businesses should be inspected in accordance with the established ordinance cycle. If inspections are short cycles, it will lead business owners to believe that you are picking on them because they are being inspected more frequently than other establishments. Fire inspectors should also be aware that they are not the only agencies conducting inspections. Owners may see housing inspectors, elevator inspectors, health inspectors, and property maintenance inspectors as well. Most fire codes require some level of interagency code coordination to avoid the issue of businesses dealing with so many inspection agencies. While some redundancy in inspections is unavoidable, coordination among agencies can help and make for much better community relations with businesses.

SCHEDULED INSPECTIONS

Scheduled fire inspections are usually reserved for large facilities that may be time consuming to inspect. The fire inspector should contact the building owner a week before the inspection to allow the owner time to gather information necessary for review by the inspector during the inspection. Informing the building owner of documents and test reports that the fire inspector must review makes the inspection go far smoother. The inspector can establish the date and time of the inspection, whom the inspector should meet with, and ensure that keys will be available to secure areas and that documentation will be gathered for review at the property. Another benefit to scheduled inspections is that building owners will tend to begin correcting obvious violations before the inspection date. Some fire officials do not like this because they will not see normal conditions of operation in the facility. Many believe that if code violations begin to be corrected before the fire inspection occurs, then fire inspectors have served their purpose well. The property will be better maintained, fewer violations will be cited, and fewer re-inspections will have to be performed. Remember, the goal of an inspection is fire code compliance. If building owners begin to correct violations because the fire inspector called to schedule an inspection date, that is a good thing!

FIRE PERMIT INSPECTIONS

Fire permit inspections differ from routine fire inspections in that they are performed to inspect a specific system or hazardous process. Fire inspectors must understand the difference in a permit inspection versus a routine inspection. First, the fire prevention code requires permits because the activity or process being performed is not permitted by the code. Permits are usually relegated to hazardous materials and processes that are prone to start or contribute to a fire. As an example, fire safety permits are required when welding or cutting operations take place or when hazardous materials are stored, handled, or used over specific amounts identified in the fire code. These quantities are referred to as the amounts requiring approval by the fire marshal or fire official. The second difference is that fire officials may place specific special conditions on the approval of a fire permit before its issuance. If a condition is not complied with or is violated by the permit holder, the permit can be immediately revoked and the process or procedure stopped. This differs from a normal inspection where adequate time to correct violations is required to be given. During the inspection process, the fire inspector may not issue the permit if violations of any permit condition exist.

Fire permits require the owner to complete an application and pay an application fee before the inspection. The payment of the fee does not guarantee the application will be approved, and the fire official may in fact deny the permit based on the application. Typically, the permit application has an approval or denied check box and lines on which to provide the reasons for denial. The fire marshal or fire official should maintain a record of all denied permit applications to maintain consistency in the program.

Fire inspectors assigned to permit inspections must review the conditions of the permit and the applicable fire code requirements before conducting the inspection (see Figure 5.3). In some cases, such as a public fireworks display, the inspector may be required to monitor the activity until it is completed. Fire permits have a specific time duration of approval, meaning that they usually can be valid for a maximum of 12 months or less, and may be specifically for the time of the event. Fire permits are not transferable and require reapplication if revoked. When the fire safety permit expires, the applicant must reapply for the permit, pay the appropriate application fee, and be reinspected by the fire inspector.

SPECIAL INSPECTION

special inspection ■ The inspection of a system or piece of equipment to determine code compliance, usually in addition to a routine fire inspection.

Sometimes special inspections are required to enforce the fire prevention code. **Special inspections** may be conducted for life safety purposes such as checking exits in large assemblies during operating hours, or they may be directed to specific target problems such as kitchen range hood and duct inspections. As an example, many states that have casinos or an entertainment industry require fire inspectors to conduct life safety inspections in the hotels, casinos, or entertainment complexes every week. These fire inspections usually occur during nights and weekends and are usually unannounced. The inspections involve mostly the examination of the elements of means of egress and overcrowding of public assemblies. This type of life safety inspection program may also include smaller bars, restaurants, and nightclubs on a periodic basis.

Another type of special inspection is kitchen range hood and duct systems, as the fire prevention code requires annual inspection of the entire kitchen range hood and duct systems. These fire inspections are conducted off normal hours when equipment is cold to the touch so that inspectors will not get burned. The inspection requires all duct access doors to be opened for examination of grease buildup. Some fire departments use a special piece of camera equipment called a bore scope to inspect the inside of the ducts for any buildup of grease. These inspections usually take several hours to complete because of the size of the systems and the difficulty of accessing the ductwork. The inspection requires assistance from the owner's contractors to gain access to ducts as well as for the testing of the fire suppression systems.

task force ■ A group of agencies that conduct coordinated inspections, usually on difficult properties.

A third type of special inspection is a **task force** or multiagency inspection. Buildings that are identified by inspection agencies as problematic may be referred to a special building task force to coordinate code enforcement efforts. The task force consolidates code issues and conducts a coordinated inspection to gain code compliance. Each participating agency, which usually includes fire, health, property maintenance, building, electrical, and plumbing inspectors, deals with its code authority but is coordinated in effort through the task force to gain legal compliance. These types of inspections are usually conducted on hard-core problem properties that create many quality-of-life issues in their communities. As an example, fire prevention bureaus often receive citizen complaints about smells from buildings and accumulation of trash on a property. When the fire

FIRE SAFETY
PERMIT
CONDITIONS

	REGISTRATION #:

LOCATION INFORMATION	
NAME:	STREET ADDRESS:
MUNICIPALITY:	COUNTY:
STATE: ZIP CODE:	AREA CODE & PHONE #:

APPROVAL OF PERMIT ON PRECEDING PAGE IS CONTINGENT ON ADHERENCE WITH THE FOLLOWING CONDITIONS:

_____
_____
_____
_____
_____
_____
_____
_____
_____
_____
_____
_____
_____
_____
_____
_____

_____   _____
FIRE OFFICIAL   DATE OF APPROVAL

Fire safety permits grant permission for activities that the code traditionally prohibits. The fire inspector may place conditions upon the permit for compliance.

FIRE SAFETY PERMIT

Inspector:_____

LOCATION INFORMATION	
NAME	STREET ADDRESS
MUNICIPALITY	COUNTY
STATE	ZIP CODE AREA CODE & PHONE #

Mailing Address

This certificate must be posted in a conspicuous location at the above premises.

Permit Type _____
Permit Fee _____
Registration No. _____
Date of Event _____
Expiration Date _____
Annual _____
Seasonal _____

_____
FIRE OFFICIAL

_____
Date of Approval

[ ] If box is checked, approval is contingent on adherence with the following conditions:
_____
_____

5:71-3.7(b)13.

FIGURE 5.3 Example of fire safety permits. *Courtesy of the NJ Division of Fire Safety*

FIGURE 5.4 Fire inspections must address attractive nuisances such as vacant buildings. *Courtesy of J. Foley*

inspector conducts the inspection, they often encounter health code, building code, and property maintenance code issues beyond the scope of the fire prevention code. Because of the extreme multiple hazards at the property, the coordination of the task force helps focus the enforcement effort into a single court action as opposed to multiple actions by each individual agency. This ensures that all of the issues are addressed promptly.

The last type of special inspection is the identification of **attractive nuisances**. This would include vacant lots, weed, grass, and trash abatement as well as vacant and abandoned buildings. These inspections often involve absentee landowners or no owners at all. Inspections of vacant lots and buildings can be very time consuming, are often difficult to provide notice of violations, and take considerable time to abate. Municipalities may have to accept some of the burden in these nuisance properties to board them up, cut the grass, and remove trash (see Figure 5.4) These costs may then be liened against the property owner. The fire inspector's time and effort must be tracked in this process so the municipality can recover the costs of abatement.

attractive nuisance ■ A vacant or dilapidated property that attracts vermin, crime, or creates public nuisances, lowering the quality of life.

Tools and Equipment

To properly conduct a fire inspection, the fire inspector must have proper tools and equipment for each specific type of inspection. First and foremost, the inspector should have a professional appearance with a clean and pressed uniform, proper footwear, and credentials. Owners will size you up within the first ten seconds of meeting you, and if your appearance is sloppy or disheveled, they will be more likely to challenge your authority and your inspection results. Under certain conditions where the inspection requires you to get dirty, then coveralls or battle dress uniforms may be appropriate.

The fire inspector should have a properly marked vehicle to carry equipment and should have at minimum the following items:

- Fire department radio
- Flashlight with spare batteries
- 25-foot retractable tape measure
- Measuring wheel for large areas
- Hard hat
- Safety glasses

- Clipboard, pens, pencils, and appropriate forms
- Handheld computer (if this technology is used)
- Spare computer batteries
- Small digital camera

In addition to these inspection items the fire inspector should also carry fire turnout gear and SCBA or other firefighting tools such as portable fire extinguishers and first aid equipment, hydrant and gas wrenches, caution tape, and other safety equipment in the vehicle. Because fire inspectors often are out on the streets, they may encounter fires in progress, vehicle accidents, and other medical emergencies. Fire inspectors may also be helpful to incident commanders by providing building intelligence during fire ground operations. If handheld computer tablets are used in the field, they should be in protected cases so that accidental dropping of the equipment will not damage it.

Professionalism

Fire departments are viewed by the public as professional organizations and are usually one of the most respected departments in city or county government. In surveys conducted by *U.S. News and World Report* in 2007, the public considered firefighting the most respected career. This poll of 1,010 adults placed firefighters ahead of scientists and teachers as being prestigious careers.[1] Fire inspectors must realize that they benefit from this public perception and must present themselves in a professional manner. Think of the old saying, "One bad banana can spoil the bunch." A customer or businessperson will carry away a quick impression of the inspector upon his or her entering the property. If the inspector's appearance is sloppy and unorganized, the customer will lose respect for the inspector and question everything the inspector identifies as a code violation. Additionally, the customer may view the fire inspector as less than competent. If the fire inspector is neat, clean, and professional in his or her appearance and approach, he or she will gain respect from the customer and be viewed as a professional. It is important to be personable and friendly to the customer at the first meeting. People sometimes get apprehensive at the appearance of uniformed personnel, but if your approach is friendly, it reduces this apprehension. It is also important to be honest in answering questions that the customer may have about a code violation or the inspection process itself. If you are unsure of the correct answer, do not make something up. Explain to them that you will research the correct information and get back to them. If you make an appointment to conduct the fire inspection or to meet a customer, keep the appointment. Keeping appointments shows professionalism and respect for the customer's valuable time. If you cannot make an appointment or are running late, call and let the customer know you will be late or perhaps reschedule a time convenient for them. As a fire inspector, using a professional approach will make your job easier and the customer more satisfied about the inspection process.

Record Review and Files

In the inspection process, the starting point for all fire inspectors should be examination of the building record or file. Fire inspections can be complicated, and a complete file review will provide the fire inspector with all of the important information regarding the property (see Figure 5.5). The inspection file provides information on the proper street address, **lot and block**, the owner and agent's name, telephone numbers, and mailing addresses. The file will have a general description of the building including basic information on the building's height, area, type of construction, utilities, fire protection features such as fire alarms or automatic sprinkler systems, the location of fire department standpipe connection, and many other items. When reviewing the file, the fire inspector should

lot and block ■ Method by which properties are identified and described in tax records and deeds.

FIGURE 5.5 Fire inspectors should review the building record before conducting the inspection. *Courtesy of J. Foley*

examine the last notice of violation to determine what types of fire code violations were encountered during the previous fire inspection. The file also contains any owner correspondence, fire safety permits, building registrations, certificates of approval for fire protection systems, test records for fire protection systems, and hazardous material inventories from the previous inspections. All of this information is vital to fire inspectors so that they are properly prepared to conduct the fire inspection. Records may also contain fire code variances that have been approved in the past and will explain the conditions of the variances so the fire inspector can confirm continued code compliance.

Upon completing the file review, the fire inspector should also review the applicable chapters of the fire code and the NFPA standards that may apply to the building. Many fire prevention bureaus may have checklists for types of violations based upon occupancy hazard. The fire inspector can use these checklists as a guide for fire code compliance. Checklists are good references but must be supported by reviewing the applicable fire codes and standards; this is extremely important in technical inspections involving hazardous processes or hazardous materials. Reviewing the file before the inspection is often overlooked by fire inspectors because they generally have a heavy workload, but all inspectors should take the time to review the file and building record; you will be surprised as to how it enhances your ability to conduct a thorough and professional fire inspection.

Right of Entry

right of entry ▪ The right of a government agency to enter private property.

Entering a private property for the purpose of conducting a fire inspection requires permission from the owner. The fire inspector must present his credentials, state the nature of his request, and make an entry demand to conduct the inspection. Demanding entry is a legal term and does not mean demand in the customary sense; it is simply stating the **right of entry** to the owner, who may refuse entry if he or she chooses. As a fire inspector, your authority to inspect is derived from the regulations and law adopted in your jurisdiction. Fire inspectors are agents of the local government with police powers dictated by the United States Constitution. The Tenth Amendment states: "The powers not delegated to the United States by the Constitution, nor prohibited by it to the States, are reserved to the States respectively, or to the people."[2] The federal government defers the rights of police powers to the states. Each state in its constitution may provide certain powers to local government as long as the state has legislated those powers and enacted them into laws. This usually happens with the adoption of a state fire prevention code. The states, however, cannot trample on the rights of any citizen, who has protection under the Fourteenth Amendment of the Constitution, which states: "All persons born or naturalized in the United States, and subject to the jurisdiction thereof, are citizens of the United States and of the state wherein they reside. No state shall make or enforce any law which shall abridge the privileges or immunities of citizens of the United States; nor shall any state deprive any person of life, liberty, or property,[3] without due process of law; nor deny to any person within its jurisdiction the equal protection of the laws." Fire inspections mandated by government constitute police powers for which citizens are protected. The Fourth Amendment protects citizens from unwarranted and unreasonable search and seizure of their private property and states: "The right of the people to be secure in their persons, houses, papers, and effects, against unreasonable searches and seizures, shall not be violated, and no warrants shall issue, but upon probable cause, supported by oath or affirmation, and particularly describing the place to be searched, and the persons or things to be seized."[4] The Supreme Court has ruled many times on unreasonable search and seizure in fire inspection and code enforcement cases including *Franks vs. Maryland* (359 US 360), *Camara vs. San Francisco* (357 US 523), and *See vs. Seattle* (387 US 541). All of the cases dealt with the violation of protected Fourth Amendment rights of individuals by government inspection agencies. The courts have ruled that to conduct a fire inspection, you must have the legal permission of the owner, and when an owner refuses entry, you must obtain an administrative search warrant based on probable cause to continue your inspection. While probable cause in an **administrative warrant** is not as specific or detailed as required in a criminal search warrant, the fire inspector must convince the judge in a compelling fashion that the fire inspection is warranted and that failure to conduct the inspection could cause harm to the public or occupants of the building.

administrative warrant ■ A order issued by a court to grant permission to an inspector to enter private property for the purpose of inspection.

Whether the fire inspection is mandatory by permit or statute, the owner still has protected rights and can refuse the inspection or revoke permission at any time. The fire inspector must be legally on the property in order to cite any fire code violations.

Legal entry on a private property occurs only under the following conditions:

1. An *exigent circumstance* exists, such as fire, toxic fumes, and explosive gases, or other imminent hazard. In these cases, warrantless entry is permitted to mitigate the emergency.
2. You are permitted to inspect building areas *open to the general public* during public operating hours. You would not, however, be able to enter areas where the public is prohibited such as storage areas or offices or back hallways unless they are part of the public's egress.
3. The *owner or agent gives you permission* to conduct the fire inspection. Keep in mind, however, that permission can be revoked at any time and once revoked you must leave the premises. It is also important to understand that unless the owner openly refuses entry, the courts will assume permission is implied.
4. The *owner or agent refuses entry* and you obtain an administrative search warrant from the court of competent jurisdiction.

The rules for right of entry are generally defined in the building and fire codes or local or state adopting ordinances, and fire inspectors should examine their local requirements for right of entry. Fire marshals, fire officials, and fire inspectors should discuss right of entry rules and issues with their legal representatives. They should know the proper procedure on how to obtain an administrative search warrant, if necessary. In a typical procedure, the municipal attorney would prepare an order outlining the owner's refusals of entry, the legal authority allowing the fire inspection to take place, the types of violations that may exist or have existed in the past, the areas of the premises to be inspected, and the time frame for the inspection to take place. The fire inspector would then meet with the judge so that the judge can ask any questions about the issuance of the search warrant. The judge, if satisfied, will then sign the warrant and issue it to the fire inspector to conduct the inspection. It is wise in these cases to execute the warrant with the assistance of the police in case the owner still refuses entry. Typically, most fire inspectors never have to go through this process; however, a small percentage of owners with a history of being problematic may require the issuance of an administrative search warrant before they allow the fire inspection to be completed. As a professional, you must remember that it is the owner's constitutional right to refuse entry, and you should follow the correct legal entry procedure so as not to have your effort negated by an improper fire inspection.

Starting the Fire Inspection

Upon your arrival at the property to be inspected, you should prepare your equipment and make sure you have a working pen, your agency identification, necessary forms, a clipboard or computer tablet, your radio or phone, and a camera, if necessary (see Figure 5.6). You may not be taking all of these items with you during the inspection, but you should have them available and ready as needed. Upon exiting your vehicle, make some initial observations from the public way, such as the street address, main entrance location, fire department connections, utility locations if visible, the nearest fire hydrant, exposed buildings, and general condition of the building and surrounding grounds. After you enter the building, request to see the owner or the person in charge of the property. Be patient, as the owner or agent may be involved in other important business activities at the time you arrive; this is especially true on cold calls and block inspections. Follow the correct rules of entry by showing your agency identification, stating the nature of the fire inspection, and requesting the owner's permission. You should also request that the owner or his or her designate accompany you on the inspection tour and that they have the appropriate access keys to all areas of the building. Depending on whether the fire inspection was prescheduled or a cold call, you may get several different responses to your request. Sometimes, the owner may have other pressing issues to deal with, and so it may be a really bad time to conduct the inspection. If this

FIGURE 5.6 Inspectors may use handheld computers to record information during the inspection tour. *Courtesy of J. Foley*

is the case, request a date and time within a week to return and conduct the fire inspection. If it is a scheduled fire inspection and it turns out to be a bad day for the owner again, provide a return date but emphasize that the fire inspection must be completed. If the owner cannot accompany you, request that a manager or other designee accompany you, but make sure it is someone who has access to all areas of the building. At the beginning of the fire inspection, it is a good time to request any documents that you may need to see before you leave the property, as this allows the building owner time to gather them for review by the inspector. These documents would include any building permits, fire suppression system certification reports, fire alarm test records, evacuation and fire plans, training records, HVAC test records for smoke control, hazardous materials inventories, and management plans or any other documents required by the code. Sometimes, a list of documents can be provided to assist the owner in maintaining records. It is also important to schedule a few minutes with the owner at the end of the inspection to review your findings. The closing interview is a critical part of correcting violations and gaining fire code compliance.

The Inspection Tour

The fire inspector will conduct an inspection tour of the building for two distinct purposes. The first and most important reason is to identify fire safety violations under the fire code and to have them corrected. The fire code violations must be recorded in an organized fashion and will be presented to the building owner as a correction list with time provided to make the necessary repairs. The fire inspection report must be clear, concise, and complete to ensure code compliance. The second purpose of the fire inspection tour is to gather building intelligence for use by the fire department during emergencies. The building intelligence that the fire inspector gathers will include the type of building construction, the building's occupancy and use, the building's height and area, and the types of fire protection systems present. The fire inspector will also identify special hazards such as flammable liquids or hazardous materials in storage or use areas of the building to provide important information to firefighters to protect their safety and welfare.

IDENTIFYING CONSTRUCTION TYPES

Identifying an existing building's type of construction can be challenging for the fire inspector. This is especially true with buildings that were renovated and may have exterior veneers such as brick facing, expanded polystyrene foam–Dryvit systems, or other types of claddings. Sometimes the best location to determine the construction type is the building's basement or attic space. In these areas the interior usually is not finished, and structural elements may be exposed.

There are five main categories of construction:

Type 1 – Noncombustible and fire resistant (most fire resistant)
Type 2 – Noncombustible
Type 3 – Ordinary
Type 4 – Mill (heavy timber)
Type 5 – Combustible frame (least fire resistant)

The factor that determines the type of construction is the structural elements of the buildings such as columns, girders, fire walls, party walls, and floor and roof framing and the ability to resist fire exposure. In Types 1 and 2 construction, all of the building's supporting elements are noncombustible, but Type 2 is not necessarily fire resistant. Type 3 construction is typically noncombustible walls or columns (masonry) and combustible floor joists and roofs. Type 4 is combustible frame, but all supporting elements are a minimum of 6 inches nominal width. Type 5 is typical combustible frame.

EXTERIOR

The inspection tour should be conducted in a systematic approach. Generally, it is easiest to begin with a 360-degree view of the exterior of the property to see the exits' discharge locations, the utility locations, and any obstructions to the fire department's access. The inspector should also observe the location of remote buildings and any other exterior hazards including material storage or vegetation. The fire inspector should make a quick sketch of the property to assist in orienting the buildings. This drawing can be refined at a later point.

ROOF COVERINGS

After the building's exterior orientation is identified, the fire inspector should enter the building and proceed to the roof. The roof inspection will reveal many fire protection features, from the location of firewalls, to hazardous exhaust systems and ventilators. If the building is a high-rise or mall, the smoke management system should also be examined, as well as general roof condition and firefighter access. Once on the roof, ensure that it is sound and identify the type of roofing materials used by code classification, as Class A, B, and C roofing. ASTM E-108 and UL 790, *Test for Fire Performance of Roofing Materials* classify roof coverings. The Class A roof is generally effective against high-fire exposure; it produces no flying brands and will not readily communicate fire. Class A is permitted on all five types of building construction. Examples of Class A roofing materials would be metal roofing, clay tiles, concrete and stone, and fiberglass asphalt composition shingles. Class B roofs afford moderate fire protection and are the minimum roofing material permitted in Type 1 fire-resistant construction. Examples of Class B roofing materials are chemically treated wood shingles. Class C roofs afford low fire exposure protection and are permitted on structures of construction types 2, 3, 4, and 5A. Typically, roll roofing and asphalt felt are Class C materials. It is also important to identify how the roof structure is supported. If the roof is supported on steel or wooden trusses or steel or wooden I-beams, it may require identification signs at the main entrance for firefighter safety. For example, the states of New York, New Jersey, Florida, and Virginia and the city of San Francisco all require firefighter warning signs for truss construction. The signs must be of a specific size and marking in each state, such as the one shown in Figure 5.7. The fire inspector should also identify any serious leaks in the roof, as these tend to cause structural damage and may result in early collapse of the building in a fire.

ROOF STRUCTURES

Identify and inspect all rooftop structures, such as elevator machine rooms or air handling rooms. In the elevator machine room, examine the shaft ventilation and fire alarm and extinguisher equipment. The room should be clean and free of waste rags and combustibles, and the elevator controls and motors should also be identified for lockout and tagout of each individual elevator control, motor, and car for rescue purposes. Air handling rooms should be free of combustible materials that could spread smoke to other parts of the building via the air ducts. The fire inspector may note other rooftop systems including lighting protection, kitchen exhaust, and chimneys as seen in Figure 5.8. These systems must have the proper rooftop clearances and be structurally sound and properly maintained and free of grease accumulation. Kitchen exhaust areas should be free of grease buildup and should be inspected and cleaned regularly.

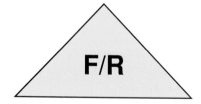

FIGURE 5.7 Many states require the installation of a roof/floor truss sign at the main entrance to truss constructed buildings.

EXAMINING THE BUILDING FLOORS

After the completion of the roof inspection, the fire inspector should begin to inspect each floor of the building. It is important at this point to use a

FIGURE 5.8 Roof structures such as kitchen exhaust ventilators must be inspected for grease accumulation. *Courtesy of J. Foley*

systematic inspection pattern. The pattern can be clockwise or counterclockwise, but it should be as consistent as possible. On each floor, the fire inspector should orient himself or herself to any landmarks or streets as a point of reference to their location in the building. Sometimes, the stair enclosures can act as an anchor point on each floor for orientation. The fire inspector will need to enter every room and closet on the tour to ensure that all hazards are identified. If the building is a residential apartment building or condominium, the fire inspector will need the permission of each occupant to enter dwelling units. The fire prevention codes usually exclude the inspection of the dwelling unit the same as they do for single-family residences. The building manager may have right of entry under the apartment lease agreement or condominium by law, but this is not always the case and should be checked. It is better practice to ask permission of the occupant before entering the unit for inspection. Fire inspectors should check their local code statute for right of entry requirements. If the fire inspector comes upon a locked room, he or she should inquire who could open it for inspection purposes. If it cannot be accessed at that time, a note on the inspection report should be made and the owner should be told that access to the area will be required on the follow-up inspection.

Upon entering each room, the fire inspector should make observations again in a systematic pattern. The inspector should scan the room in layers such as the ceiling, the walls, the floor, the location of the exits, the types of fire protection systems, and equipment that is present. A mental or written checklist and consistent observation pattern will improve your level of observation.

Ceiling Inspection

When scanning the ceiling, look for holes and penetrations or missing ceiling tiles, as these provide passageways for smoke and fire to attack the building's structural components. Smoke detectors should be inspected for proper spacing to HVAC supply or return ducts. Air supply will move smoke away from the detector, so placing detectors too close to a return may cause them to become dirty very quickly. Sprinkler head spacing and sprinkler head type should be inspected. The inspector should look for obstructions to the sprinklers such as heads too close to beam soffits or ceiling coffers, or light fixtures below the sprinkler head, as well as other obstructions. The inspector should note the type and general temperature range of the sprinklers. Temperature range is determined by the

systematic inspection pattern ■ A pattern of movement and recording information to permit better organization of information in a fire inspection report.

sprinkler frame color listed in NFPA 13, *Standard for the Installation of Sprinkler Systems*. The inspector should note the general condition of the ceiling. Inspectors should also note any ceiling beam and pocket that may affect heat and smoke passage. These areas must be properly protected by smoke detectors.

Wall Inspection

During the wall inspection, the fire inspector should note the interior finish material, any hanging decorations, exit locations, exit signs, and emergency lighting units. If there are no emergency lights, then the normal lighting must be on emergency power from a generator and should be tested in the presence of the fire inspector. The walls should be free of holes and penetrations that could spread fire and smoke. The fire inspector should look for electrical system violations, such as open junction boxes, exposed wire conductors, and the use of extension cords. Finally, the fire inspector should note the type and location of manual fire alarm pull stations, speakers, strobes, abort switches, and portable fire extinguishers.

Floor Inspection

The fire inspector should make observations on the floor area, which should include any storage, the exit access pathways, obstructions to the exits, the type of flooring and finish, any trash, portable space heaters, and general housekeeping condition of the room. Proper housekeeping ensures that exit pathways are clear and that storage of goods is compact and orderly, reducing the ability of materials to burn rapidly. Poor housekeeping may also indicate inadequate maintenance personnel or management of the facility.

Fire Protection System Inspection

Inspection of the fire protection equipment requires a close examination of the portable fire extinguishers. Questions to ask yourself include the following: Have the fire extinguishers been properly serviced? Are they the right type and size? Are they hung at the proper height and locations? The proper distribution of fire extinguishers is found in NFPA 10, *Standard for Portable Fire Extinguishers* and is based on the hazard classification of the occupancy. The fire alarm equipment needs to be examined for proper location of smoke or heat detectors and manual pull stations, and the inspector should identify any unprotected areas or any equipment that is obstructed, damaged, or improperly installed. All of the fire alarm spacing rules come from NFPA 72, *National Fire Alarm and Signaling Code*. During the examination of the automatic sprinkler system, the inspectors need to identify the temperature rating of the sprinkler heads. The inspector should note any obstructions or commodities located within 18 inches of the sprinkler head, as storage must be below that level. The fire inspector should also examine the sprinkler system design hazard classification plate. Sprinkler system design information is located on the riser by the main control valves in accordance with NFPA 13, *Standard for the Installation of Sprinkler Systems* (see Figure 5.9). The fire inspector needs to ensure that the commodities stored within each building space match the sprinkler system hazard classification. As an example, a sprinkler system designed for light hazard residential or light commercial use cannot adequately protect a room with appreciable flammable liquids, as this would create a mismatch in the system design and could result in the system's failure to control a fire.

Equipment and Storage

The last area of observation is what equipment or storage is in the room. The fire inspector should examine the storage, equipment, processing machinery, and any other hazards or potential ignition sources within the room. The fire inspector should ensure that all equipment is properly wired to the electrical systems and not running on temporary extension cords. If the equipment produces heat, the specification plate should be checked for minimum clearance to combustible materials. Equipment should also be listed for the application it is used for and should show recognized testing laboratory

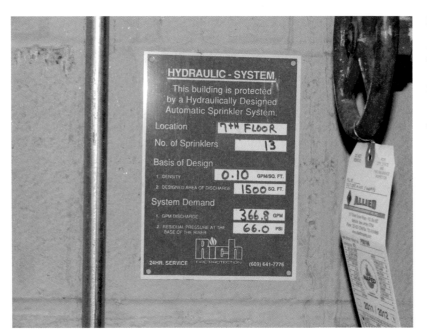

FIGURE 5.9 The automatic sprinkler system design information can be found on the design plate located on the sprinkler system riser above the system main control valve.
Courtesy of J. Foley

identifications. Motors should be examined for dust and grease as well as any signs of friction in moving parts.

By using a check list to make observations, the power of identifying violations is strongly increased. Identifying fire hazards, no matter how small, helps to reduce the likelihood of a fire starting. Remember, the goal of the fire inspection is to see that the building, equipment, processes, and occupant behaviors comply with fire safety standards. Fire inspectors must be mindful that the fire prevention code is a minimum fire safety standard, so any violations will place the building below minimum acceptable standards and could place the building's occupants at risk.

INSPECTION OF PORTABLE FIRE EXTINGUISHERS

Portable fire extinguishers are considered first aid appliances and are not generally required by building codes. The installation, maintenance, and inspection requirements are found in the fire prevention codes. Portable fire extinguishers are selected, installed, and maintained as specified in NFPA 10, *Standard for Portable Fire Extinguishers*. The fire prevention code requires the installation of extinguishers in A, B, E, F, H, I, M, S, R-1, R-2, and R-4 buildings. The IFC also has 40 other locations based on hazardous processes that require dedicated portable fire extinguishers to be present. The use of portable fire extinguishers is divided into five major classes of fire and achieves a rating based on the effectiveness of the extinguisher on each fire class. The classes of fire are as follows:

- Class A – Ordinary combustibles, wood, paper, cardboard, etc.
- Class B – Flammable and combustible liquids
- Class C – Energized electrical equipment
- Class D – Flammable metals
- Class K – Kitchen with grease and vegetable oils

Portable fire extinguishers may be rated for a single class of fire, such as a 2A 2-1/2 gallon pressurized water extinguisher, or multiple classes, such as a 2A 20BC dry powder fire extinguisher. The key for the fire inspector is to make sure the extinguisher is proper for the hazard it is supposed to extinguish. The general placement and travel distance for portable fire extinguishers is established in NFPA 10 based on the building's hazard classification. NFPA 10 bases the classification on the type of fuel present as follows:

- **Light Hazard:** Class A fuels with small amounts of Class B; typically schools, residential, business buildings
- **Ordinary Hazard:** Moderate Class A and Class B fuels or Class I and II commodities; typically small manufacturers, mercantile business storage facilities
- **Extra Hazard:** Appreciable Class A and B fuels with Class I and II commodities; factories, industrial buildings, flammable liquid warehouses

The placement requirements for class A portable fire extinguishers are as follows:

	LIGHT HAZARD	ORDINARY HAZARD	EXTRA HAZARD
Minimum "A" rating	2A	2A	4A
Max floor area in square feet	3,000 sq. ft.	1,500 sq. ft.	1,000 sq. ft.
Max floor area per extinguisher	11,250 sq. ft.	11,250 sq. ft.	11,250 sq. ft.
Max travel distance	75 feet	75 feet	75 feet

Class B fuels require the following extinguisher ratings and travel distances:

	EXTINGUISHER RATING	TRAVEL DISTANCE
Light hazard	5B or 10B	30 or 50
Ordinary hazard	10B or 20B	30 or 50
Extra hazard	40B or 80B	30 or 50

As you can see from these tables, the rating of the fire extinguisher increases when it is further away from the hazard. The fire code will also require the placement of specific fire extinguishers such as class K portable extinguishers within 30 feet of cooking appliances or 40BC rate extinguishers within 25 feet outside or 10 feet inside a flammable liquid storage room. In metal turning shops class D fire extinguishers may be required. Class D extinguishing agents are not universal like ABC or dry powder and are usually specific to the particular flammable metal being used. These fire extinguishers are over and above those needed for general occupancy. When conducting the fire inspection the inspector should look for the following:

1. Is the fire extinguisher the proper size and type for the hazard class?
2. Is the fire extinguisher properly mounted and within 3.5 to 5 feet depending on the extinguisher's weight with 4 inches of floor clearance?
3. Is the fire extinguisher visible and unobstructed by doors or equipment?
4. Is the extinguisher charged and pressurized, and does it have a current service tag? The inspector should give the extinguisher a visual check for pressure on the gauge, lift the extinguisher to feel if it is full, examine the hose and nozzle for defects, and ensure that the safety pin is inserted and secured for use.
5. If the extinguisher is 5 or more years old, it should have an internal inspection verification ring around the neck of the cylinder indicating that the inspection was performed.
6. The inspector should examine the hydrostatic testing date of the extinguisher cylinder. Generally water, foam, wet chemical, and halon extinguishers must be tested

every 5 years and dry powder extinguishers every 6 years. CO_2 fire extinguishers are pressure tested every 10 years.

7. Fire extinguishers may be placed in locked cabinets to prevent vandalism. The cabinet must have a vision panel and information on how to access the fire extinguisher such as break glass.

8. The operating instructions and use of the extinguisher should appear on the label of the unit.

The fire inspector should also see whether the building owner provides training to employees on the proper selection and use of portable fire extinguishers. Almost all extinguishers in the market today can be operated safely by remembering the acronym PASS, where P = pull the pin, A = aim the extinguisher at the base of the fire, S = squeeze the two handles together at the proper distance, and S = sweep the nozzle back and forth until the fire is extinguished.

Portable fire extinguishers are a key element in addressing small fires within buildings. The successful use of these devices depends on the proper type and distribution of the extinguishers, the proper maintenance of the units, and the operator's knowledge of how to use the equipment. Fire inspectors must make sure that all of these elements are addressed during the annual fire inspection.

Document and Certification Review

At the beginning of the fire inspection process, the fire inspector should request that the owner show all of the necessary documents required to be inspected under the fire prevention code. Both the *International Fire Code* and NFPA 1 specify documents to be provided for inspection to determine fire code compliance. These documents may include the following information:

- Fire alarm system testing reports meeting the requirements of NFPA 72, *National Fire Alarm and Signaling Code*
- Automatic fire sprinkler system inspection test reports in accordance with NFPA 25, *Standard for the Inspection, Testing, and Maintenance of Water-Based Fire Protection Systems* (These records will reflect quarterly, biannual, and annual testing.)
- Fire department standpipe test records required in NFPA 25, *Standard for the Inspection, Testing, and Maintenance of Water-Based Fire Protection Systems*
- Fire booster pump test records – NFPA 25
- Certification of range hood fire protection systems, including replacement of fusible links every 6 months
- Kitchen system cleaning schedules
- HVAC smoke control systems balancing reports
- Testing of emergency generators and automatic transfer switches – NFPA 110, *Standard for Emergency and Standby Power Systems*
- Treating of draperies and decorative material flame proofing – NFPA 701, *Standard Methods of Fire Tests for Flame Propagation of Textiles and Films*
- Hazardous material safety data sheets or MSDS
- Hazardous materials management plan
- Fire safety plan
- Fire evacuation plan
- Employee training records, where required
- Building permits or certificates of approval on new equipment or construction within the facility

All of these records shall be maintained by the building owner and are subject to inspection to determine code compliance. Often, these inspection and maintenance service records are overlooked by the owner until the fire inspector requests to review them. In some states, the regulations require that contractors forward copies of fire protection test records to the fire marshal or fire official for review. In many states, contractors that work on fire equipment or fire protection systems must be certified and licensed by the state and must comply with all the requirements established under the adopted fire prevention code. The reporting of testing may also require the use of standardized form and formats by contractors. These forms should comply with the requirements of the applicable NFPA standards.

Closing Interview

The closing interview with the building owner or agent is probably the most critical part of the fire inspection process. The closing interview gives the fire inspector the opportunity to discuss both the positive fire safety aspects observed during the fire inspection and any fire code violations that were identified or that may pose a serious fire risk. The fire inspector should develop a checklist of serious violations that require correction or that may be expensive to correct as he or she conducts the fire inspection. These salient points should be the subject of the closing interview. The inspector should avoid discussing minor violations, as this may be viewed as nit picking by the owner. The smaller violations, if properly identified, will get corrected. It is the more serious and more expensive items that should be the subject of the closing discussion with the owners.

It is always a good practice to begin discussion on a positive note. Tell the owner the things you observed that had a positive effect on building safety. Let the customer know if he or she is doing things correctly; this reinforces good attitudes and safety behaviors. The fire inspector should discuss the important aspects of the fire inspection, and these should be the points that are addressed quickly, such as inoperable fire alarms or automatic sprinkler systems, obstructed or locked egress, and elimination of toxic gases or flammable vapors, among others. The inspector should discuss the nature of the violation, the risk involved if it remains uncorrected, and the options for correcting the violation. As an example, if the violation was keeping a fire door chocked open, the options for correction could be closing the door or connecting the door to an automatic releasing device operated by the fire alarm system or a smoke check door closer. This provides the building owner or agent with several options to explore to see which will accommodate their needs the best.

Often, fire inspectors are unwilling to suggest methods of correction of fire code violations. Making suggestions is not correcting the violations, it is presenting viable options to solve or mitigate a problem. The fire inspector must be a salesperson for fire safety and must be capable of assisting the building owner or agent to comply with the fire code by educating them. If the fire inspector has knowledge of a possible solution, he or she should discuss the options with the customer. Fire inspectors should not be the adversary of customers but should be their partner in helping them to comply, thus providing greater public safety. The closing interview is the opportunity for the fire inspector to share his or her knowledge and build a relationship with the customer. A good professional relationship will reduce fire code violations and ensure compliance into the future. A good closing interview will leave the customer less apprehensive about what needs to be accomplished to gain fire code compliance and will make the owner more confident that the fire inspection was a worthwhile experience (see Figure 5.10).

The Fire Inspection Report

The fire inspection report is the vehicle by which compliance is gained.

A well-written fire inspection report begins with well-written field notes. If your field notes are organized and were systematically taken during the inspection tour, your fire inspection report will be well organized and easy to follow.

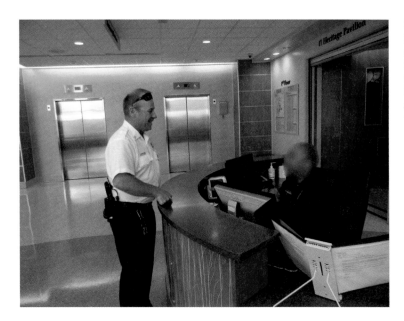

FIGURE 5.10 The fire inspector should conduct a closing interview with the owner to discuss any violations identified in the inspection tour. *Courtesy of J. Foley*

The beginning of the fire inspection report should contain all of the "need to know" information such as date of the inspection, building address, building use, group, and business name. The names of the owner, agent, and/or manager of the company; the business mailing address or PO box; and the telephone numbers of the premises and of any emergency contacts also should be included. In some cases, an agent must be specified for service of all notices, so the fire inspector must identify who that agent is and their address and telephone number. The inspector must check local code requirements on service of notices for this information. The fire inspection reports may also provide nice-to-know information that will include but not be limited to building construction type, height, and area; the types of utilities and emergency shut-off locations; and fire department–specific information, such as the location of fire alarm control panels, fire sprinkler system control valves, fire department connections, system maintenance contractors, 24-hour emergency contact numbers, and location of lock or "Knox Box" for "right to know" information. All of these pieces of information should be gathered and verified at the beginning of the fire inspection and listed on the report as shown in Figure 5.11a and b. This basic information is helpful to firefighters in the preparation of fire department preplans.

Fire inspection reports may be generated automatically by computer programs or typed into computer forms from the inspector's field notes. Fire inspectors should never send handwritten reports to owners, as they appear very unprofessional and may not always be legible or easy to read. Fire inspection reports must be clear, concise, and complete in order to ensure that fire code violations will be corrected. The fire inspection report must comply with the legal requirements of the regulation, especially if additional enforcement actions are required, such as the issuance of fire code penalties, or if the code violations are appealed. In some states, the fire code may require fire inspections to be in a specific format on a state-approved fire inspection form. The state of New Jersey, as an example, has 14 mandatory fire inspection forms that all inspection agencies within the state must use. There are three general types of fire inspection report formats: narrative types, checklist types, and combinations of narrative and checklist types. If checklists are used, the fire inspector must be sure to check every block as applicable or nonapplicable on the form. Open or nonchecked items can lead to violations not being corrected or confusion on code violations. Nonchecked items can also be a basis for appeal and can lead to difficulties in appeal and court actions for the fire inspector. If all items are not checked

Need to Know Information

INSPECTION REPORT

Type of Inspection: [ ] Annual [ ] Quarterly [ ] Complaint

Municipal Code	Occupancy Load	LHU Class	BOCA Use	Seasonal	Registration Number

Business Name	Street Address

Municipality	State	Zip Code	Telephone

Owner's Name	Street Address

Municipality	State	Zip Code	Telephone

Mailing information (if different from above) Business Name	Street/P.O. Box

Municipality	State	Zip Code	Telephone

Nice to Know Information

Attic [ ] yes [ ] no
Basement [ ] yes [ ] no
Roof Hatches [ ] yes [ ] no
Skylights [ ] yes [ ] no

Exit Signs [ ] yes [ ] no
Emergency Lights [ ] yes [ ] no
Fire Escape [ ] yes [ ] no
Elevators [ ] yes [ ] no
Elevator Recall [ ] yes [ ] no

Stories _____
Exits Per Floor _____

Area (in Sq. Ft.)
Building: _____
LHU _____
Basement _____

Extinguishers ( ) yes ( ) no
Test Records ( ) yes ( ) no

Cooking Protected ( ) yes ( ) no ( ) n/a
Test Records ( ) yes ( ) no

Sprinklers: ( ) yes ( ) no ( ) n/a
( ) full ( ) basement
( ) partial ( ) spray booth
Test Records: ( ) yes ( ) no

Fire Pump: ( ) yes ( ) no
Test Records: ( ) yes ( ) no

Fire Department Connection: ( ) yes ( ) no

Sprinkler Alarm: ( ) yes ( ) no
( ) local ( ) central

Date of Last Inspection: _____

Fire Detection System
( ) yes ( ) no

Test Records
( ) Smoke Detectors - Hard Wired ( ) yes ( ) no
( ) Smoke Detectors - Battery ( ) yes ( ) no
( ) Heat Detectors ( ) yes ( ) no
( ) Manual Pull ( ) yes ( ) no
Alarm: ( ) Local ( ) Central Station

Standpipes
( ) yes ( ) no ( ) wet ( ) dry
Test Records: ( ) yes ( ) No

Fire Department Connection
( ) yes ( ) no
Alarm: ( ) yes ( ) no
( ) Local ( ) Central

Date of Last Inspection _____

PERMITS
[ ] Annual
[ ] Temporary

Type: _____
_____
_____
_____
_____

Date issued: _____
Date of Expiration: _____
Permit Number: _____

Emergency Generator: ( ) yes ( ) no ( ) N/A   **Records:** ( ) yes ( ) no

Valid C.O. (if known): [ ] yes [ ] no
[ ] not available
Date issued: _____

Floor Construction	[ ] Concrete	[ ] Wood	[ ] Trusses			
Bearing Walls	[ ] Concrete	[ ] Block	[ ] Wood	[ ] Brick	[ ] Metal	[ ] Other
Ceiling	[ ] Plaster	[ ] Sheet Rock	[ ] Wood	[ ] Acoustic	[ ] Metal	[ ] Other
Roof Construction	[ ] Concrete	[ ] Reinf. Concrete	[ ] Wood	[ ] Trusses	[ ] Metal	[ ] Other
Heating [ ] Oil	[ ] Gas	[ ] Electric	[ ] Other	[ ] Hot Water	[ ] Hot Air	[ ] Steam
Electric	[ ] Fuses	[ ] Circuit Breakers		**Wiring** [ ] Good	[ ] Poor	

Number of Violations
Maint: _____   **Retro:** _____

_____ _____ _____
Inspector (print)   Certification #   Inspection Date

5:71-3.7(b)1.   Report Reviewed By: _____   Comments on Back: [ ] yes

FIGURE 5.11a Example of fire inspection form, including general information. *Courtesy of the NJ Division of Fire Safety*

FIRE CODE VIOLATIONS

Owner/Agent: _____

Premises: _____

_____

_____

[ ] *If box is checked, a New Jersey State Uniform Construction Code Permit is required.*

Page ____ of ____

Registration No: _____

Date: _____

Print Name: _____

_____

Inspector's Signature

U/A1 _____

U/A2 _____

U/A3 _____

[ ] Maintenance
[ ] Retrofit
[ ] Continuation Sheet

Inspector Name & Reinspection Date>

Violations cited in the above premises are as follows:

No.	Location	Nature & Description of Violation & Action Required for Abatement	Code Reference N.J.A.C. 5:70-3 / 4	Abate By Time/Date	U/A1	U/A2	U/A3

[ ] If box is checked, see additional page(s) for violations _____ to _____

Key: "A" Violation Corrected
      "U" Violation Uncorrected

NOTE: The numbering of violations is for identification purposes only and shall not be construed as bearing in any way on the seriousness of any violation.

N.J.A.C. 5:70-3, 2006 International Fire Code New Jersey Edition.

N.J.A.C. 5:70-4 New Jersey Fire Safety Code

I-102 Revised 7/09

FIGURE 5.11b Example of fire inspection form violation report. *Courtesy of the NJ Division of Fire Safety*

as to their applicability, the inspection has the appearance of being incomplete or haphazard. If a narrative report format is used, you must ensure that all violations are individually cited and numbered. Generalized statements about violations such as "Repair all exit lights" may lead to some exit lights being missed, especially in large buildings. All fire code violations need to be specifically linked to a location.

INSPECTION CERTIFICATE

Registration No.:
DATED:

Premises:

Take Notice:

This certifies that the referenced property has been inspected pursuant to the Uniform Fire Safety Act and satisfies minimum requirements of the New Jersey Uniform Fire Code.

This certificate is valid for one year unless inspections are required on a quarterly basis, in accordance with N.J.A.C. 5:70-2.5. In such case, this certificate shall not be valid unless the quarterly blocks are completed.

Commissioner,
Department of Community Affairs

1st Qtr	2nd Qtr	3rd Qtr	4th Qtr

By: **Fire Official**

This certificate must be posted in a conspicuous location at the above premises.

5:71-3.7(b)11.

FIGURE 5.12 The inspection certificate identifies compliance with minimum fire code requirements. *Courtesy of the NJ Division of Fire Safety*

Last and most importantly: The fire inspection report must tell the building owner or agent the date on which the fire inspector will return for reinspection. It should explain the owner's legal rights of appeal and how to request additional time for correction. Appeals usually have to be filed within a specific time frame by the customer once the report is received. The fire inspection report should explain the outcome that can result if the owner fails to comply, such as monetary penalties, revocation of permits, or further court actions by the enforcing agency. These appeal and extension rights also should have been pointed out to the customer during the closing interview.

Certificate of Inspection

certificate of inspection ▪ A certificate issued to an owner upon satisfactorily complying with the fire prevention code.

Most fire prevention codes require that the fire inspector, at the completion of the inspection process, issue a **certificate of inspection** (see Figure 5.12) or a final inspection report abating all of the fire code violations. Realize that the inspection is not complete until all code violations have been corrected and any outstanding monetary penalties, inspection fees, or appeal hearings have been properly resolved. Inspection compliance certificates usually must be posted for public inspection and verify to the public that the facility meets the minimum fire safety standards. Inspection certificates may also be updated quarterly in large places of public assembly to ensure compliance with life safety requirements. This is established by the inspection cycle specified in the jurisdiction's adopted codes. Typically, fire prevention bureaus issue inspection compliance certificates annually.

Fire Safety Permits

Fire safety permits will require additional special fire inspections related to the permit application. The difference between a normal fire inspection and a permit fire inspection is that permit inspections usually are more technical in fire code requirements and are smaller in scope as far as regulations that apply. Fire safety permits are required because the fire prevention code prohibits the specific activity. Typically, permits require an application fee to be paid before issuing the permit and conducting the fire inspection. Codes require that for all permits issued, an inspection must occur. The payment of a permit fee is not tacit approval for the owner to operate or conduct a process, and in fact, the permit may be denied in certain circumstances based on the application information. Fire safety permits are issued for processes that involve a high potential for fire incidents. These hazard areas may include welding and cutting, use of explosives, discharge of fireworks, storage of hazardous materials, bonfires, torch down roofing, use of open flames in places of public assembly, and similar types of activities. All permits must have fire safety inspections before the permit's issuance, and the fire inspector may revoke the permit when any conditions of the permit are violated. Fire safety permits are issued for a fixed duration of time. The time may range from a few hours for a fireworks display to 12 months for a welding and cutting shop. The fire marshal or fire official may place additional conditions on permits that may include the presence of a fire inspector during the activity. As an example, a public fireworks display generally requires the fire inspector to be present, and the inspector makes the final determination as to the firing of the display and may stop it at any time for safety concerns. Fire safety permits are different from normal fire inspections because the fire inspector is not required to provide any time for correction of violations. If the operator violated a permit condition, the activity's permission is either revoked or suspended until compliance is reestablished. Without a valid permit, the activity must cease.

Summary

The fire inspection process is a vital part of a code enforcement system. An effective fire inspection educates the customer, reduces fire risk, and inspires greater awareness of safety in the community. Fire inspectors and the inspections they conduct provide an invaluable service to the community, reducing the loss from fire. Fire inspectors must demonstrate professionalism and confidence, but must also be personable and compassionate to owners' concerns. Treating the owners as customers increases respect for the fire department, enhances professionalism, and assists the fire inspector in gaining trust and,

ultimately, code compliance. During the fire inspection tour, the inspector should educate the building owner or their designate and build customer relations. Remember that fire inspectors are not the enemy of the business community, and we are all on the same side of the problem—against fire. Conducting a fire inspection is more than just writing code violations; it is an opportunity to build relationships and trust with the business community and to change the attitudes and behaviors of each owner and improve fire safety within their buildings and the community as a whole.

Review Questions

1. Using your fire prevention code book, list the types of certifications and test records that must be reviewed during a fire inspection.
2. List and describe the five main components of a complete fire inspection.
3. List the types of information that are necessary to review before conducting a fire inspection.
4. List and describe the methods of scheduling fire inspections and address the benefits and the problems with each method.
5. List the methods of complying with legal right of entry for fire inspections. Develop a procedure or guideline for fire inspectors to follow to ensure proper right of entry.
6. Develop a list of property types in your community and develop an inspection schedule to be

submitted for ordinance approval based on these types of buildings.
7. List and describe the important aspects of the fire inspection closing interview.
8. Conduct a mock inspection of a facility and describe how you would conduct the fire inspection. Prepare a graphic of the building to demonstrate your method.
9. Conduct a mock fire inspection of a facility and complete a fire inspection form from your jurisdiction. Describe the method used to prepare the report and any changes you would make to the inspection form.
10. List the critical elements necessary for an effective fire inspection report.

Endnotes

1. http://www.usnews.com/usnews/biztech/ articles/070802/2prestige.htm
2. U.S. Constitution, Tenth Amendment.
3. U.S. Constitution, Fourteenth Amendment.
4. U.S. Constitution, Fourth Amendment.

Courtesy of J. Foley

board of appeals, *p. 117*

Dillon rule, *p. 122*

discovery, *p. 128*

due process, *p. 122*

expert witness, *p. 128*

fact witness, *p. 127*

fiduciary interest, *p. 118*

fruit of the poisonous tree, *p. 128*

judgment, *p. 122*

notice of violation, *p. 110*

order to show cause, *p. 126*

overrule, *p. 127*

preliminary hearing, *p. 126*

sustain, *p. 127*

OBJECTIVES

After reading this chapter, the reader should be able to:

- Understand of how a fire inspection report is prepared.
- Know the key elements of a notice of violation.
- Know the different tools available to a fire inspector to gain fire code compliance.
- Understand appeals and court proceedings.
- Understand the fire inspector's role as a witness in court testimony.

Professional Levels of Job Performance for Fire Inspectors as Cited in NFPA 1031 and NFPA 1037

- NFPA 1031 Fire Inspector I *Obj. 4.2.1 Prepare inspection reports*
- NFPA 1031 Fire Inspector I *Obj. 4.2.6 Participate in legal proceedings*
- NFPA 1031 Fire Inspector I *Obj. 4.3.15 Determine code compliance*
- NFPA 1031 Fire Inspector II *Obj. Investigate complex complaints*
- NFPA 1031 Plan Examiner I *Obj. Prepare reports*
- NFPA 1037 Fire Marshal *Obj. 5.2.6 Direct the development, maintenance, and evaluation of department record keeping system*
- NFPA 1037 Fire Marshal *Obj. 5.6.5 Manage the appeal process*
- NFPA 1037 Fire Marshal *Obj. 5.6.6. Manage the process of record keeping*

- NFPA 1037 Fire Marshal *Obj. Manage a program for alternative compliance measures*
- NFPA 1037 Fire Marshal *Obj. Manage a program to coordinate with other agencies*

Introduction

The fire inspector must obtain the building owner's or agent's permission to have proper and legal entry to a building. Fire inspectors must state the nature of the visit, present their credentials, and request to proceed before they can move to the next step in the inspection process. The next step in the process is the identification and correction of fire code violations. Because fire code enforcement may require legal action, the fire inspector must have knowledge of the legal requirements necessary to properly enforce the fire prevention code.

The fire inspection process is a law enforcement action. The process is not voluntary to the business operator and, in fact, usually is mandated by municipal ordinances or state regulations. The duty of the fire inspector is to identify and correct any fire code violations that they observe or encounter. The fire inspector, however, is not the final judge or jury in the process. Like any other law enforcement activity, identifying fire code violations requires following correct legal protocols and affording the business owner his due process. We will discuss the correct methods for preparing notices and attaining correction. The fire inspector must understand the legal process necessary to force compliance if violations are not corrected voluntarily. We also must examine violations that create an imminent hazard to the occupants and must be immediately corrected. Fire inspectors must also prepare inspection cases for appeal hearings or court actions and will be required to testify in those cases.

Record Keeping and Technology

Before we discuss the preparation of notices of violation and fire inspection reports, we should discuss the two main methods by which records are maintained. Traditionally, fire prevention bureaus have kept written or typed records in property files for each inspection they conduct. Each state's archive laws require specific time frames for which particular reports must be retained by the enforcing agency. Fire officials and fire marshals should check their local archiving requirements. Hard copy files still need to be maintained for public inspection purposes and court of board of appeals cases. While computer technology reduces paperwork, it will not totally eliminate it in public inspection agencies.

In today's modern fire prevention bureau, files usually are maintained on computer database systems. The number and types of computer information management systems is extensive but they generally fall into one of three categories: off-the-shelf software solutions, custom designed program software, and Web-based application software. The off-the-shelf solutions are generally used by small to medium jurisdictions, as they usually are low cost and affordable. The disadvantage of these programs is that you may be required to modify your fire inspection practices to fit the software, and they may not provide all of the information management requirements you desire. Custom designed software may be utilized in larger jurisdictions that have specific information technology departments and must integrate record keeping with other agencies. However, these systems are expensive and may not totally provide all of the functionality necessary in fire inspections. The other disadvantage is the IT department will have total control over the technology, making changes and permissions difficult. The third and growing option—Web-based inspection applications—generally are more flexible in design but may have security issues. These Web-based systems have a distinct advantage in that information is housed at multiple storage locations, so if a local system goes down, the information is still retrievable. Many of the Web-based systems also have the building and fire codes already in the system, making them easier to use. The fire official, fire marshal, and fire inspector should all be included in the evaluation of any software application. Ultimately, an end user who is not satisfied with the application's ease will not use the technology effectively.

The second part of the technology equation is the hardware necessary to support the system. Today, handheld computers are being replaced by iPads and tablets. These devices are smaller, lightweight, and relatively easy to use. The inspector must be careful when using these technologies, however. As an example: While the inspector could e-mail a fire inspection report immediately from an iPad, the courts may not consider this properly served, as the fire code has no provision for e-mail service. Considerations for handheld devices should include rugged protective cases, sufficient battery life or spare batteries to get through the workday, avoiding outside lighting conditions that may make the device hard to see, and ease of use (see Figure 6.1). Some departments may also provide capability for printing reports in the field from portable vehicle printers. Fire officials and fire marshals should think about field printing, as it doesn't allow supervisory review of the work product. The field of fire inspection and fire department management software is vast, and generally no one size fits all solutions. It is a good practice to review vendor information and to speak with actual fire inspection users before making any software and hardware decisions.

FIGURE 6.1 Fire inspector using a computer tablet for fire inspection. *Courtesy of J. Foley*

Types of Corrective Orders

Corrective orders given by fire inspectors can take different formats depending on the jurisdiction's fire code requirements. Fire inspectors should not use verbal orders to owners to correct code violations. Verbal orders are not documented and can be misunderstood by the person receiving the order. Any violation that is explained verbally must also be issued in writing on the fire inspection violation report. Fire inspectors should also note the violation, even if it is corrected in their presence. Fire inspectors must realize that verbal orders can be ineffectively communicated, not clearly understood, or in some cases, may be denied as ever having been said. The critical point here is that if it is a code violation, it must be documented.

Corrective orders may vary in the way they are prepared. Some fire departments may use specific fire inspection forms for correction of violations, while others may use a letter format outlining the specific violations to be addressed. In some cases, a combination of the two methods may be used to provide extra detail or information on the correction of the violation. Some states require specific mandated fire inspection forms and notices to be used by all fire marshals or fire officials within the state. The benefit to the state is uniformity in communications, and it also ensures that all fire inspectors and fire officials are providing the specific legal components and requirements of the state fire prevention code to every building owner. Uniformity streamlines the inspection process and eliminates confusion and conflict in enforcement methods.

notice of violation ■ A notice issued to a building owner with regard to open code violations specifying correction dates, appeal rights, and statutory penalties for noncompliance.

A **notice of violation** order is a document used in conjunction with the fire inspection report to identify fire code violations and gain code compliance. The notice of violation, or NOV, must specify who is receiving the notice, whether it is the owner, the occupant, or the registered agent for service, or all three (see Figure 6.2). Fire inspectors must identify the correct person to receive the order, and that person is the one who has the power and ability to correct the violation.

In most cases, multiple copies of the orders may be issued to all parties to ensure compliance. Fire prevention codes have administrative requirements that NOVs be served upon the person either causing or responsible for correcting the actual code violation. An order should not be served upon occupants who have no power or authority to correct the violation. As an example, you could inspect a mercantile store, which has an egress violation due to placement of stock in front of the exit door and a sprinkler system violation for maintenance of the system. The egress violation may be addressed to the occupant or tenant of the store and the sprinkler violation to the building owner to correct. Each of these individuals may or may not have the power to correct the given code violations. In these cases, it is best for the fire inspector to direct the NOV to both parties to expedite the correction process. Fire prevention codes usually establish that if a tenant fails to correct the code violation, then ultimately, it becomes the owner's responsibility to resolve. This is why the owner should always be served with a notice of violations. Fire inspectors must also understand that when owners rent properties, they surrender certain property rights to the renter; however, most building leases have clauses allowing the owner access for the purpose of complying with legal requirements of government, such as complying with fire codes. In situations of leased properties, fire inspectors must still be cautious and ensure they have proper consent from both parties to perform the fire inspection in case certain areas of the building are retained under the owner's control. This may occur in buildings of mixed occupancy.

The notice of violation (NOV) also provides the owner with information on follow-up actions by the fire inspector, such as how to request additional time to correct the violations, and it explains the owner's appeal rights and how to request a hearing. The NOV may also include the imposition of potential fines or the possible closure of the property. Most importantly, the order states the compliance dates by which the violations must be corrected or additional enforcement actions will occur. The NOV form

NOTICE OF VIOLATIONS
and
ORDER TO CORRECT

Page _____ of _____

Date: _____ Inspector: _____ Registration No: _____

(Name of Business, Structure, Premises)		
(Address)		

OWNER	AGENT	TENANT/OPERATOR
(Name)		
(Address)		
(City, State, Zip)		
(Telephone Number)		

YOU ARE HEREBY NOTIFIED THAT an inspection of the above referenced property by the _____

page(s).

YOU ARE HEREBY ORDERED by the COMMISSIONER to correct the violations listed on the accompanying "violations" page(s) within the time, or by the date specified. If a reinspection discloses that violations have not been corrected, you will be subject to penalties of up to $5,000.00 per violation per day or as otherwise authorized by the Act and Department Regulations. IN ADDITION, the ACT imposes liability on the owner for the actual costs of fire suppression where a violation directly or indirectly results in a fire.

Commissioner, Department of Community Affairs

By: _____
FIRE OFFICIAL

I hereby acknowledge receipt of a copy of this NOTICE OF VIOLATIONS and ORDER TO CORRECT.

_____	_____	_____	_____
Signature	Printed Name	Title	Date

APPEAL RIGHTS-EXTENSIONS

5:71-3.7(b)2   >See reverse side for information concerning your administrative appeal rights<

FIGURE 6.2 Example of Notice of Violation Form. *Courtesy of the NJ Division of Fire Safety*

should describe how a building owner can file for appeal and should include the filing fees and legal time limits on filing the appeal. In most cases, the owner must file the appeal in writing within 15 days of receiving the notice of violation. If the owner fails to request an appeal within the required time frame, the appeal rights are considered waived and he or she no longer has a right to appeal the code violations. As an example, in some states a building owner who requests additional time to correct code violations automatically waives the right to appeal the notice of violation. The concept is that if additional time is needed to correct, the violations must exist. Fire prevention codes may also establish minimum correction times that the fire inspector must provide before any additional enforcement actions can be taken. Fire inspectors must be aware of these time frames when establishing reinspection dates to ensure they have provided the minimum necessary time to correct the code violation.

Preparing Notices of Violation and Fire Inspection Report

Buildings that have violations of the fire prevention code must be given a formal notice of violation along with the fire inspection report (see Figures 6.3 and 6.4) that outlines the owner's rights as well as any consequences for failure to comply. Each code violation must be cited correctly on the fire inspection report and should be in a format that meets the legal requirements of the fire prevention code. Violations should also be in a format that the owner can easily understand and therefore correct without extensive code knowledge. Simple violation formats such as action, item, location, violation, and statute citation will help make the violation easy to understand while still meeting the legal citation requirements of the code regulation. Owners may also have access to the fire prevention code for review through municipal or state government websites, and most municipal governments retain a hard copy of the code with the city clerk for public viewing.

Action

The action is what you need the owner or tenant to do to correct the code violation. Action can usually be described in single words: repair, replace, install, remove, open, close, test, certify, relight, obtain, relocate, and so on. These single action words direct the owner or tenant on what action is to be taken to resolve the code violation.

Item

The item follows the action and is the fire code violation itself. Items should be explained in simple phrases that describe the item of violation such as the following examples:

- Remove *curtains in front of the exit door*
- Test *automatic fire pump*
- Install *UL-listed door opener on stairway door*
- Replace *damaged manual fire alarm pull station*

The item should be described in its simplest terms and not fire code jargon that the building owner may not readily understand.

Location

The location is the specific area where the code violation exists. This may require a room number, a room or area name, a floor level in the building, or any other identifier that will assist the person finding and correcting the code violation.

- Remove curtains in front of exit door *in the theater at stage left*
- Test automatic fire pump *in parking garage of third floor*
- Install UL-listed door opener on stairway door *on fourth floor, north stair #4*
- Replace damaged manual fire alarm pull station *at rear exit from kitchen*

The clarity of these first three elements of action, item, and location are the most critical to getting the owner to properly identify and correct the code violations.

Violation and Code Citation

The last element—the language of the fire code citation itself that has been adopted by statute or ordinance—is for the purpose of legal enforcement. The violation and code citations provide the specific language and code section that is in violation. Often, the code language can be confusing to laymen as to what the requirement means. The violation language and citation must be included for appeal purposes should the violation be challenged in a board of appeals or in a court. Without the actual language and code citation

INSPECTION REPORT

Type of Inspection: [ ] Annual   [ ] Quarterly   [ ] Complaint

Municipal Code	Occupancy Load	LHU Class	BOCA Use	Seasonal	Registration Number

Business Name	Street Address		
Municipality	State	Zip Code	Telephone

Owner's Name	Street Address		
Municipality	State	Zip Code	Telephone
Mailing information (if different from above) Business Name	Street/P.O. Box		
Municipality	State	Zip Code	Telephone

Attic   [ ] yes   [ ] no
Basement   [ ] yes   [ ] no
Roof Hatches   [ ] yes   [ ] no
Skylights   [ ] yes   [ ] no

Exit Signs   [ ] yes   [ ] no
Emergency Lights   [ ] yes   [ ] no
Fire Escape   [ ] yes   [ ] no
Elevators   [ ] yes   [ ] no
Elevator Recall   [ ] yes   [ ] no

Stories _____

Exits Per
Floor _____

Area (in Sq. Ft.)

Building: _____
LHU _____
Basement _____

Extinguishers   ( ) yes   ( ) no
Test Records   ( ) yes   ( ) no

Cooking Protected ( ) yes   ( ) no   ( ) n/a
Test Records   ( ) yes   ( ) no

Sprinklers: ( ) yes   ( ) no   ( ) n/a
                      ( ) full   ( ) basement
                      ( ) partial   ( ) spray booth
Test Records: ( ) yes   ( ) no

Fire Pump: ( ) yes   ( ) no
Test Records: ( ) yes   ( ) no

**Fire Department
Connection:**   ( ) yes   ( ) no

Sprinkler Alarm: ( ) yes   ( ) no
                      ( ) local   ( ) central

Date of Last Inspection: _____

Fire Detection System
( ) yes   ( ) no

Test Records
( ) Smoke Detectors - Hard Wired   ( ) yes   ( ) no
( ) Smoke Detectors - Battery   ( ) yes   ( ) no
( ) Heat Detectors   ( ) yes   ( ) no
( ) Manual Pull   ( ) yes   ( ) no

Alarm: ( ) Local   ( ) Central Station

Standpipes
( ) yes   ( ) no   ( ) wet   ( ) dry

Test Records: ( ) yes   ( ) No

Fire Department Connection
( ) yes   ( ) no

Alarm: ( ) yes   ( ) no
        ( ) Local ( ) Central

Date of Last Inspection _____

PERMITS

[ ] Annual
[ ] Temporary

Type: _____

_____
_____
_____
_____

Date
issued: _____

Date of
Expiration: _____

Permit
Number: _____

Emergency Generator: ( ) yes   ( ) no   ( ) N/A   **Records:** ( ) yes   ( ) no

Valid C.O. (if known): [ ] yes [ ] no
[ ] not available
Date issued: _____

Floor Construction	[ ] Concrete	[ ] Wood	[ ] Trusses			
Bearing Walls	[ ] Concrete	[ ] Block	[ ] Wood	[ ] Brick	[ ] Metal	[ ] Other
Ceiling	[ ] Plaster	[ ] Sheet Rock	[ ] Wood	[ ] Acoustic	[ ] Metal	[ ] Other
Roof Construction	[ ] Concrete	[ ] Reinf. Concrete	[ ] Wood	[ ] Trusses	[ ] Metal	[ ] Other
Heating [ ] Oil	[ ] Gas	[ ] Electric	[ ] Other	[ ] Hot Water	[ ] Hot Air	[ ] Steam
Electric	[ ] Fuses	[ ] Circuit Breakers		**Wiring**	[ ] Good	[ ] Poor

_____   _____
Inspector (print)   Certification #

Number of Violations

Maint: **Retro:**

_____
Inspection Date

5:71-3.7(b)1.   Report Reviewed By: _____   Comments on Back: [ ] yes

---

FIGURE 6.3 Example of Fire Inspection Form—General Information. *Courtesy of NJ Division of Fire Safety*

FIRE CODE VIOLATIONS

Owner/Agent: _____

Premises: _____

_____

_____

[ ] *If box is checked, a New Jersey State Uniform Construction Code Permit is required.*

Violations cited in the above premises are as follows:

[ ] Maintenance
[ ] Retrofit
[ ] Continuation Sheet

Inspector Name & Reinspection Date>

Page ____ of ____

Registration No: _____

Date: _____

Print Name: _____

_____
Inspector's Signature

U/A1 _____

U/A2 _____

U/A3 _____

No.	Location	Nature & Description of Violation & Action Required for Abatement	Code Reference N.J.A.C. 5:70- 3 / 4	Abate By Time/Date	U/A1	U/A2	U/A3

[ ] If box is checked, see additional page(s) for violations _____ to _____

NOTE: The numbering of violations is for identification purposes only and shall not be construed as bearing in any way on the seriousness of any violation.

I-102 revised 7/09

Key: "A" Violation Corrected
"U" Violation Uncorrected

N.J.A.C. 5:70-3, 2006 International Fire Code New Jersey Edition.

N.J.A.C. 5:70-4 2006 New Jersey Fire Safety Code

FIGURE 6.4 Example of Fire Inspection—Violations Form 2. *Courtesy of NJ Division of Fire Safety*

being provided, the fire inspection report is technically flawed and most likely will be vacated in an appeal or court hearing. The action, item, and location should provide sufficient information to correct the code violation for most people. The specific violation and code citation provide the legal requirements of identifying the specific section of the fire prevention code and the statutory code section where it can be found.

- *Violation #1: Remove curtains in front of exit door in the theater at stage left.*
  IFC 1027.4 Furnishings and decorations. Furnishings, decorations, or other objects shall not be placed so as to obstruct exits, access thereto, egress there from, or visibility thereof. Hangings and draperies shall not be placed over exit doors or otherwise be located to conceal or obstruct an exit. Mirrors shall not be placed on exit doors. Mirrors shall not be placed in or adjacent to any exit in such a manner as to confuse the direction of exit.
- *Violation #2: Test automatic fire pump in parking garage of third floor.*
  IFC 913.5 Testing and maintenance. Fire pumps shall be inspected, tested, and maintained in accordance with the requirements of this section and NFPA 25.
- *Violation #3: Install UL-listed door opener on stairway door on fourth floor, north stair #4.*
  IFC 703.2.2 Hold-open devices and closers. Hold-open devices and automatic door closers, where provided, shall be maintained. During the period that such device is out of service for repairs, the door it operates shall remain in the closed position.
- *Violation #4: Replace damaged manual pull station at rear exit from kitchen.*
  IFC 901.6 Inspection, testing and maintenance. Fire detection, alarm, and extinguishing systems shall be maintained in an operative condition at all times, and shall be replaced or repaired where defective. Nonrequired fire protection systems and equipment shall be inspected, tested, and maintained or removed.

The fire prevention code violations should be listed on the inspection form established by the jurisdiction. Some fire departments utilize check-off forms to assist the fire inspector in identifying the specific code violations. If your department uses this type of form, make sure that all appropriate blocks are marked or, if not applicable, they are identified as such. Incomplete forms may become a technical issue if the violations are later appealed. If the fire inspector does not complete the inspection form properly, it raises issues of his professional knowledge as well as the thoroughness of the fire inspection. Fire inspectors need to perform their job professionally each and every time and cite violations correctly. Fire code violations usually fall into two general categories: those that are specific to a fixed single location, and those that reoccur throughout a property in many different locations. An example of a specific item would be a fire pump required to be tested annually and certified by documentation; in most cases there are only one or two fire pumps at the property, and they usually are located in a fire pump room. An example of a single reoccurring violation may be numerous exit signs that are not lit throughout the entire building. The fire inspector may write these violations as one particular type of code violation, but they must specify every specific location where the violation reoccurs. It is not a good fire inspection practice to identify the violation as throughout the building or the entire building. If you do this, it is very likely that not all exit signs will be repaired because some will be missed.

The fire inspector must identify a date for reinspection of the property; this is referred to as the time to correct. These reinspection dates may vary based on the gravity of the specific violation and in some instances, multiple notices of violations may have to be issued to avoid confusing the building owner as to when specific violations must be repaired. One method that can be employed is to use a separate inspection form for violations that require a short period of time to correct but are not necessarily imminent hazards. These forms are sometimes referred to as *Field Correction Notices* and stand as their own legal notice of a code violation. As an example, these notices can be used for items such as fire alarm system trouble, minor conditions obstructing egress, failure to obtain a fire safety or building permit, or removal of trash from the premises. Field correction notices are usually issued to the owner or occupant and have a correction time of 24–48 hours. The regular fire inspection report and notice of violation contains the code violations, which the owner may have 30–60 days or longer to correct based on the need to obtain contractors, architects, engineers, and plan reviews by building code officials.

REINSPECTIONS

It is equally important that the fire inspector return on or immediately after the reinspection date to determine whether violations have been corrected. While the inspector's goal should always be to return on the appropriate date, in the fire department, emergencies happen that may affect the fire inspector's routine on any given day. If you miss the reinspection date, you should make that reinspection a priority for the next available workday. Note also that the fire inspector cannot return before the reinspection date cited on the notice of violation unless the owner requests an early reinspection. The fire inspector can take no further actions, such as instituting fire code penalties on any notice of violation until the reinspection date on the notice of violation has passed.

Upon reinspection, the fire inspector will encounter one of three possible outcomes. The first outcome is that all violations are corrected or abated properly completing the fire inspection. This occurs in approximately 20 to 25 percent of all annual fire inspections. The second outcome is that violations are partially corrected by the owner. In this case, the fire inspector must make some value judgments. If the owner is making a diligent effort to comply but is running into contractor or parts problems, additional time to correct the violations is warranted unless they immediately affect life safety. If the owner corrected a few easily correctable violations, like changing a light bulb, in 30 days, but has left the more difficult items uncorrected, the fire inspector should be more inclined to institute other enforcement actions to motivate the owner to code compliance. Such actions could be the institution of monetary penalties or fines based on what the fire prevention code prescribes. In most of these cases, compliance is usually gained after the initial penalties are issued. These partially corrected violation situations are common and occur in 30 to 70 percent of annual fire inspections. The last scenario is that upon reinspection of the property, no violations have been corrected. This occurs about 25 to 30 percent of the time. In this case, the next step must be the institution of a monetary penalty or other action that the fire prevention code may prescribe including court action, if necessary. The fire inspector must always prepare for this last outcome when writing his fire inspection reports. Even in these difficult enforcement cases that account for 25 to 30 percent of fire inspections, only about 1 to 3 percent end up in a board of appeals or other court action. This is not to say that gaining code compliance will be easy, as the last 25 to 30 percent of inspections will dominate a large portion of the fire inspector's and fire official's time to gain compliance. Fire inspectors must assume that every inspection may go to court; therefore, the inspector must make sure to properly perform each step in the process so as to have a successful outcome in court. The fire inspector also must realize that conducting the fire inspection creates a special duty to the inspector to follow up as the regulation requires. Failure to reinspect promptly may make the inspector and municipality liable should anyone be hurt in a fire.

Fire Code Variations

The fire prevention codes generally allow building owners to request a variation to any code requirements. Fire inspectors should check their local codes for variation request requirements. Typically, variations must be requested in writing and cannot be considered for financial reasons alone. The request must demonstrate some level of not being achievable due to physical constraints or configurations within the building. The variation must provide a viable code alternative that will provide equal or greater safety to the occupants than the original code requirement. Variations must be approved or denied by the authority having jurisdiction and may be the subject of appeal if they are denied. Variations do provide an opportunity for fire inspectors to attain greater protection within the building.

A number of existing rooming houses over three stories were required to enclose the interior stairways with fire doors and fire barriers in accordance with specific retro-fit requirements in the state fire prevention code. These buildings were originally constructed as single-family homes and generally had open stairs that led to narrow 36-inch hallways. From a practical standpoint, by installing fire doors and fire barriers in the halls, the owner would no longer be able to move furniture into the rooms without great difficulty. The solution was to eliminate the fire doors and barrier requirements contingent on the owner's installing a compliant NFPA 13R automatic sprinkler system. The residential sprinkler installation was actually more cost effective, provided greater public safety, and allowed the owner better use of the building. The variance was a win/win for all.

Serving Notices of Violation

After the notice of violation and fire inspection report is prepared, it must be properly served upon the owner of the building or the person responsible for correcting the code violation. Fire inspectors should check their local fire code enabling ordinance or state statute for the requirements on service of notices. Generally, these include personal service, registered mail with return receipts, or posting of notices.

The first method is personal service upon the owner, as this implies that the fire inspector hand delivered the notice and the owner signed a receipt that they received it. The signed document satisfies the **board of appeals** or court requirements that the owner has received the notice of the violation and fire inspection report on a specific date. Hand-held computer tablets with fire inspection software may allow the inspection report and notice of violation to be printed in the fire inspector's vehicle, and the owner's signature can be captured by the computer. Personal service also requires that the notice be delivered to an adult or individual over a specific age, usually 14. Fire inspectors should review their local requirements for personal service of notices and make sure the person being served is a legal adult according to their statute.

board of appeals
■ A body of members consisting of non-judicial persons who are stakeholders in the field appointed to resolve appeals.

Most fire prevention bureaus have fire inspectors prepare their reports and send them out by mail after a supervisor reviews the report. If the notice and report are to be mailed, they should be sent with a return receipt requested to ensure an accurate record of delivery. Boards of appeals and courts require reasonable proof that the notice was properly delivered to the defendant. In any appeal or court action, the first hurdle for the fire inspector is proof of service. If NOVs and fire inspection reports are typed and mailed, the fire inspector must also allow sufficient time for paperwork preparation when considering the reinspection date. In most cases, that should be a minimum of 3 days to a maximum of 7 days. Many fire prevention codes require that a minimum time be provided to correct the code violations, which is usually 15 days. If you allow 2 days for typing and preparing the notice, 3 days for mailing, and 15 days to correct the violations, at minimum, the reinspection cannot occur until the twenty-first day after the initial inspection. Typically, 30 days to correct violations provides adequate time for all of these processes to occur, and this ensures proper service and adequate time to correct violations.

In cases of absentee landlords or abandoned properties, it may be more difficult to locate a property owner. The courts usually allow reasonable efforts to be made in delivering the notices of violation. These methods include mailing the notice to the owner's last known address and posting the notice at the main entrance of the building. You may also be required to file a copy of the notice with the state attorney general or the state secretary of state. Typically, after a short period of time, such as 3 days, the courts will consider the notice legally served upon the owner.

Another important point to consider with regard to service is who is considered the owner under the definition of the fire prevention code. Fire prevention codes generally have liberal definitions of owners or agents for the purpose of service. These definitions often take into consideration employees that work for the owner, any agent of the owner, or any person reasonably in charge of the physical property or entities having a **fiduciary interest** in the property such as banks and mortgage companies. A good example would be the foreclosure process of a business. If the owner walks away from the property, the mortgage holder will become the owner by default and therefore be responsible for code corrections and the abatement of the violations. The fire inspector must be aware that properties can never go without an owner. If no one claims ownership, the municipal government will end up being the owner due to unpaid tax liens. Last, once the fire inspector has attained the proof of service, he or she must ensure that all proofs are properly filed with the corresponding NOV and fire inspection report so that they have the required paper trail should a property owner request an appeal or other court action.

UNSAFE STRUCTURE ORDERS

The unsafe structure order is used for emergent situations that may eventually create an imminent hazard situation. Unsafe structures are buildings, structures, or equipment that pose a significant hazard to life, safety, or health and require immediate attention from the owner or occupant to abate the hazards. Typically, unsafe structure orders are issued to allow a short period of time to remedy the code violations. During that time, the fire inspector may have to seal a hazardous appliance or piece of equipment such as a defective heater, water heater, or cooking appliance. The process of sealing an appliance is sometimes referred to as *red tagging*, where the inspector places a tag on the appliance requiring that its use be suspended until corrections are made. Once sealed, it is unlawful for anyone to operate the equipment until it is inspected and approved by the fire inspector. Often, an unsafe structure order leads to the issuance of an imminent hazard order, causing the removal of occupants from a building until the code violations are corrected. Unsafe structure conditions may include a structural defect with the building; illegal or improper occupancy; hazardous or defective equipment, such as an oil heater or gas-fueled appliance; or an open, vacant, or abandoned structure. Typically, the fire prevention code requires the fire inspector to also notify construction or building officials anytime an unsafe structure order is to be issued, as this may impact the owner's certificate of occupancy.

ABANDONED BUILDINGS

The securing of vacant buildings is a critical part of the fire inspector's duties. Vacant structures are attractive nuisances and are a breeding ground for arson and other crimes. The NFPA reported that from 2003 to 2006 over 31,000 structure fires occurred in vacant, unsafe structures. These fires resulted in 141 civilian deaths, 15 firefighter deaths, and 4,500 firefighter injuries. That translates to 13 percent of all firefighter injuries during that time period. The NFPA also reported that 43 percent of these vacant structure fires were intentionally set and were proliferating the crime of arson in communities.[1] Fire prevention inspectors must not overlook vacant structures in their community, as they are breeding grounds for larger problems and crimes that reduce the public's safety.

Most municipalities have local municipal code provisions for the demolition of abandoned structures in a state of extreme disrepair. Fire inspectors may find it difficult to locate owners of abandoned structures for the purpose of conducting an inspection. Most courts will permit a warrantless inspection of the property if the building is open and vacant. The court assumes that the owner at that point has no reasonable expectation of privacy; however, it is still better in these cases to get a inspection warrant, especially if the municipality is going to demolish the structure because it is unsafe.

```
┌─────────────────────────────────────────────────────────────────────────┐
│                      NOTICE OF IMMINENT HAZARD                            │
│                              AND                                          │
│                 ORDERS TO TAKE CORRECTIVE ACTION                          │
│                                                                           │
│  ISSUED TO:                              FOR PREMISES:                     │
│                                                                           │
│                                                                           │
│  PAGE _____ OF _____        DATE:        REG. #:                      │
│                                                                           │
│  You are hereby notified that an inspection by the _____ disclosed violations of the │
│  Uniform Fire Code (N.J.A.C. 5:70-1 et. seq.) promulgated pursuant to the New Jersey Uniform Fire Safety │
│  Act (N.J.S.A. 52:27D-192 et. seq.) which constitutes an IMMINENT HAZARD TO THE PUBLIC HEALTH, │
│  SAFETY OR WELFARE. The violations are specified on the accompanying "Violation" page(s). │
│                                                                           │
│  Due to the violations listed on the accompanying page(s), YOU ARE HEREBY ORDERED, to take corrective │
│  action regarding the above referenced premises as specified adjacent to the box(es) indicated below. │
│                                                                           │
│       [ ] VACATE the above referenced building, structure, or premises    │
│             by _____ am/pm on _____,                │
│                                                                           │
│       [ ] CLOSE the above referenced building, structure, or premises     │
│             by _____ am/pm on _____,                │
│                                                                           │
│       [ ] REMOVE _____ of / in the above referenced building,      │
│             structure, or premises by _____ on _____.           │
│                                                                           │
│                    Commissioner, Department of Community Affairs           │
│                                                                           │
│           By: _____                    │
│                              Fire Official                                │
│                                                                           │
│                          Appeal Rights                                    │
│      See reverse side for information concerning your administrative appeal rights. │
│                                                                           │
│      5:71-3.7(b)7.                                                        │
└─────────────────────────────────────────────────────────────────────────┘
```

FIGURE 6.5 Example of order of imminent hazard. *Courtesy of NJ Division of Fire Safety*

IMMINENT HAZARD ORDERS

The nature of an imminent hazard is that if the fire inspector does not take some form of immediate action, a tragedy is likely to unfold. The term "imminent" means ready to take place or threatening. The issuing of an imminent hazard order will always require that the fire inspector evacuate the occupants from the imminent hazard situation until such time as the violations can be corrected or mitigated to reduce the life safety risk (see Figure 6.5). Examples of imminent hazards include buildings that have any of the following conditions:

- Unsafe structural condition, or potential building collapse
- Locking or blocking of means of egress
- Presence of explosive fumes, gases, or materials in violation of the fire prevention code
- Presence of toxic gases, flammable vapors, or flammable and combustible liquids in violation of the fire prevention code
- Inadequacy of fire protection systems, broken and disabled fire alarms, broken sprinkler systems, etc.

The key to identifying an imminent hazard is that if the fire inspector fails to act, the outcome is all but certain to have a negative consequence. The format of an imminent hazard order will require removal of the occupants, closing of the structure, and removal of the violations before the owner can reestablish occupancy or use. In the cases of imminent hazards, owners can usually request immediate appeal hearings to the fire inspectors verbally, and a hearing must be conducted within 24–48 hours of the imminent hazard action. The owner may also file a court action for injunctive relief on an order to show cause. Unlike other fire code enforcement actions, under an imminent hazard order the enforcing agency may still take action to remove occupants from the building or may institute other precautionary measures even though the matter is under appeal. The courts do not stop or stay the fire inspector's orders during the appeal process because of the potential risk to the occupants. Often, a fire inspection and violation notice may expand into an unsafe structure order or imminent hazard because the owner has failed to abate fire code violations at the property. As an example, in Paterson, New Jersey, the local fire official inspected 56 public schools in 2007 and cited numerous violations of the fire prevention code over the summer vacation. The schools opened in September but failed to correct the fire violations and were fined $13,000 by the local fire official. The school board continued not to address the violations and in January, the fire official eventually had to close 42 of the 56 local schools under imminent hazard order until the serious code violations were corrected.[2]

Because imminent hazards can be cross-jurisdictional, the fire inspector should communicate with all building inspection agencies to gather support for the order. This includes the building and construction officials, electrical inspectors, health department officials, and any other agencies needed, especially the solicitor and the engineer's office. The issuance of imminent hazard orders can create a firestorm in a community and can become a highly politically charged event. The fire inspector must perform his duty, but must also be prepared for the pressures that may follow. In the case described above, 36,000 students were affected, and generally this will cause tremendous political pressure to reopen the closed schools as quickly as possible. The fire inspector must ensure that the imminent hazards are properly inspected and rectified before allowing reoccupancy. Having good interagency support and communications assists the fire inspector by ensuring that the imminent hazards are properly rectified. Fire inspectors can be assured that building owners who allow their properties to fall into such a state of disrepair may attempt to paint any imminent hazard order as an overreaction by an overzealous fire inspector. Good interagency support edifies the fire inspector's assessment of the building and makes it more difficult for owners to raise such allegations. Fire inspectors must ensure that all code violations are properly cited, identified, explained, and supported by other agencies in the jurisdiction, including the city administration. Before such orders are issued, elected public officials should be briefed. This is because of the political nature of displacing building occupants. Briefing government officials, including the mayor, avoids miscommunication and embarrassment to the public official when approached by the media. Fire inspectors must remember that taking someone out of their building or home will generate news media questions and community concerns to public officials. Having officials properly briefed by the fire department presents a more unified public appearance and better support to the fire prevention effort.

Fines and Penalties

Another tool that fire inspectors may have at their disposal is the imposition of monetary fines or penalties. Typically, the fire prevention code establishes minimum and maximum fine structures within the administrative regulation chapter of the fire code. Fines usually may be instituted as single fines for a violation or daily fines for every day the violation continues to go unabated after the compliance date. Fire prevention codes also may have rules for fire officials to reduce or waive imposed fines based on compliance with the notice of violation and the owner's commitment not to have the violation reoccur (see Figure 6.6).

CITY OF ATLANTIC CITY
NEW JERSEY
FIRE PREVENTION BUREAU

ORDER TO PAY PENALTY

DATE ISSUED: ______
PREMISES: ______

ISSUED TO: _____

Take notice that on _______, you and/or your agent suffered a fire at _____, at which Code Violations were discovered.

This Investigation revealed that violations existed that were uncorrected. According to our records a permit was not obtained or required Safety Regulations followed.

Therefore, pursuant to the Uniform Fire Safety Act, N.J.S.A. 52:27 D-192 et. Seq., you are Hereby Ordered To Pay A Penalty Of $_______.

In accordance with the above referenced law, you **MUST PAY** the total amount due within 30 days after issuance of this order. If you fail to make full payment within 30 days, this matter will be referred to the Office of the City Solicitor for summary collection pursuant to the Penalty Enforcement Law (N.J.S.A. 2A:58-1 et. Seq.).

Take notice That Pursuant to N.J.A.C. 5:70-213, penalty assessments can be $150,000.00 or the actual cost of fighting the fire, which ever is greater.

If you believe the public interest in fire safety supports compromise of any penalty assessed, promptly send a written, detailed explanation to the Bureau, Attention: Fire Official. Include your Registration Number and daytime telephone number.

Take Notice that this Penalty is in addition to all penalties previously assessed.

Make Check Payable to: City of Atlantic City, City Hall – Suite #406, 1301 Bacharach Boulevard, Atlantic City, New Jersey 08401-4603

James M. Foley
Fire Official

INC: Appeal rights

FIGURE 6.6 Example of order to pay penalty

An important aspect of imposing monetary penalties as a fire code enforcement tool is the agency's ability to actually collect the fines. Penalties that are issued and never collected send a signal to the community that the fire prevention effort is a paper tiger and eventually the penalties will be ignored. Penalty collection usually happens by one of two methods. The first method is that the fire official may send an unpaid penalty notice to the owner demanding payment of the fine after the specified payment date has passed. Often, the second notice encourages the recipient to pay the fine and avoid entanglement with the code enforcing agency. The second method is to file a claim in court for summary judgment or a lien on the property in the amount of the unpaid fine. These actions, depending on the amount of money due, may go into small claims court or a special civil court and usually follow an established administrative procedure for the court to collect the fines. The key element in these collections again is proper proof of service to the owner. In these summary proceedings, there is generally no testimony from the defendant with regard to the violations. The fire inspector will be required to testify to the facts leading up to the fine and present copies of the paperwork and proofs of service for each document placed in front of the court. The judge will then render a decision on whether the spirit or intent of the law was followed and will order compliance and payment, including the addition of court costs. The court then will issue a **judgment**. The enforcing agency may then use the judgment to obtain either a property lien on real estate or they may exercise a sheriff's sale of physical property. In most cases when the sheriff's officers show up, the fine gets paid.

Some states employ other monetary tools to gain fire prevention code compliance. For example, some states may recover firefighting costs from property owners that have uncorrected fire code violations. The imposition of fines is a tool that fire inspectors may use to gain code compliance in more difficult inspections. A key element that the fire marshal and fire official must deal with is the collection of these fines in order to keep fines as a viable component of their code compliance program.

Closing Properties

Fire prevention codes may include provisions for closing properties for repeat serious code offenses. Typically these requirements are aimed at places of public assembly where there is a high life safety risk. The fire prevention code will be specific as to the types of violations where these provisions may be applied. Usually, these cases are often imminent hazards such as locking exit doors with the building occupied. The closing rule generally can be applied only if the violations, which have to be substantially the same violation, are repeated within a specific period of time. For example, a nightclub chain locking a side exit door would constitute an imminent hazard and the building would be vacated until the door was opened. If a year later a different exit door was locked, it would also be an imminent hazard, but because it is a repeat violation that is substantially the same under the fire prevention code, a closing order could be instituted. Punitive closings of buildings may run 30–60 days, depending on the circumstances of the hazard (see Figure 6.7). As an enforcement tool, because the outcome is so severe to the owner or operator, there is tremendous incentive to not repeat past mistakes. Fire inspectors need to check their local code administrative section to establish what compliance tools are available to them in difficult code compliance cases.

Legal Procedures and Due Process

We have discussed the different types of notices and orders that fire officials and fire inspectors may issue during the fire inspection process. A critical key element of the issuing notices is to ensure that the legal process has been properly followed. All orders by the fire inspector must follow the rules of **due process** provisions in the fire prevention code. The state must respect all legal rights of the property owner that are due to a person under the state or federal constitution. When the government takes any action that deprives a person of their fundamental rights they are in violation of the Fifth Amendment of the U.S. Constitution. Not providing proper due process to an aggrieved person creates harm to that person and offends the rule of law that this country is founded upon. This is why it is essential for the fire inspector to follow the fire prevention regulation and provide sufficient time to correct the fire code violations based on their effect on the occupant's life safety. The fire inspector must obtain proper service of the notice of violation and must also provide the building owner with appeal rights if the owner disagrees with the inspector's findings.

Fire inspections fall under a legal requirement referred to as the **Dillon rule**. This rule states that a municipal or county government has only those powers expressly granted to them by charter or state legislation.[3] All powers and authority granted to conduct fire inspections and issue notices of violation must be traceable to the enabling legislation either directly or indirectly. If no legislation can be found to support the authority, then the government lacks the ability to address the situation.

TO: PREMISES:

DATE: REG. #:

Pursuant to the Uniform Fire Code (N.J.A.C. 5:70-2.18) the violations cited on the attached page(s) are found to the WILLFUL or GROSSLY NEGLIGENT and in VIOLATION of Orders dated _____.

You are therefore ORDERED to IMMEDIATELY VACATE and CLOSE the above referenced premises until the violations are ABATED and the time period specified below has passed. Should you fall to do so, a penalty of $ _____per day shall be assessed for each day you remain open for business.

The violations will NOT be deemed abated until verified by the State Division of Fire Safety. A reinspection will be undertaken within 48 hours of receiving written notice from the owner/agent stating that the subject violations have been abated or removal completed.

As a result of the above finding, you are further ORDERED to keep the specified premises vacated and closed for a period of _____ days (not to exceed 60 days) following the Division's verification of abatement and until a certificate of continued occupancy issued pursuant to the State Uniform Construction Code Act is obtained by the owner.

Commissioner, Department of Community Affairs

By: _____
                 Fire Official

Appeal Rights
See reverse side for information concerning your administrative appeal rights.

5:71-3.7(b)8.

FIGURE 6.7 Example of closing of property order. *Courtesy of NJ Division of Fire Safety*

Typically, the due process clauses in fire prevention codes provide that from the date of service of a notice of violation or any other orders, the owner may file an appeal of the order within an established time, usually 15 days. If the owner files an appeal, all further enforcement actions related to the violations are stayed until a hearing is conducted before the appropriate hearing body (see Figure 6.8). The only exception to a stay of action is in imminent hazard cases where the enforcing agency may keep the building vacated until the hearing is completed. The hearing venue may be a municipal court, state court, or (usually) a local or county board of appeals. State statute or enabling legislation will establish the rules that apply to boards of appeals. The state legislation will establish the makeup of members of the board as well as procedures on timeliness, evidence submission, testimony, and the decision-making processes. Boards of appeals usually represent a cross section of the community stakeholders. Typically, boards have professional members like architects, engineers, and attorneys; building trade members such as electricians, carpenters, masons, and plumbers; public safety members like fire inspectors and building code officials; and members of the public to represent the general community. Each board will have specific rules on the quorum required to conduct official business as well as provisions on conflict of interest of board members. Fire inspectors should review and understand the legal rules that apply in a board of appeals hearing for testimony, evidence, cross examination, and final decisions.

ADMINISTRATIVE APPEAL RIGHTS

The owner of the premises or of the use, or an authorized agent of the owner MAY CONTEST THIS ORDER at an Administrative Hearing. The request for a hearing must be in writing within 15 days after receipt of this order and addressed to:

In accordance with N.J.A.C. 5:70-2.19 an appeal shall be signed by a proper party and shall include:

a) The date of the act, which is subject of the appeal;
b) The name and status of the person submitting the appeal;
c) The specific violations or other act claimed to be in error; and
d) A concise statement of the basis for the appeal.

You are advised that only matters deemed to be CONTESTED CASES, as defined by the Administrative Procedures Act, will be scheduled for a Hearing. If a hearing is scheduled, you will be notified in advance of the time and place.

EXTENSIONS

If a specified time has been given to abate a violation, YOU MAY REQUEST AN EXTENSION OF TIME by submitting a written request to the _____. To be considered, the request must be made before the compliance date specified and must set forth the work accomplished, the work remaining, the reason why an extension of time is necessary and the date by which all work will be completed.

TAKE NOTICE THAT, pursuant to N.J.A.C. 5:70-2.10(d)2, an application for an extension constitutes an admission that the violation notice is factually and procedurally correct and that the violations do or did exist. In addition, the request for an extension constitutes a waiver of the right to a hearing as to those violations for which an extension is applied.

PENALTIES

Pursuant to N.J.A.C. 5:70-2.12, a violation of the Code is punishable by monetary penalties of not more than $5,000 per day for each violation. Each day a violation continues is an additional, separate violation except while an appeal is pending.

ALSO TAKE NOTICE THAT, pursuant to N.J.A.C. 5:70-2.12A, when an owner has been given notice of the existence of a violation and has not abated the violation, that owner shall, in addition to being liable to the penalty provided for by N.J.A.C. 5:70-2.12, be liable to a dedicated penalty in the like amount.

A violation that is recurring justifies imposition of an immediate penalty without the necessity for an interval in which corrections can be made. A violation shall be deemed to be a recurring violation if a notice has been served within two years from the date that a previous notice was served and the violation, premises and responsible party are substantially the same

Claims arising out of penalty assessments can be compromised or settled if it shall be likely to result in compliance. Moreover, no such disposition can be finalized while the violation continues to exist.

Any penalties assessed are in addition to others previously assessed. Penalties must be paid in full within 30 days after an order to pay. If full payment is not made within 30 days, the local enforcing agency may institute a civil penalty action by a summary proceeding under the Penalty Enforcement Law (N.J.S.A. 2A:58-10 et seq.) in the Superior Court or municipal court.

NOTICE:

If you require guidance or advice concerning your legal rights, obligations or the course of action you should follow, consult your own advisor.

FIGURE 6.8 Example of appeals rights. *Courtesy of NJ Division of Fire Safety*

The Board of Appeals Hearing

When a matter is appealed, the enforcing authority will be requested to submit all related reports to the board of appeals as well as any photographs or other supporting documents necessary to hear the case. These documents often are requested at a preliminary hearing or work session, which is held before the formal hearing. At the preliminary hearing, the board may ask any questions of the enforcing agency and will request submission of any other information or evidence for examination by board members before the hearing. Owners may or may not be present at the preliminary hearing, as the board's first order of business is to determine that the enforcing agency acted within the requirements of the code. The date of the hearing will then be established by the board. Boards of appeals usually have hearings once or twice per month unless it is an imminent hazard case.

The fire inspector should arrive at the hearing site early and make sure he or she has a professional appearance; a fire inspector who looks professional is less likely to be challenged by the board or the defendant. The enforcing agency generally presents their case first after being sworn in. When testifying, the fire inspector should speak clearly to any questions and maintain eye contact with the board members. When questions are asked, they should be answered in a succinct manner without getting off the topic. Frequently, board members may drift away from a violation issue; part of the fire inspector's duty is to politely refocus the discussion back on the code violation. It is not uncommon in appeal hearings for the building owner to bring attorneys along to do the questioning. These attorneys may object to certain answers given by the fire inspector. The board chairman usually addresses these objections, as appeals hearings are less formal than court hearings. In any case, if an objection is raised, stop talking until the chairman addresses the objection and tells you to continue your testimony. As you speak about each code violation, stay with the facts as you see them; do not get into personal attacks or make any derogatory comments about the owner. Remember that ultimately decisions of a board of appeals must be based on facts and how the fire prevention code applies. If you can guide the board down that road, you will win your decision.

The fire inspector must be prepared for any questions the board may ask at the hearing. The board members may ask questions on specific violations or technical references related to the particular violation such as a referenced NFPA standard. At the conclusion of the enforcing agency's presentation, the building owner then presents his side of the case and may ask any questions of the fire inspector pertaining to the notices of violation and fire inspection report. Finally, the board members may cross-examine both parties or ask for any public comments from interested parties before closing the testimony. Once testimony is closed, the board will discuss the facts of the case, and any board member may offer a resolution or recommendations in the form of motions to the board chairman. The board members then will vote on each resolution or recommendation with the winning vote being a simple majority. The board then will render a final decision in writing to all parties in the matter. The decision of the board of appeals is binding on the enforcing agency and the owner. If there is disagreement with the final decision, either party may appeal the matter to a court of competent jurisdiction. In these situations, the courts usually remand the case back to the board of appeals for a second review before bringing the case into the court system. The courts rely on the boards to resolve these code matters, as they are more knowledgeable in the regulations than are the courts, and this keeps the courts from being backlogged with small disputes.

INJUNCTIVE RELIEF AND COURT ACTIONS

Owners and enforcing agencies also have the right to take any case into a court of competent jurisdiction to obtain injunctive relief without filing an appeal. This action is a legal

ATLANTIC CITY SOLICITOR'S OFFICE
1301 Bacharach Boulevard
Atlantic City, New Jersey 08401
(609) 347-5540

CITY OF ATLANTIC CITY A Municipal Corporation of the State of New Jersey Plaintiff, vs Defendant(s)	SUPERIOR COURT OF NEW JERSEY ATLANTIC COUNTY LAW DIVISION–SPECIAL–CIVIL PART DOCKET NO: COMPLAINT STATUTORY PENALTY

DEMAND: $ ___

PRELIMINARY STATEMENT

This is a summary action brought against defendants for enforcement and recovery of penalties pursuant to the Penalty Enforcement Law, N.J.S.A.2A: 58-1 et seq., and N.J.A.C. 5:18-1 et seq., for defendants' violations of the State Uniform Fire Safety Act, N.J.S.A.52:27D-192 et seq.

Count One – "Against Owner"

1. Defendant, __________, is the "owner" (as such term is defined in the Uniform Fire Code, N.J.A.C. 5:18-1.5) of a building located at ____, Atlantic City, New Jersey 08401.

2. Defendant, ____, has a business, or a legal interest in said property.

3. Plaintiff, through the Fire Safety Division of the Atlantic City Fire Department, conducted an inspection of the aforementioned building on ____. A re-inspection was made on____.

4. Pursuant to said inspection, defendant was issued and served with a Notice of Violations and Order to Correct (See Exhibit "A").

FIGURE 6.9 Example of order to show cause.

order to show cause ▪ A judge's order mandating that a person appear in court at a specific date and time for a given legal or factual reason.

preliminary hearing ▪ An initial hearing before a judicial official or nonjudicial board for the purpose of gathering facts in determining issues between public authorities and property owners.

proceeding called an **order to show cause**, which is a preliminary action filed with a judge for a hearing by either the owner or the enforcing agency. The owner may seek relief from an enforcing agency's order while the enforcing agency may be filing to request the courts to enforce the order (see Figure 6.9). Filing an order to show cause requires the fire inspector to appear and testify in a **preliminary hearing** before the judge to review the order and the enforcement actions. After hearing testimony, the judge may direct questions to both parties involved with regard to aspects of the case. In these cases, attorneys for both the plaintiff and defendant will be present. The judge has the authority to stay, uphold, modify, or deny any of the enforcing agency's orders if it is believed that the actions taken are not in compliance with the rule of law as stated in the Dillon rule. The judge can also independently enforce corrective actions upon the owner on behalf of the enforcing agency. Fire inspectors must apply due diligence and be prepared in all code enforcement cases, as any fire inspection may end up in the court system.

Court Hearing Preparation

We have reviewed the basic concepts of a board of appeals hearing and the legal process involved in preparing for board of appeals cases. The rules for a court hearing are different and are more formal than those of a board of appeals. The fire inspector must be aware of these differences when preparing for a hearing in court as opposed to a board of appeals hearing.

Before the court hearing, the fire inspector should review the report and review all of the pertinent fire code requirements that were cited as violations. The fire inspector should have a firm understanding of the code intent behind each code section so as to appropriately and professionally address all questions related to each violation. A fire inspector can say nothing worse at a hearing than "The fire code says you have to do it." This shows a lack of understanding of the foundation of the code requirement and reflects poorly on the fire inspector's preparation. The fire inspector should review the *International Fire Code* fire prevention code commentaries and any appropriate NFPA handbooks to attain additional knowledge on the code background. These books provide the foundations for understanding the code regulations and provide illustrations and insight into the proper application of the fire prevention code.

After you have prepared and reviewed all of your documents, place them in chronological order and prepare a time line of events from the initial fire inspection to each reinspection and the subsequent results of those inspections. Times and dates as well as what is or is not corrected during each reinspection are very important to your success in court. Your timeline should be in a bullet list fashion for easy reference. Unlike appeals hearings, in a court hearing, you should have the representation of an attorney. Remember that the opposition will have a lawyer prepared to refute your report and testimony. Review your case with your attorney, as he or she will frame the correct legal questions to ask you during your testimony. The attorney also will look for any weakness in your case that may be challenged by the opposition in a cross-examination. Your attorney must have copies of all the documents in your file and will determine which documents to place into evidence during the hearing. Your attorney should go over the questions he or she will ask you and the way in which you should answer them.

THE COURT HEARING

As a fire inspector, you should arrive at the courthouse early and meet with your attorney to discuss any strategies. You should also be properly dressed in uniform and have your file for reference with you. In the court room, the plaintiff and the defendant will sit at tables on opposite sides of the judge. The plaintiff (enforcing agency) will go first, and your attorney will place you on the witness stand to testify. In most cases, the fire inspector will be a **fact witness**, meaning that they will testify only to what they have personally observed firsthand. Do not make references to what others have told you or things that you have not witnessed as it will be considered hearsay, and the opposition most likely will object. Anytime an attorney objects to your answer or a question being asked, stop talking and wait for the judge to give you direction as to whether the objection is **overruled** or **sustained**. The key element is to always tell the truth based upon your knowledge and first-hand observations. The defendant's attorney will have an opportunity to cross examine you and identify any inconsistencies in your testimony. Another key element of importance is to answer each question succinctly, never providing more information than requested. Be aware that the opposing attorney may ask trap questions

fact witness ▪ A witness who testifies to facts of which he or she has personal knowledge.

overrule ▪ To disallow the argument of a person by a higher authority.

sustain ▪ To affirm the validity of an argument by a higher authority.

in a "yes or no" fashion that do not allow you to explain your answer. Defense attorneys practice these techniques and are generally very good at them. It's ok if they trap you in a "yes or no" question, as your attorney will allow you to answer it properly on a redirect or may object to the relevancy of the line of questions being asked. Never answer a question unless you understand what is being asked. Be careful with leading questions such as "Inspector Smith, wouldn't you agree that . . .". Do not allow the opposition to put words in your mouth. If you don't understand the question, say so and have it rephrased. Be polite and don't get angry. Sometimes an attorney's entire technique is to get the witness angry to demonstrate some malicious intent against the attorney's client. Speak clearly and loudly to the judge or jury when giving your answers. Most importantly, be professional.

expert witness ■ A witness with special expertise and knowledge who can testify on complex technical issues.

In some cases, your attorney may wish to classify you as an **expert witness**. If the opposing attorney objects, you will be required to be qualified as an expert. The reason an attorney may wish to qualify you as an expert is that a fact witness cannot render opinions whereas expert witnesses can. This qualification is based on your experience, education, and training in the fire prevention field. It is always good to have a copy of your curriculum vitae to address the question of whether or not are you an expert. After the judge is satisfied that you have expert knowledge in your field, you may then render opinions to hypothetical questions. Juries tend to give more credence to expert witnesses than to fact witnesses, but they also expect the expert to provide complex information in a way that it can be easily understood. The use of visual aids to illustrate concepts in a simple fashion often can help. These aids can be diagrams, photos, or material tests that illustrate the fire prevention concept. Keep in mind that often a picture is worth a thousand words in illustrating points.

RULES OF EVIDENCE

discovery ■ A procedure allowing both sides of a case to obtain all factual issues related to the case.

While we may not think of a fire inspection report as a piece of evidence, it is. The fire inspection report documents the fire code violations that existed at a specific date and time. Each reinspection and other actions taken illustrate the owner's effort to comply or failure to do so. Each report and follow-up report becomes a significant piece of evidence. Before the hearing, all records and evidence must be provided to the opposition in a process called **discovery**. Any evidence not provided in discovery most likely will not be permitted in the case. During the hearing, all evidence must be submitted together with proofs of service to the court as well as to the opposition to ensure that all have the same evidence against them and due process has been properly given. Typically, each document is labeled or identified by the plaintiff's attorney as it is entered into testimony, and each item is issued an evidence number. The evidence will also be shown to the opposition for any objections that they may have to its being submitted. The judge rules on whether to allow the evidence into the case or make it inadmissible. Fire inspectors may also have other physical evidence to submit to the board including photographs, videos, e-mails, drawings, approvals, or certifications from other agencies that are not part of the initial fire inspection report. All of these documents must also be submitted to the opposition before the hearing as a matter of discovery. If the fire inspector has taken photographs of violations and is submitting them as evidence, they need to make sure the owner's consent to photograph the property was obtained. The photograph should be date and time stamped as to when the inspection was made and the photo was taken. Fire inspectors can request permission to take photographs during the inspection tour. If the inspector has no permission, the court may disallow the photos as **fruit of the poisonous tree**, making them inadmissible. This permission may be implied if they were taken during the inspection tour without objection of the owner. If photographic evidence is used, each picture must be explained as to what it reflects in relation to the fire inspection report. In some cases, a photo log

fruit of the poisonous tree ■ Evidence that has been obtained without the owner's permission or a warrant.

and diagram will assist the court in determining where the photos were taken, including the direction and angle with respect to a floor layout diagram. If you have doubts about permission on photographs, either do not use them or get an administrative warrant to allow you to take new photographs. Other evidence that the fire inspector may wish to submit may come from other inspection agencies, such as stop work orders or communications between inspection agencies on referral forms or citizen complaints. These also must be provided to the opposition during discovery to be admissible in court. The fire inspector must also keep in mind that during the course of the inspection, he or she cannot seize or remove any type of equipment from the property such as extension cords, kerosene heaters, or chains and locks unless the fire prevention code expressly gives the fire inspector such authority or he or she has consent from the owner to do so. Evidence is an important part of proving a case in court and must be properly collected and documented in order to be admitted. While code enforcement cases do not have the same standards of evidence collection that a criminal court case does, the fire inspector still needs to consider proper evidence collection and presentation to be successful in court.

PROFESSIONAL DEMEANOR

The fire inspector must present himself or herself in a professional manner at all times during a fire inspection and subsequent hearings. In cases that go to court, one of the best defenses that commonly is pursued is that the fire inspector has singled out or is treating the owner differently than other building owners and is selectively enforcing the fire prevention code. The inspector must be prepared to demonstrate that the actions taken were not selective and that other buildings of similar occupancy also are inspected. If the case relates to an open and abandoned property, the inspector should make sure that other open and abandoned properties have been noticed in the area, as this negates the claim of selective code enforcement.

In either a court or a board of appeals case, the defense attorney, who often does not have a great defense, will try to place the fire inspector on the defense for improper inspections or technical flaws in the report. Often, the best defense is to create confusion for the board or the jury so that the testimony is disregarded or not given adequate credibility in the decision process. Attorneys do this by asking questions to paint the fire inspector into a corner, to place words in his or her mouth, or to evoke some emotional reactions like anger. Attorneys may ask rapid-fire questions to confuse the fire inspector in his testimony, especially with dates if the paperwork is not correct or if the fire inspector shows a lack of confidence in testifying. By no means is testifying in court easy or enjoyable; however, it is an important aspect of fire code enforcement. The following methods can be employed by the fire inspector to control the situation:

1. Keep your cool; never get angry or insult the opposition's attorney. Keep in mind that like you, the attorney is just doing his job.
2. Be prepared. If there are problems with paperwork be aware of them, and if asked, explain the error. Always tell the truth.
3. If you do not understand the question, have it rephrased. This slows the tempo of the questioning down and keeps you from answering incorrectly.
4. If you make a mistake, it is okay. The prosecuting attorney will redirect questions to address mistakes and get the correct answer.
5. When asked a question, pause before you answer to collect your thoughts and slow the questioning tempo.
6. Speak clearly and maintain eye contact with the board members, the judge, or the jury. Ultimately they are making the final decision.
7. Don't let the attorney put words in your mouth using statements like, "Would you agree" or "Isn't it true that . . .". Look for these trap phrases and answer carefully.

8. Don't talk too much. If asked a question, answer it, but do not volunteer information, as it will likely get you in trouble.

9. Sit up straight and tall and try to relax. If asked a question about a specific item on a document, ask to examine the document before answering; don't go off the top of your head.

10. If you are a fact witness, speak only to what you know. Avoid what others have told you.

11. If you are an expert witness, you may give opinions; however, make sure they are based on factual information and fire science. Do not render opinions on things outside your field of expertise.

12. If objections are raised again, remain silent until the judge directs you to continue or sustains the objection.

Court appearances and testimony can be harrowing experiences, even for seasoned fire inspectors. New inspectors should attend hearings of senior inspectors to get a flavor of the environment and to see how the process works. Unfortunately, the only way you can learn these lessons is by testifying under actual conditions, but don't be intimidated: Your professionalism and knowledge of your case will get you through successfully.

Summary

We have examined the legal instruments of the fire inspector, which are the fire inspection report and notice of violation that are issued to the property owner or agent. Reports must be clear, concise, and complete if code violations are to be properly repaired. The fire inspector must make sure that all reports and notices are properly served to the owner, the occupant, or the agent who has the power to correct the fire code violations in the property. Inspectors need to follow required compliance schedules for reinspection and must be fair handed by providing adequate time to correct outstanding code violations. Fire inspectors must advise owners of their rights of appeal and explain the appeal process if they wish to exercise those rights. The fire inspector must ensure that they do not selectively enforce the fire prevention code and

that they practice due diligence throughout the enforcement process. In courts and boards of appeals, the inspector must act professionally, be knowledgeable, and be prepared to give factual and truthful testimony in the case. Before a hearing, the fire inspector should work with an attorney to prepare the case and review the strengths and weaknesses of the case. Lastly, the fire inspector should not take winning or losing a code enforcement case to heart. Remember that justice is blind and that sometimes the guilty go free so that the system can avoid prosecuting the innocent. The main goal in courts and board of appeals hearings should be to determine whether the fire inspector performed his or her duty to the best of his or her ability and offered the best defense possible to maintain the public's safety.

Review Questions

1. List the critical elements of information when writing fire prevention code violations.
2. Review the types of forms used by your fire prevention office. Describe the purpose and use of each type of form.
3. What types of certifications must be reviewed by the fire inspector during a fire inspection. How are these certifications obtained?
4. Examine the numbers of fire inspections and reinspections that are performed by your fire prevention bureau. What is the average percentage of violations that are corrected during initial fire inspections and reinspections?
5. List and describe the different methods of service for notices of violation in your jurisdiction. What steps must the fire inspector follow in each method?
6. Examine your fire prevention code and list the requirements for the following:
   a. Service of notices
   b. Filing appeals
   c. Correcting fire code violations
7. Develop an inspection program for dealing with abandoned structures. What are the key elements for the following:

   a. Inspection of abandoned properties
   b. Service of notices
   c. Methods of correction
8. What are the elements that must be present under your fire prevention code to issue an order of imminent hazard?
9. List and describe the types of fines that may be issued for violations of your fire prevention code. How are these fines collected?
10. Describe the due process requirements of your fire prevention code. What are the critical elements and how is the process conducted?
11. You are preparing to go to court with an important fire code enforcement case. Describe the steps you should take in preparing for court.
12. Describe the types of testimony that each of the following witnesses may provide in a court hearing:
    a. Fact witness
    b. Expert witness
13. List some of the rules that a fire inspector should follow when testifying in a court of law.

Suggested Readings

Local fire prevention code.

International Code Council. 2002. *Legal Aspects of Code Administration*. Falls Church, Virginia: International Code Council.

Sanderson, Richard L. 1975. *Readings in Code Administration*. Chicago Illinois: Building Officials and Code Administrators International.

Varone, Curt. "Fire Law." http://firelawblog.com/

Endnotes

1. Marty Aherns, *Vacant Building Fires* (NFPA, Fire Analysis & Research Division, 2009), 5.
2. Paul Cox, "Patterson Schools are Shut for Serious Fire Safety Violations," *The Star Ledger,* September 9, 2007.
3. International Code Council, *Legal Aspects of Code Administration* (Falls Church, Virginia: International Code Council, 2002), 12.

Courtesy of J. Foley

KEY TERMS

automatic transfer switch, *p. 154*

bonfire, *p. 140*

flambé, *p. 143*

Knox Box, *p. 152*

plenum, *p. 149*

seal the equipment, *p. 154*

treadle, *p. 152*

UL-300, *p. 149*

OBJECTIVES

After reading this chapter, you should be able to:

- Understand the importance of the authority, scope, and purpose of the fire prevention code.
- Know the common type of fire hazards found in all structures.
- Know the fire service requirements for access and hazard identification.
- Know the critical building systems and the points of inspection required for safety maintenance.

Professional Levels of Job Performance for Fire Inspectors as Cited in NFPA 1031 and NFPA 1037

- NFPA 1031 Fire Inspector I *Obj. 4.2.4 Investigate common complaints*
- NFPA 1031 Fire Inspector I *Obj. 4.3.6 Determine the operational readiness of fire detection and alarm systems*
- NFPA 1031 Fire Inspector I *Obj. 4.3.7 Determine the operational readiness of portable fire extinguishers*
- NFPA 1031 Fire Inspector I *Obj. 4.3.8 Recognize hazardous conditions involving equipment, processes, and operations*
- NFPA 1031 Fire Inspector I *Obj. 4.3.12 Code compliance for incidental storage*

- NFPA 1031 Fire Inspector II *Obj. 5.3.6 Evaluate hazardous conditions involving equipment, processes, and operations*
- NFPA 1031 Fire Inspector II *Obj. 5.3.12 Code compliance of heating, ventilating, and air conditioning*
- NFPA 1037 Fire Marshal *Obj. 5.8.1 Codes, standards, and jurisdictional requirements*

Introduction

Fire inspectors learn as they gain experience that the fire prevention code and the National Fire Protection Association standards contain hundreds of thousands of code requirements. It is important for fire inspectors to gain knowledge of the more commonly used standards within the fire prevention code as well as a familiarity with other technical standards that may apply to specific properties in their jurisdiction.

Fire codes are legally adopted by statute; standards are adopted in the fire prevention codes by reference. What this means is that only the standard that the fire code identifies by year and edition may be enforced as part of the regulation. Every fire code has a list of applicable standards in the fire code's appendix. The appendix will list the standards for each standard-making organization from American Society for Testing and Materials (ASTM) to the National Fire Protection Association (NFPA). Fire inspectors should also be aware of the scope, authority, and the interrelationship between the building code and fire code as well as any other adopted NFPA standards under those codes. Often, building and fire codes may be from different adoption years and may be applying different standards. The fire inspector should determine which standard is going to apply. For example, if a state has adopted the 2006 *International Building Code* and the 2003 *International Fire Code*, the IBC references NFPA 13-2007 and the IFC references NFPA 13-2002. In a new building constructed under the IBC, the 2007 standard is applicable, and in older buildings the 2002 standard would be applicable. This distinction is important because buildings must be maintained in accordance with the code requirements under which they received their certificate of occupancy. The exception is the fire prevention code edition, which is the minimum requirement for maintenance regardless of a building's age. Another important point to consider is that when enforcing the fire prevention code, a code requirement will always override any requirement specified in a standard if they are in conflict. This is because the code is adopted by legislation and the standard is adopted by reference.

All fire prevention codes provide additional guidance to the fire inspector in the administration code section on the proper application of standards. Each state regulation will establish a hierarchy of how building and fire codes relate and are to be applied. For example, if the fire prevention code does not address the fire code requirement on a specific process or hazard, the inspector may be directed to the building code. If the building code is silent and has no provision, the inspector may be directed to any other applicable safety standard as *prima facie* evidence for compliance with the intent of the regulation. What *prima facie* means is that the evidence presented, unless it is disputed, would stand as fact for code compliance. It is not uncommon for fire inspectors to encounter situations

A new fire inspector had to inspect an older, seven-story seashore hotel. The hotel had one interior exit stairway and an exterior metal fire escape at the other end of the building. The fire escape was accessed by a door on the second through sixth floors, but not the seventh. On the seventh, residential sleeping floor, the fire escape was accessed by a window and required the occupants to descend a vertical unprotected ladder to the sixth-floor fire escape platform. As the fire inspector looked down from the seventh floor, he knew that an older person or a young child could not possibly descend this ladder to the fire escape below safely.

As the fire inspector researched the fire prevention code, the only requirements were to paint the fire escape and conduct a load test if required. The fire code did not address the use of a ladder on the seventh floor of a fire escape. The fire inspector then examined the building code, but that didn't address fire escapes at all because new buildings may have only interior or exterior stairways. The state fire prevention code then directed the fire inspector to any applicable NFPA standard that can be used as prima facie evidence for code compliance. NFPA 101, Life Safety Code®, includes a section that applies to existing fire escapes. The standard requires that the access to the fire escape for multiple units be through a door, and that each fire escape landing be served by a platform and stairway. In fact, the only place that a ladder may be used is on the fire escape to access the roof, or from the last landing to the street. In the case of the roof access, a fall guard must also be provided. This case went to a construction board of appeals hearing, as the owner believed that the fire escape pre-existed the code and was grandfathered. The fire prevention official was successful in having the owner correct the code violation based on the Life Safety Code requirements. These requirements were presented as prima facie evidence in this case even though NFPA 101 was not adopted by reference in the fire code.

that may not be clearly defined in the fire code. Being able to navigate the codes and standards and "connect the dots," so to speak, is an important aspect of fire code enforcement and understanding the interrelationship of building and fire codes and standards.

Now let's examine the basic types of fire code violations that are often encountered by fire inspectors. The fire prevention code generally addresses common fire code violations in the first ten chapters under categories such as "General Precautions Against Fire," "Fire Service Features," "Building Services and Systems," "Fire Protection Systems," "Means of Egress," "Interior Finish and Decorations," and "Fire Resistive Rated Construction." These chapters make up approximately the first third of the fire prevention code book. Inspectors may encounter these common fire code violations in almost every building that they inspect.

Purpose, Scope, and Applicability

The first and most important section of the fire prevention code is the administrative chapter's section titled "Scope & Applicability." This section provides direction to the fire inspector on the purpose of the fire prevention code and the types of building uses it applies to. The purpose of the fire prevention code is generally to establish the minimum safety standards that apply to buildings, based on nationally recognized fire safety practices to reduce the fire and explosion incidents and also to protect life and property. The scope of the regulation identifies which buildings will be inspected and which processes will be regulated. Generally, the fire prevention code's scope includes provisions for structures, processes, and human behaviors related to occupancy; the storage of combustible or hazardous materials; hazardous occupancy conditions; and the maintenance of fire protection systems. The applicability of the code also addresses the boundaries or limits of what can be inspected under the fire code and its relationships to other applicable codes. The fire prevention code is maintenance oriented and applies to all new and existing buildings within a jurisdiction. The fire prevention code does not override building code requirements for the construction or renovation of a building with few exceptions. These exceptions generally relate to either a retro-fit fire code requirement or the changing of a fire protection standard such as UL-300 for kitchen exhaust hoods. In either case, fire inspectors must check their local code requirements for proper application. The scope also limits the inspection of one- and two-family dwellings, although this too may vary from state to state.

Some states require fire inspections on first occupancy or when the property is resold, and others may require inspections if the residence is not owner occupied. Some fire departments may conduct home fire inspections as a courtesy service to the owner. In either case, home inspections are geared more to educating the occupant on minimum safety measures such as working smoke detectors, two means of planned escape, and minimizing storage hazards around heaters or garages. The applicability provisions of a fire code also reference outdoor activities that may be subject to violation or other actions under the fire prevention code such as accumulation of combustibles, hazardous materials storage, or discharge of fireworks.

The administrative section also contains a "severability and validity" clause that protects the entirety of the code should a requirement be determined to be unconstitutional or illegal by the courts. Severability means that only those sections identified by the court as improper would be affected and not the entire fire code document. This is important because from time to time, courts may issue decisions on certain aspects of the regulation, either changing or nullifying the particular fire code requirement.

TECHNICAL ASSISTANCE

Fire inspectors often are required to make judgments on fire protection systems or methods that they may have limited knowledge about when enforcing the fire code. In these complex cases, the fire marshal or fire official may request a report or evaluation of the process or equipment in question by a technical expert. The technical opinion may be from an architect, engineer, testing laboratory, or other qualified professional that is appropriate and can offer an unbiased opinion in the evaluation. The cost of the expert is the responsibility of the building owner. Typically, this requirement may be helpful when an owner requests a deviation or variance from a fire code requirement. The appropriate professional will be able to address the variation and how it will provide equal or better protection for the building. This provides better information to the fire inspector so he or she can make an informed and reasonable decision.

MAINTENANCE OF FIRE PROTECTION

The fire prevention code requires maintenance of all of the safety features installed under the building code to establish a safe occupancy. The fire prevention code provides the maintenance safeguards that are applicable for the life of the building, including regular scheduled fire inspection and fire system testing. The fire code requires that the fire inspector be furnished with testing and inspection reports on each specific fire protection or building system covered by the regulation. The fire code identifies the types of approvals the fire inspector may accept for compliance with the code including equipment listings and labels.

General Fire Code Provisions

The general provisions of the fire prevention code are established in the administration section of the regulation and often include specifics on unsafe structures, imminent hazards, fire safety permits, and fire inspection schedules. Each state or local jurisdiction may adopt additional specific rules and requirements in the administration section that are unique to that state. The general provisions of the fire prevention code also usually contain a "general precaution clause" that applies to all buildings. These general provisions are intended to cover hazardous situations that may arise where there is no specifically identified fire code requirement.

Some codes define these in ten general hazard categories as follows:

1. Dangerous conditions that may cause or contribute to the spread of fire or endanger occupants.
2. Conditions that interfere with the operation of a fire protection system or equipment.
3. Obstruction of exits and fire escapes that may restrict egress or interfere with the fire department's operation.
4. Accumulation of dust or waste materials in HVAC systems or grease in kitchen or other exhaust ducts.
5. Accumulation of grease on kitchen equipment or under and around other mechanical equipment.
6. Accumulation of rubbish, boxes, trash, or other materials or the excessive storage of combustible materials.
7. Hazardous conditions arising from improperly installed electrical equipment or appliances.
8. Hazardous conditions arising from defective or improperly installed equipment for handling combustibles or hazardous materials.
9. Dangerous or unlawful amount of combustibles, explosives, or hazardous materials.
10. All equipment, processes, or operations that are in violations of the intent of the fire prevention code.[1]

As you can see, just about every hazard that can be encountered may fit into one or more of these ten categories. The code writers' intent, however, is not to have the fire inspector cite every violation under these general provisions. These provisions are intended to be used as a tool by the fire inspector for those hazardous conditions that are encountered where no specific code requirements exist. Fire inspectors must use a degree of discretion when citing violations under the general section and make sure that there is not another, more applicable fire code requirement or an exception to the requirement in the other chapters of the code book. The key element is that each fire inspector must learn the interrelationship of the codes used in their jurisdictions so they may properly apply the regulations.

RECORDS

One of the most challenging tasks required during the fire inspection for the fire inspector is to review maintenance records that the owner is required to keep . Records will fall into three general categories: equipment maintenance, emergency preparedness, and safety training of employees. The equipment records will include both test and inspection reports on active fire protection systems such as automatic fire alarms and fire suppression systems, as well as passive fire protection certifications on treatments of draperies or other interior finish materials. The owner may also be required to keep inspection log reports on specific systems as required by NFPA 72, *National Fire Alarm and Signaling Code* or NFPA 25, *Standard for the Inspection, Testing, and Maintenance of Water-Based Fire Protection Systems* for both the fire alarm and sprinkler systems. Every active system will require annual maintenance testing as well as some special inspection reports on long-term maintenance requirements, such as internal pipe inspections. The fire inspector should request copies and review the following records during the fire inspection:

- Automatic sprinkler systems test report
- Automatic fire alarm systems test report
- Private fire hydrant systems test report
- Smoke control and stair pressurization systems test report
- Fire pumps test report
- Generators test report
- Fixed fire suppression systems test report
- Elevator recall systems test

- Portable fire extinguishers service verification
- Flammability certifications

Owners must also keep records on the following employee training and emergency preparedness requirements:

- Portable fire extinguisher training
- Fire and evacuation drill records
- Emergency fire safety procedures
- Evacuation plans
- SARA hazardous materials inventories sheets
- MSDS sheets

OBSTRUCTION OF EGRESS

The code violation most often encountered by fire inspectors is an obstructed or blocked means of egress. The most critical building system is the components of egress. The fire prevention code established the maintenance rules for egress in each of the three components of the egress system. Exit access is the egress component within a room or space that allows free travel to the room doorway. Corridors may also be part of exit access and generally will have fire resistance rating requirements based on the limitation of travel in two directions. Storage, furniture, and other items are permitted in exit access provided they do not encroach on the minimum required egress width of the corridor. The second component is the exit itself, and this is a protected element with unlimited travel distance. Exits are fire rated and are not permitted to be used for any other purpose but exiting. The exit can be an exterior door, or an interior or exterior stairway. The third component is the exit discharge, which connects the exit to the street or an area of refuge away from the building. On the inspection tour, the fire inspector should examine each room and space for adequate and multiple egress paths. Rooms that have an occupant load of fifty or more persons must have multiple exit doors, and they must have doors swinging in the direction of exit travel. Corridor doors should be self-closing or on automatic hold-open devices attached to the fire alarm system. The doors should be listed and labeled under NFPA 80 based on the hourly fire resistance rating of the corridor wall. Doors should fit in the frames tightly and not bind or have space around them other than at the sill. All elements of egress must have emergency lighting and exit signs that provide the required illumination established in the building codes. Emergency lighting should be tested and witnessed by the fire inspector to ensure its proper operation. Battery pack units will have test buttons that activate the lights and ensure that the battery is charged. Fire codes require illuminated or self-luminous exit signs meeting UL 924 requirements. Exit signs are required at the entrance to exit stairs, at all exit doors, and directional exit signs are required when the path of the exit is long or changes direction. Fire inspectors must check illuminated exit signs for both primary and secondary power, and self-luminous signs from proper UL listings. The fire inspector must ensure that all means of egress are unobstructed, well lit, safe for travel, and unlocked at the point of discharge. When a fire inspector encounters means of egress violations, they are required to be addressed immediately, as they may pose an imminent hazard to the occupants. If the egress cannot be easily remedied, then the fire inspector should reduce the occupant load to a safe acceptable level or remove the occupants from the area until it can be properly addressed and restored. Egress is a very complex code requirement with many rules that fire inspectors must remember.

IGNITION SOURCES

An important aspect of controlling fires is the control of potential ignition sources. In order to have a fire start, there must be a source of ignition. Typical ignition sources are either electrical, chemical reactions, mechanical energy from friction, or direct flame or radiant energy contact with a combustible material. The fire inspector should be looking for these types of ignition sources during the course of the fire inspection. The goal of fire prevention is to identify and to separate the fuel source from any potential sources of ignition. Heat sources, such as space heaters, are required to be maintained at a safe distance from combustibles, usually about 3 feet. It is not uncommon in business offices to find portable space heaters under desks or next to files, waste cans, and combustible furnishings. The fire inspector must be vigilant for these types of violations.

The fire code also requires that any material that can undergo spontaneous combustion should not be disposed of in combustible trash receptacles or within 10 feet of a building. Spontaneous combustion is an exothermic chemical reaction that occurs as rags wet with finishing oils dry. If the heat cannot be dissipated because of insufficient ventilation, the rags eventually will get to their ignition temperature and start a fire. There have also been other cases of spontaneous ignition of non–talcum-powdered latex surgical gloves caused by chlorine gas interacting with oxygen in the air, creating a heat buildup in cartons of gloves that cannot dissipate through the packaging. Spontaneous ignition of finishing oil rags was cited as the cause of the high-rise fire that killed three firefighters at One Meridian Plaza in Philadelphia, Pennsylvania. Rags soaked with chemicals that may spontaneously ignite should be placed in a metal container with a proper lid and should be kept remote from the building.

COMBUSTIBLE WASTE

Combustible waste has a very broad definition under most fire prevention codes. Combustible waste encompasses all trash, wastepaper, litter, wood, and straw as well as high weeds and vegetation. These types of materials are not permitted to accumulate either inside or outside of a structure where they can become a fire hazard. The fire code also specifies that trash cannot become a public nuisance or health hazard, even in small quantities, and must be properly stored in trash receptacles with proper fitting lids. Traditionally, trash was disposed of in 60-gallon metal trashcans; today, however, most trash receptacles are made of combustible plastics. Fire inspectors must ensure that these receptacles are not stored under fire escapes or exterior fire stairways. Trash receptacles should be stored at least five feet from the building for reasonable safety. Often, trash accumulation may become a target for arsonists or juvenile fire setters. Many buildings have been destroyed and people have been killed in fires that have started on the exterior of the building in a trash receptacle. Commercial dumpsters are also used to remove trash in commercial buildings, multifamily dwellings, and during property renovations. These dumpsters should be located so as not to reduce access to the fire department and should be 5 or more feet from the building. Dumpsters should never be located under the building's eaves or near a fire hydrant. Most municipalities have local permits and regulations on dumpster placement. Dumpsters should be monitored and removed promptly once full.

OPEN BURNING OUTDOORS

The fire prevention code recognizes that in some cases, open burning may be required for industry or occupations, such as farming, as well as recreational cooking and other recreational activities. These fire activities require obtaining a fire safety permit from the local fire marshal or fire official, as well as approvals and permits from the state department of environmental protection or division of forestry services. Fire inspectors should check their jurisdiction for the process and preapprovals required under the Federal Clean Air Act.

Open burning is not permitted if it creates a public objection due to levels of smoke or odors of smoke in an area. Open burning is also not permitted in conditions that may spread the fire such as high winds. Farmers are usually prohibited from open burning to clear land without proper preapprovals from environmental agencies. Open burning may not be used for trash disposal and may not contain items that produce air pollution effluents like tires or flammable liquids, as burning these are violations of the Federal Clean Air Act and possibly other state or local pollution regulations.

Recreational open burning usually falls into one of three categories: cooking, heating, or entertainment. Examples would be campfires, barbeque pits, heating barrels, outdoor fire places, and bonfires. Open burning for cooking must occur in an approved camping location and should not be within 25 feet of a building, structure, or other combustible materials. All combustibles should be cleared from the area so that fire cannot spread. Fires in an approved noncombustible container such as a burn barrel may not occur within 15 feet of a structure. The fire codes also require that outdoor fires must be properly attended to until they are extinguished. The person tending the fire must have a method of extinguishment in the form of a portable fire extinguisher, a water bucket, sand, or a hose. Fire inspectors have the authority to require that any fire be extinguished if it presents a hazard to any structures or causes complaints from the public.

bonfire ■ A fire of dry wood materials used to excite or rally a crowd before a large event such as a holiday or sporting event.

Bonfires also are restricted in the fire prevention code and may not be ignited without a fire safety permit and proper inspection. All bonfires must have a minimum clearance of 50 feet to structures as well as the elimination of combustibles that may spread the fire to within 50 feet of a structure. The International Fire Code restricts the amount of fuel to be used in a bonfire to 5 cubic feet of clean combustible materials. The bonfire's burning duration may not exceed 3 hours unless approved by the fire marshal or fire official. Only seasoned dry wood may be used in bonfires to avoid spreading the fire beyond the permitted area.

Bonfires can be very dangerous events, with one of the worst bonfire accidents occurring at Texas A&M in 1999 when a 5,000-log bonfire being constructed collapsed and killed 12 students and injured 27. Many smaller accidents have occurred around the country with people improperly building bonfires. Often, the culprit is the use of flammable liquids to ignite the fire. The fire prevention code requires that bonfires be ignited with tinder, and flammable liquids are prohibited. In 2009, a violent bonfire explosion happened in southern New Jersey in preparation for a high school football game. The fuel was a dry wooden pallet to which firefighters added 15 gallons of diesel fuel. The heavier-than-air fuel vapors collected in the pile, and when firefighters ignited the bonfire, the pile exploded, injuring at least one firefighter and sending shards of the burning pallet into the surrounding crowd. Flammable and combustible liquids should never be used in the preparation of bonfires.

OPEN FLAMES

The use of open flames in public activities covers a wide variety of requirements under the fire prevention code. These activities include the discarding of ignited smoking materials, the use of candles, fires used in theatrical performances, the use of torches for paint removal or building repair, and the preparation of flaming foods such as flambé. The fire code additionally prohibits the discarding or throwing away of any form of burning materials that may start an unwanted fire. Every year, hundreds of fires are started by careless discarding of smoking materials. These items must be placed in a noncombustible container or ashtray. With the changes in public opinion on second-hand smoke, many states have declared public buildings and specific public use buildings such as restaurants and theaters as prohibited from smoking. These regulations have forced most occupants to smoke outdoors, resulting in an increase in exterior building fires caused by carelessly discarded cigarettes. Fire inspectors should be aware of any state or local restrictions on smoking and should ensure that those establishments provide a safe area outdoors where people can smoke and properly discard smoking materials. The use of mulch or combustible planting materials should be restricted in these smoking areas.

Cooking Appliances

The use of portable outdoor barbeques and fireplaces also has significantly increased throughout the country as people create outdoor living spaces. In multiple-family dwellings, the use of these appliances may be restricted or completely prohibited. Generally, these devices should not be installed within 5 feet of combustible construction or stored on combustible decks. Grills used next to or under the eaves of a building may cause a structure fire. Reasonable clearance to combustibles will apply whether the appliance is fueled by solid fuel or liquefied petroleum gas (often referred to as LP or LPG). The fire inspector should check the local fire code requirements for cooking with grills on building balconies. In many jurisdictions, grills on building balconies are prohibited unless the balcony is protected by an automatic sprinkler system.

Candles

Candles and candelabras often may be used in places of public assembly, religious ceremonies, and special events and celebrations. The use of candles is a major source of fire ignitions and must be controlled to prevent accidental fires. If the candle device uses liquid fuel, it must be equipped with a self-extinguishing tipover device that also restricts the fuel leakage. The base of the candle must be stable and must return to its upright position when tilted to 45 degrees and released. All liquid fuels must be properly stored in flammable liquid cabinets, and any shades or other decorative devices on top of the candle must be noncombustible. If tall candelabras are used, they must be securely fastened to a stable base so they may not be tipped over accidentally. Candles require adequate clearance to combustibles and should not be permitted in exit access aisles or exits where people may have to move rapidly and they can be knocked over or set clothing on fire. Balloons or other decorations such as tall, dry flower arrangements should not be permitted where candles are being used. These items may be moved by air conditioning, wind, or breezes or people inadvertently and become ignited by the open candle flame. Fire inspectors should be aware of the seasonal use of candles for holidays such as Christmas, Hanukkah, Halloween, and Thanksgiving and remind the public about the safe use of open flames. This can be accomplished by public safety announcements by local media outlets. The use of candles is also restricted in R-2 residential use groups, specifically in residential dormitories such as student housing on college campuses.

Theatrical Fires

Magic shows, concerts, plays, or other types of theatrical performances often use fire and open flames in their productions or events. Open flame or pyrotechnic devices may be used only after proper permits have been obtained and inspections have been conducted. The use of open flames and flame-producing appliances specifically designed for theatrical effects must comply with NFPA 160, *Standard for the Use of Flame Effects Before an Audience* for specially constructed theatrical props. Many of these types of props are one-of-a-kind devices and require very careful inspection for proper listings of components and parts. All materials, pipes, valves, cylinders, and electrical switches must be compatible with the fuel used in the prop. Keep in mind that the fire prevention code prohibits open flames in places of public assembly unless they meet specific requirements of the code. The requirements include the location of the prop behind the fall-line of the stage's fire curtain. All combustibles should have a clearance of ten feet from the open flame or pyrotechnic device. The NFPA standard also requires clearances of 10 feet to the stage entertainer and 20 feet to the proximate audience. This is the minimum distance from audiences and entertainers. Most importantly, the entertainer or facility owner is required to obtain a fire safety permit and have a fire inspection conducted before the event. In 1984, singer Michael Jackson received severe burns when pyrotechnics ignited his hair. Pyrotechnics also provided the ignition source for the Station Fire in Warwick, Rhode Island, that led to the deaths of 98 patrons.

Torches

Many buildings have been lost to fires as a result of paint-removing torches, torch down roofing, and cutting or welding too close to combustible materials. According to the

The Station Nightclub Fire—West Warwick, Rhode Island

On February 20, 2003, one of the worst nightclub fires in United States history took place at the Station night-club. The club was occupied by 440 persons of whom over 100 died and 200 were injured by a fire that became untenable in under five minutes. The fire was ignited when a pyrotechnic gerb, a small spark-producing firework, ignited the egg crate foam plastic that was attached to the walls at the rear of the stage to reduce sound. The fire quickly spread on the foam plastic and reduced the possibility of exiting safely within a matter of minutes. The club owners had not filed with the local jurisdiction for a fire safety permit for the display of pyrotechnics indoors. Pyrotechnic displays require a fire safety permit, and a fire inspection must be conducted under NFPA 1126, Standard for the Use of Pyrotechnics Before a Proximate Audience. A complete permit inspection most likely would have prohibited the use of pyrotechnics in the nightclub and at a minimum, would have brought into question the flammability of the wall coverings next to the pyrotechnic devices.

The grand jury indicted the club owners and the tour manager for the band and during hearings with the prosecutor were extremely concerned with the lack of a proper fire inspection by the fire inspector from the Warwick Fire Department. They believed that the inspector should have detected the foam plastic material that led to the rapid spread of the fire. The prosecutor explained that under Rhode Island law, the fire inspector was given sovereign immunity unless he acted maliciously or in bad faith in conducting the fire inspection. The grand jury decided that there was not sufficient evidence of either bad faith or malicious intent for the fire inspector not recognizing the potential hazard of the foam, and the inspector was not indicted. The jury most likely would have indicted the fire inspector had the sovereign immunity law not been in effect. The results of the trial placed the business owner in jail for 15 years on wrongful death charges. The band tour manager pleaded to 100 counts of manslaughter and received a 10-year sentence, and many of the businesses and several industries, including the state, had to settle civil suits from the victim's families in the following amounts:

WPRI-TV settled for 30 million for blocking the exit with its camera crew.

JBL Speakers settled for $815,000.00 for foam in their speaker systems.

Anheiser Busch settled for $5 million for sponsoring the concert.

The local beer distributor settled for $16 million for sponsoring the concert.

Clear Channel Radio settled for $22 million as a concert sponsor.

Home Depot settled for $5 million for selling the foam.

Polar industries settled for $5 million for distributing the foam.

Sealed Air Corporation settled for $25 million for manufacturing the foam.

The City of Warwick and State of Rhode Island settled for $10 million.

The Great White Band settled for $1 million.

In turn, the results of this horrific fire helped to make improvements in the code enforcement procedures for fire marshals in the State of Rhode Island, including certification and better training in fire code enforcement. An important takeaway lesson from this incident is that fire inspectors must conduct through and complete fire inspections and pay attention to detail. In many states sovereign immunity laws are being altered based on the level of industry knowledge that exists and the training that fire inspectors are required to receive to perform their jobs.

NFPA, from 2003 to 2006, misuse of propane torches and acetylene cutting equipment resulted in 5,580 fires, 22 deaths, and 223 civilian injuries in nonresidential structure fires. Many major fires like the loss of a Universal Studios back lot in Hollywood, California; the Monte Carlo Hotel fire in Las Vegas, Nevada; and a $2 million school fire in Washington, DC, all in 2008, were the result of the improper use of open flame torch and cutting devices.[2] Anytime open flame devices are utilized, a fire safety permit is required by the fire prevention code, and the operator must follow the required safety rules in NFPA 51B, *Standard for Fire Prevention During Welding, Cutting, and Other Hot Work*. The fire prevention code requires the presence of at least one portable fire extinguisher with a minimum 4A rating as well as the institution of a fire watch for at least one hour after the torch or cutting device has been used. Operators of torches must have access to surrounding combustible areas to ensure that a fire cannot extend into concealed spaces.

Open flame devices may not be used in hazardous areas unless the area can be rendered safe by separation, protection, or removal of combustibles or the containment of the flame to prevent it from contacting combustible materials. The NFPA reports that of all the welding and torch-related fires that occurred from 2003 to 2006, 43 percent were caused by welding, 12 percent by paint removal, and 7 percent by plumbing repairs.[3] Fire inspectors must be vigilant on the use of torches and flame-producing devices within their jurisdictions. Inspectors should follow up on hot work permits to make sure that the operators are following safety rules and fire watch procedures. The fire inspector should identify all of the rules as conditions of the fire safety permit.

Food and Beverage Preparation

Many restaurants prepare special foods in a **flambé** cart or may have a signature beverage or dessert that is ignited before being brought to the customer's table. The codes generally prohibit the moving of flaming food items by the restaurant staff as they may be accidentally spilled or dropped and can ignite other materials. Every restaurant that prepares foods such as flaming steaks, desserts, or drinks must follow certain fire code regulations. These regulations restrict the amount of burning flammable liquid to no more than 1 fluid ounce. The foods should not be transported but must be prepared at the table on an approved flambé cart. The cart should have a secure bottle holder and also a wet towel or other extinguishing agent to smother flames in an emergency. The flame effect should not extend beyond 8 inches nor create an extremely high flame. A fire safety permit for such activities should be required and should include a training requirement outlining all of the safety considerations and code restrictions that the operator must comply with.

flambé ▪ A term used in the restaurant industry to describe food that is presented to the customer with an open flame. The source of the flame is usually liquor poured around the plate and ignited for a dramatic effect.

SMOKING MATERIALS

Smoking materials such as cigarettes, cigars, and matches have been a leading cause of fires in the United States for years, especially in residential fires. In the last ten years, however, smoking trends have been significantly decreasing because of societal pressure regarding the negative health effects of smoking and second-hand smoke. We have also seen smoking material fires reduced by changes in technologies such as smoke detectors and fire alarms as well as ignition-resistant furniture and bedding and fire-safe cigarette advances. According to the National Health Service, Centers for Disease Control and Prevention, in 1965, 42.4 percent of adults smoked cigarettes. In 2010, the number of adult smokers decreased to 19.3 percent, or almost a 50 percent decrease over the 45-year period.[4] While the fire code doesn't regulate smoking materials in most residential buildings, the effects of careless smoking in commercial buildings have significantly declined. The United States Fire Administration identified smoking as a cause in 2 percent of residential fires but 14 percent of residential fire deaths.[5] The fire inspector's role in reducing smoking-related fires is to identify areas where the use of smoking materials will create a fire hazard. Those areas of the building should be properly posted as no-smoking areas. Typically, these hazardous locations are where combustible or flammable materials are being stored, handled, or used. The fire inspector should ensure that these hazardous areas are posted as no smoking and the owner enforces the regulation. The removal of any posted "no smoking" sign or smoking in a prohibited area is a fire code violation. The owner may designate smoking areas within the guidelines of the law in the particular state. All permitted smoking areas must have ashtrays and proper disposal receptacles for cigarettes. Smoking areas outside should be free of combustible landscaping materials. While fire inspectors are not the "smoking police," they should observe where these activities occur. Many serious fires that have caused the deaths of firefighters were the result of smoking in nonsmoking areas. The use of smoking materials in a nonsmoking area was determined to be the cause of the Super Sofa Furniture Store fire in 2007 in Charleston, South Carolina, that took the lives of nine Charleston firefighters.

One day while driving through an industrial area of the city, a fire inspector saw a man dragging a fuel hose from an exit door of the building to his laundry truck parked at the curb. The fire inspector stopped and asked the gentleman what he was doing, and he told the inspector that the owner had installed a fuel tank for the vehicle to make it more convenient to refuel. When the inspector entered the emergency exit door to speak with the owner, he was confronted with a 1,000-gallon tank resting on the floor of the laundry facility. The tank was improperly installed without building permits and was located in a hazardous area, not to mention it obstructed the secondary means of egress. The tank constituted an imminent hazard, and all sources of ignition and personnel had to be secured until the tank could be defueled and removed from inside the building. Fire inspectors must stay vigilant in their communities. During difficult economic periods, business operators may develop creative solutions to reduce overhead cost. This can include situations like the illegal storage tank or even making their own bio-diesel fuel to operate their vehicles. Unfortunately, many of these green energy solutions may violate the requirements of the fire prevention and building codes.

REFUELING EQUIPMENT

Anytime flammable or combustible liquids are being transferred or stored, the potential for a fire increases dramatically. Fire inspectors must carefully evaluate refueling operations and ensure they are performed within the fire code requirements. Fire inspectors should question how generators and other types of combustion-powered equipment are fueled. The fire prevention code prohibits the storage, operation, and repair of fueled vehicles inside buildings unless they are specifically designed for such use. When vehicles are displayed inside buildings, they must have their batteries disconnected and contain a limited amount of fuel. The fuel tank is required to be secured, and the vehicles may not be fueled or defueled inside the building. Fuel-powered vehicles also may not be driven inside buildings other than those designed for automobiles, as there can be a buildup of carbon monoxide gas within the building. Frequently, fire inspectors encounter fuel-fired appliances such as portable kerosene heaters or kerosene salamanders during fire inspection tours; these are especially prevalent in buildings undergoing renovation or construction. In many states and local jurisdictions, liquid fuel–fired appliances like kerosene heaters are prohibited from use in commercial and multifamily residential buildings. The maximum allowable quantities of flammable and combustible liquids stored in a building are 10 gallons of a flammable liquid and 60 gallons of combustible liquids. Fuels exceeding these allowable quantities require a permit and approval from the fire marshal or fire official. The flame-producing appliance may also require a permit from the fire department to operate inside a building. Fire inspectors may prohibit the use of any portable heating equipment if it presents a hazard due to the location, the quantity of fuels, or other safety concerns such as carbon monoxide production or proper clearances to combustibles.

VACANT STRUCTURES

Vacant structures create community blight and are often havens for illegal activities in communities. Vacant buildings frequently are targets of both juvenile fire setters and arsonists because they have easy access and getting caught by police is not likely (see Figure 7.1). Fire inspectors will spend most of their time inspecting occupied buildings; however, they should not overlook the vacant structures in their communities. Vacant structures also present many cross-jurisdictional hazards, especially between the building, health, police, and fire agencies. Such hazards include drug abuse and medical waste in the building, prostitution, dead animals, feces, and rotting trash accumulation. Additionally, there may be structural issues with walls, floors, or roofs, and in some cases, illegal position of the property. In the city of Philadelphia in 1985, a group of people calling themselves "MOVE" barricaded a vacant home that they illegally occupied and turned it into a fortress. The property blighted

FIGURE 7.1 Open and vacant buildings create urban blight and destroy neighborhoods. These buildings are frequently havens for criminal activity and targets for arsonists. *Courtesy of J. Foley*

the neighborhood and generated many police, fire health, and building department complaints from the neighbors. When the mayor of the city, Wilson Goode, finally had to act on the situation, it turned into a major disaster, causing a fire that destroyed 65 homes and eventually a lawsuit of $1.5 million that the city had to pay to "MOVE." To avoid these situations, inspectors need to actively address vacant structures and make sure they are boarded and secure. Many municipalities may take an interagency or task force approach to managing vacant buildings. These task force units usually include code enforcement agencies, the police, and local citizen group representatives such as town watch. The public works departments and administrators from city government are also involved in these taskforces in a support role. The difficulty in addressing vacant structures is locating the owner and getting the owner to reinvest money into the abandoned property. Most vacant structures are vacant because a business has failed or the mortgage payments could not be made. Consider that there were 1 million foreclosures in 2010 and over 2.87 million mortgages that were prone to failure. This illustrates the scope of the problem in every community.[6] The fire prevention code requires that buildings that become vacant must be properly safeguarded from trespassers. The building owner must provide security, which includes locking, blocking, or boarding exterior openings on the property. Utilities must be disconnected, and combustibles should be removed from the property. If the building has sprinklers or a fire alarm system, the system must be maintained in operational condition unless the fire marshal or fire official approves the system to be disabled. In cold climates, a wet pipe sprinkler system must be drained or it will freeze and break if the building is not heated. Fire inspectors also should require that placards be installed on any unsafe structures or buildings that are imminent hazards or subject to demolition (see Figure 7.2). Firefighter warning placards are based on Federal Emergency Management Agency and NFPA urban

FIGURE 7.2 Fire inspectors must ensure that buildings that are imminent hazards are properly marked and identified for firefighter safety. *Courtesy of J. Foley*

Insp. 2/20/2012

Building was in normal condition at the time of inspection.

Insp. 2/20/2012

Structural hazards exist in the interior, firefighters should use extreme caution.

Insp. 2/20/2012

Extreme risks exist and firefighters shall use caution and fight fires from the exterior.

FIGURE 7.3 The IFC requires vacant buildings that are to be demolished or are imminent hazards or unsafe structures to be placarded using the FEMA hazard classification system. *Courtesy of Federal Emergency Management Agency*

search and rescue identification systems for dangerous buildings (see Figure 7.3). The intent of the placard is to identify the hazard to firefighters so they can maintain a defensive position during a fire and reduce their risk. Vacant structures are extremely hazardous to firefighters' safety, and fire prevention inspectors need to ensure that the locations of these hazards are properly identified and communicated to the first responders.

Fire inspectors addressing the vacant structure problem will find that locating a responsible party to repair the structure is a time-consuming and difficult code enforcement issue. In these cases, the municipality often is required to take action to abate the hazard, which involves utilizing city resources and money. Fire inspectors should keep records of resources deployed and any funding expended to address the code violations so that the municipality may later recover its expenditures through a property lien. Municipalities may have to remove the combustibles from the property and board and secure the premises, which requires time and labor cost. In many cases of imminent hazard or unsafe structures, the municipality may have to demolish the building, which requires court action and additional expenditures to the jurisdiction. Fire marshals and fire officials need to establish an interagency approach and develop protocols to address the vacant structure problem, reduce blight, and maintain safe communities.

VEHICLE IMPACT PROTECTION

Vehicle impact protection is a requirement of the fire prevention code for the protection of aboveground storage tanks, utility meter connections, means of egress doors, and other locations that may be impeded by parked or moving motor vehicles. Typically, impact protection requires steel post of 4- to 6-inch diameter pipe to be placed at the proper depth in the ground with adequate distance from the storage tank to keep vehicles from contacting it (see Figure 7.4). These posts or bollards must be capable of withstanding a force of 12,000 pounds at a height of at least 36 inches. Depending on the size and weight

FIGURE 7.4 Aboveground tanks must be protected by bollards to prevent vehicle impacts. *Courtesy of J. Foley*

of the moving vehicles in the protected area and footing depths, the size and height of the bollards may vary. Bollards should be painted in a highly visible color and should be spaced so that a vehicle impacts more than one in a collision, reducing the ability of a vehicle to penetrate the protected area. Fire inspectors should identify the hazard areas on a building that may require impact protection during the exterior inspection of the building. It is especially important to identify the means of egress discharges in tight parking lot areas to ensure that a vehicle does not block the exit discharge.

GENERAL STORAGE

The arrangement of storage in a building is a common fire hazard that the fire inspector will encounter. Storage must be neat and orderly and stored away from potential ignition sources such as heaters or electrical equipment. Fire inspectors must be attentive to the types of storage also; flammable and combustible liquids and aerosols require proper storage cabinets or approved locations depending on the type and quantity stored. Storage piles should be 24 inches below the finished ceiling and 18 inches below the automatic sprinkler heads (see Figure 7.5). Fire inspectors should also inspect the hazard classification of the automatic sprinkler system to ensure that storage does not exceed the permitted system design. The fire inspector may determine the hazard classification from the hydraulic information plate on the sprinkler system riser. Storage is not permitted to encroach on exit access or exit stairs and should not be permitted in electrical, mechanical, boiler, or similar equipment spaces. Fire inspectors should be especially observant in critical system locations such as the fire pump room and HVAC mechanical air-handling rooms. A storage fire in an air-handling room will spread smoke and heat throughout the building via the ductwork. Frequently, building owners store boxes of combustible air filters in these rooms. This should not be permitted, as the cartons are combustible. In larger buildings, maintenance workers often set up shops in the mechanical spaces, such as plumbing, painting, or electrical shops. Mechanical rooms are not acceptable areas for paint storage, flammable gases, welding and cutting operations, or carpentry shops where dust is generated. Flammable and combustible dust may collect in ductwork and create a vehicle for fire spread or potential dust explosion. Rooms containing fire protection equipment are required to be fire resistant rated for 1–2 hours. If combustibles are stored in these areas and a fire occurs, you cannot go out the next day and replace a fire pump or pump controller. The fire inspector must help the owner understand that without these critical fire protection systems operating, the building will be closed. General storage also should not be located in any concealed spaces unless it is protected by 1-hour fire-resistant

FIGURE 7.5 Mercantile storage must be kept 24 inches below ceilings and 18 inches below automatic sprinklers. *Courtesy of J. Foley*

During the inspection of a casino hotel, fire inspectors were examining an HVAC air-handling room. They opened an access door to the HVAC supply duct, which was approximately 10 feet high by 15 feet wide. When they looked inside, they found paint stored in 5-gallon pails and indication of paint spraying being conducted inside the HVAC duct. All of the materials were removed immediately in the inspector's presence and a notice of violation and penalty was issued. Fire inspectors must inspect all rooms and spaces, especially in buildings where work space is at a premium. Often, building maintenance shops are an afterthought in the building design or become repurposed by the owners, and these shops end up in critical areas that are not designed for such shop functions.

assemblies and automatic sprinklers. In 1988 five New Jersey firefighters were killed in the Hackensack Ford fire because of improper auto part storage in the concealed space of a bow string truss roof. Fire inspectors must be constantly vigilant for storage in improper locations in buildings. Fire inspectors should also require that wooden storage pallets not be stored or allowed to accumulate inside or adjacent to a building. Pallets present a unique fuel source that can develop into a significantly large fire very quickly.

Outdoor storage also must be inspected and should be neat and orderly and at least 15 feet from any structure or the eaves of a structure. Storage is also prohibited within 15 feet of property lines so as to not expose neighboring properties. The fire prevention code will usually require the installation of a 2-hour fire-resistant wall, 30 inches higher than the storage pile if the 15-foot requirements cannot be met. Outdoor piles of storage must be stable and should not exceed 20 feet in height.

COOKING

Cooking and cooking appliances are the leading causes of fires in the country in both residential and commercial buildings. The NFPA reported that between 2004 and 2008, there were 8,160 fires in eating and drinking establishments related to cooking. These fires resulted in three fire deaths, 100 civilian fire injuries, and $229 million in direct property damage. In commercial establishments, the fire inspector must examine the cooking and ventilation equipment for proper installation, venting, and most importantly, proper cleaning. The standards that will apply to these restaurant systems are NFPA 96, *Standard for Ventilation Control and Fire Protection of Commercial Cooking Operations* and the ICC Mechanical Code. The fire prevention codes require that owners of restaurants provide the fire inspector with a cleaning schedule for all commercial cooking systems. Cleaning should be at least quarterly, and for heavily used appliances, more frequently. Cleaning is critically important to the fire safety of the facility.

Ventilation Fan

Inspection of cooking appliances should begin on the building's roof. Ventilation fans should not be grease laden and should not be spilling grease and oil onto the roof surface. Ventilation fans are hinged and may be lifted for inspection of the duct below. The inspector should note that the electrical components of the fan are out of the air pathway. The inspector should also examine the fan specification plate to ensure that the fan is listed for grease-laden vapor. Examine the electrical connection for the fan's power supply. It should not penetrate the duct at the roof level. Fans should have proper curbing and grease collection features and should terminate away from building openings.

Appliance Cooking Line

Once the roof exhaust has been examined, the fire inspectors should inspect the appliance cooking line and the ventilation hood. All appliances should be located under the hood by at least 6 inches. The surrounding ceiling should be examined for discoloration and grease

FIGURE 7.6 Commercial kitchens must have K-class portable fire extinguishers, and all hazards under the range hood must be controlled by manual reset automatic gas valves and electrical system shunts. *Courtesy of J. Foley*

UL-300 ▪ An advanced test method for commercial cooking fire suppression systems established in 1998 because of high-efficiency cooking appliances and the increased use of vegetable oils over animal fats.

not being captured by the hood. NFPA 96 and the IBC mechanical code both require that all ignition systems and fuel sources be controlled with the activation of the fire suppression system. Electrical sources should be shunted out by an electrical relay, and natural gas should be controlled using a manual reset gas valve (see Figure 7.6). The ventilation hood should meet the requirements of **UL-300** and have the appropriate compatible fire suppression systems installed within it. A K-class portable fire extinguisher shall be located within ten feet of the cooking line as well as the manual pull station for the surface protection system.

plenum ▪ A void space behind the filter rack in a kitchen hood where grease-laden air collects before moving out the exhaust duct. It is protected by plenum nozzles from the fire protection systems and is the location of the fusible links that operate the system.

Plenum Inspection

The **plenum** of the ventilation hood is the space behind the filter rack. Plenum filters should be removed and examined for compliance with grease-laden vapor recovery. Proper filters capture the grease by baffles that allow the grease to separate from the air and run down to a grease collection area and grease cup at the bottom of the hood plenum (see Figure 7.7). Mesh-type ventilation filters are not approved for the capture of grease-laden vapors and may become hazardous over time. All grease filters should be removed during the plenum inspection. The fire inspector should next examine the fire suppression system's fusible links for proper date and installation as well as any accumulation of grease (see Figure 7.8). Links are to be replaced biannually, and the fire suppression system is required to be completely inspected annually.

Cooking ventilation hoods must be kept clean and grease free. The fire inspector can determine the cleanliness of the hood and plenum by using an NFPA 96 grease comb.

FIGURE 7.7 Filters for grease-laden vapor are baffled to allow the grease to collect in a trough at the bottom of the filter rack. The filters should be tight and undamaged. *Courtesy of J. Foley*

FIGURE 7.8 The fusible link system is located behind the filter rack. In this photo, the link is covered with grease and in need of replacement. This duct also requires cleaning. *Courtesy of J. Foley*

Grease combs have four offset teeth, with each tooth spaced 0.125 inch higher than the next one. The comb's teeth are placed parallel to the surface and the comb is dragged along the hood surface. The fire inspector will note which tooth picks up grease. This will determine whether the hood is clean or hazardous. The comb can be purchased from organizations such as the International Kitchen Equipment Cleaning Association (IKECA) and is a very useful fire inspection tool. Often, just physical observation by the inspector can determine that the hood and plenum are in need of cleaning (see Figure 7.9). In larger restaurants, multiple hoods may be interconnected to long runs of ductwork. This grease duct must be properly inspected to reduce the possibility of a major fire. The mechanical codes require that fire-rated duct access panels be placed for inspection and cleaning every 20 linear feet and at each change of direction in the duct. Fire inspectors may find that these requirements are inadequate in most large ventilation systems. Fire inspectors may require additional duct access panels to be installed to properly clean the ductwork if the code-required access panels are inadequate. A more acceptable distance for cleanout is most likely every 12 feet along the duct. In larger facilities, the fire inspector may also require specialized equipment to perform the duct inspection such as high reaches, ladders, or pole cameras for inspection of the duct interior. In large facilities, the kitchen duct inspection may have to be conducted as a separate fire inspection because of time involved and the impact on the business. These large systems are usually inspected at off hours when the restaurant is closed to limit any business interruption.

Kitchen Hood Fire Suppression Systems

Commercial kitchen systems are required to be protected by automatic fire suppression systems. The type of system will vary depending on the age of the equipment. Older systems

FIGURE 7.9 The fire inspector can observe that this hood is in need of cleaning. Grease accumulating and dripping from the plenum indicates that the duct also is in need of cleaning. *Courtesy of J. Foley*

may be dry chemical using the extinguishing agent sodium bicarbonate, while newer ones will be wet chemical systems, which use agents of potassium acetate or citrate, or they may be NFPA 13 sprinkler systems using water as the extinguisher agent. Several hood manufacturers use water washdown systems with a combination of wet chemicals or sprinklers for cooking surface protection. The fire inspector must be familiar with the different system manufacturers in their jurisdiction. The system specification can usually be obtained from the installation company or the company's website. Based on the type of fire suppression system that is installed, the fire inspector should examine each application nozzle for proper height and coverage as well as grease infiltration protection (see Figure 7.10). Dry and wet chemical system nozzles must have proper covers to keep grease from clogging and obstructing the nozzle opening. All components of the system, including the control head and agent storage cylinder, must be located where they can easily be inspected. The fire inspector should examine the control head for proper arming and tagging by a certified fire protection installer. The system cabinet door should be opened to ensure that all cables are properly attached and that safety firing pins have not inadvertently been left in the firing control. These pins are installed when service technicians change fusible links in the system to prevent accidental discharge. The inspector should also inspect the manual gas valve cable and the shunt relay for any electrical appliances or outlets under the hood. Fire inspectors should attempt to witness periodic testing of these fire protection systems to become more familiar with their operation. It is important for the safety of the fire inspector to make sure the system is cold to the touch before conducting any inspection of the equipment or hood and ductwork. Kitchen fire suppression systems should be connected to the automatic fire alarm system, if one is present in the building.

FIGURE 7.10 Fire suppression system nozzles must be protected from grease infiltration. Often, the rubber caps deteriorate or fall off, causing the nozzles to clog with grease. *Courtesy of J. Foley*

Fire Service Building Features

The fire inspector should be familiar with all of the fire service features that the fire prevention code requires in buildings. When the fire inspector is conducting the exterior examination of the building, the inspector should note all available access for fire department apparatus around the building. Fire apparatus requires all-weather roads that are stable in rain or snow and capable of supporting the fire apparatus weight. (Fire apparatus may weigh 14–35 tons.) All-weather access roads must be at least 20 feet wide and have vertical clearance of at least 11 feet 6 inches to accommodate larger aerial fire apparatus. Fire apparatus generally require a minimum turning radius of at least 25 feet, and on dead-end roads over 150 feet in length, there must be a "T" connection at the end of the road to allow the engine to turn around.

Roads should not exceed 12 percent in grade, and the angles of approach at the street should be level enough to not cause the engine to bottom out when entering or exiting the property. Any bridges located in the roadway must be capable of supporting the fire apparatus weight, and any gates or security measures placed across the access road must have

FIGURE 7.11 Knox Boxes allow the fire department access to secured locations. The box on this building is placed high to avoid tampering. Key boxes may also be equipped with alarms. *Courtesy of J. Foley*

the capability of being opened in an emergency. **Treadles** and tire punching devices should not be installed in the fire department access lanes or in locations where fire apparatus may move to connect to a fire hydrant.

All buildings must have their street address visible from the roadway with at least 4-inch letters, and any internal roads within a industrial complex shall have proper street signs to identify locations. If the fire department participates in a **Knox Box** or key box system, the appropriate type of box should be installed and the owner should provide the necessary facility keys for the fire department to permit access to the property or any specific equipment within the building (see Figure 7.11). If the property has hazardous materials on site, the appropriate hazardous materials management plan and inventory information must also be kept in a suitable key lock box.

treadle ■ A security device installed in parking areas to prevent vehicles from leaving without paying a fee, the treadle has spikes that rise up and puncture the tires of the vehicle if it enters the wrong way.

Knox Box ■ A Knox Box is a key box with master keys provided to only the fire department for access. The Knox Company is the leading manufacturer of fire department key box systems.

Firefighter Hazards

Any building roof opening, ventilation shaft, open pits, or scuttle into which a firefighter could fall and be injured or killed must be identified with signage to make the firefighter aware of the potential hazard. Any security devices intended to kill or maim intruders is prohibited. These devices include gases, razor wire, sharp or pointed objects, or other similar implements. All exterior fire department access points or ventilation windows shall be appropriately marked for fire department access, especially in windowless spaces. All storage areas or rooms containing hazardous materials shall have the appropriate NFPA 704 placards visible from the point of entry.

Water Supply

Industrial facilities with private water mains shall maintain and test those mains and hydrants in accordance with NFPA 24, *Standard for the Installation of Private Fire Service Mains and Their Appurtenances.* Fire hydrants should be properly spaced, but no more than 400 feet apart, and they should be on adequately sized mains to provide the necessary fire flows for the facility. Fire hydrants should also be protected from vehicle impact if located in areas where moving vehicles are in close proximity. Fire hydrants should have a clear space of 3 feet around them.

Inspection of Building Service and Systems

The building service systems are those systems that provide light, heat, air conditioning, and movement to the occupants of the building. Each system will require an inspection on the fire safety aspects of the system by the fire inspector.

Oil Burning and Heating Equipment

All heating equipment should be installed as described by the jurisdiction's building and mechanical codes. Fire inspectors should examine each appliance for proper clearances to combustibles as referenced on the appliance specification plate located on the equipment (see Figure 7.12). Oil-fueled equipment must also provide a diagram showing the location of the fuel shutoff valves for the main oil lines; this generally is required to be posted by the furnace. Oil burners require an emergency fire safety switch to turn the

FIGURE 7.12 Boilers and heaters must have adequate clearances to combustibles and proper draft on the chimney and vent. *Courtesy of J. Foley*

electricity off to the oil pump in the event of a fire or other emergency (see Figure 7.13). This switch must be properly identified and is located in the path of entrance to the heater room, usually at the top of the stairs. Appliance connectors, chimneys, and vents also should be inspected to ensure they are not leaking carbon monoxide and are properly installed. All vent pipes must angle up to the chimney connection; this prevents condensation from collecting in low points and corroding the pipe. The vent connection should be tightly sealed to prevent escaping carbon monoxide from backing up into the building. Oil-burning and gas-fired equipment should not use the same chimney or flue space. Chimneys and vents also must have proper clearance to all combustible surfaces. Reduction of clearances can only be made with noncombustible materials installed in accordance with clearance reduction tables found in the mechanical code. Fuel oil tanks inside or outside buildings are usually limited to 660-gallon capacity.

FIGURE 7.13 Oil burner fire safety switches must be located at the entry point to the equipment. *Courtesy of J. Foley*

Larger tanks must comply with the requirements of NFPA 31, *Standard for the Installation of Oil-Burning Equipment*, and are either underground or protected by fire-resistive enclosure of the aboveground storage tanks. The fire inspector should examine the general condition of the heating equipment and should examine any draft points for rust or effluents backing up from the vent. The fire inspector can check draft by starting the appliance and lighting a match, blowing it out, and seeing if the residual smoke is drawn into the draft hood of the vent. Most fire prevention codes either prohibit or restrict the use of portable unvented heating appliances in buildings where people sleep or gather with the exception of one- and two-family dwellings. Every open combustion heater requires combustion air to function properly. Indoor combustion air requires 40 cubic feet of volume for every 1,000 BTU of heat input. If openings or grills are provided into a room for makeup air, they should provide at least 1 square inch of opening for every 1,000 BTU input rating of the heater. Installed and approved unvented heaters may not draw combustion air from bedrooms, storage rooms, or bathrooms under most building codes.

Unsafe Heating Appliances

Every fire code provides the authority to the fire inspector to shut down or prohibit the use of any unsafe heating appliance. Unsafe conditions may include improper vents and chimneys, insufficient clearance to combustibles, leakage of carbon monoxide, or other defects in the system that create a potential life safety risk. The fire inspector may **seal the equipment** by shutting it down and tagging the appliance as not to be used until the violations are corrected. The heating appliance may not be returned to service until the fire inspector authorizes it and removes the seal or tag. The fire inspector also should notify the local gas utility or the owner's oil burner service company that the appliance is sealed. Any person who operates the appliance in violation of the order is subject to fines under most fire prevention codes.

Emergency Electrical Generators

Buildings that are required to have emergency generators must have those generators tested under a 30 percent service load for 30 minutes monthly and subjected to an annual load test for 2 hours. Fire inspectors should be present during these annual generator tests to ensure the proper operation of the emergency lighting and life safety systems including the operation of the **automatic transfer switches**, or ATS. NFPA 110, *Standard for Emergency and Standby Power Systems*, requires that the generator be tested at 25 percent of the load capacity for 30 minutes, 50 percent of the load capacity for 30 minutes, and 75 percent of the load capacity for 60 minutes during the annual test. The primary power failure or shutdown must switch emergency system electric loads to the secondary power supply in 10 seconds and standby loads within 30 seconds of power failure (see Figure 7.14). To properly witness this test at least three fire inspectors are required: one at the emergency generator, one at the ATS switch, and one to walk the affected area on emergency power to ensure that the emergency lights and life safety systems are operational. This test takes approximately 2 hours per generator to complete. It

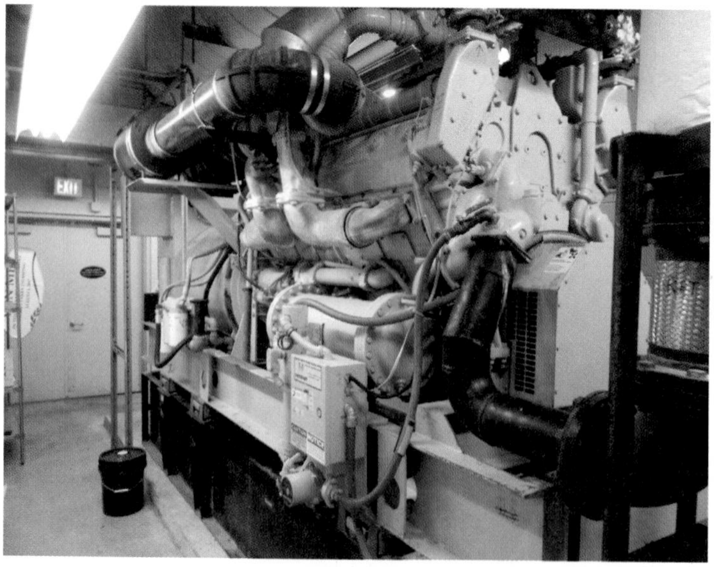

seal the equipment ■ This term is used to describe placing the equipment out of service with a tag. This is sometimes referred to as "red tagging" an appliance. Once the tag is placed upon the equipment it may not be used until it is reinspected by the authority having jurisdiction.

automatic transfer switch ■ This is an electrical switch that transfers power from primary sources to secondary sources such as emergency generators. Often referred to as an ATS switch, it must transfer power automatically within 10 seconds of primary power failure.

FIGURE 7.14 Diesel generators are used to provide secondary emergency power within 10 seconds of primary power failure. *Courtesy of J. Foley*

is critically important that the ATS switch be operated on a regular monthly basis or it may fail to operate properly under emergency conditions. The fire inspector should examine the owner's records for these tests.

Wiring and Electrical Hazards

All buildings have electricity and therefore may have electrical fire hazards. In today's environment, almost everything is electrical or has an electrical charging adapter. The most common electrical violations the fire inspector will encounter are the abuse of the electrical system with multiplug adapters and the improper use of extension cords. While fire inspectors are not electricians, they must be capable of identifying basic electrical fire hazards and their remedies. In challenging situations, the fire inspector should request the assistance of an electrical inspector to evaluate any potentially hazardous equipment or appliances.

Extension Cords

The use of an extension cord is permitted as a temporary power supply but cannot be used as a substitute for permanent electrical wiring. Inexpensive extension cords are generally constructed of 16AWG-stranded wire, which can handle approximately 13 amps of electrical resistance. Normal residential house wiring is 14/2 AWG wire and can handle 110 V/15 amps. Most commercial wiring is 14/2 12/2 AWG wire and can handle 110 V/20 amps. The problem with the use of extension cords in place of permanent wiring is that they are exposed and thus subject to potential mechanical damage as well as overload. Extension cord wire is rated lower than a normal electrical circuit breaker of 15–20 amps. This means that the extension cord can overheat and start a fire without the circuit breaker ever detecting the overload condition and tripping the electrical circuit. When extension cords are used temporarily, they may power only a single electrical device. Extension cords should not have multiplug adapters attached to them with numerous electrical devices connected like computers, refrigerators, printers, or copy machines. The fire inspector should look for extension cord abuse and have these electrical hazards removed. These cords should not be run beneath carpets or poked through walls or wired into permanent electrical systems. The National Electrical Code, NFPA 70 requires that wall outlets be spaced at 12 feet so that lights and similar appliances are within reach of a 6-foot appliance cord. If the fire inspector finds extension cords being used, it identifies a need for more permanent outlets to be installed.

Multiplug Adaptors

In today's electronic device environment, it is difficult to avoid using multiplug adapters, especially with so many low-voltage transformers being used to recharge portable devices. Any electrical plug strip must be listed to UL 1363, *Relocatable Power Taps*, and must have an internal power control and circuit breaker switch built in. Multiplug adapters may not be used with extension cords but must go directly to the wall outlet. They may not be run through any walls or other concealed spaces. Fire inspectors should check to ensure that multiplug adapters are not interconnected to additional adapters.

Open Junction Boxes and Exposed Wires

The National Electrical Code, NFPA 70, requires that electrical wiring be protected behind wall construction or within conduit on any exposed surfaces. Wiring systems should always be protected at junctions, wall outlets, and wall switches with proper electrical covers and plates. Generally, when a fire inspector sees an open junction box, it should be a red flag to closely examine the electrical system for code violations. Junction boxes are usually open because somebody has added a wire to the circuit. The National Electrical Code provides limits on the permitted number of wires entering a junction or outlet box by the size of the wire and the cubic inches of the box. Often, when additional wires are added, the cover plates will no longer fit, as the box is overcrowded with conductors. Open junction boxes are a hazard and good indicators of nonelectricians working on the electrical system. If it appears that the box is overcrowded with conductors, the fire inspector should insist that a licensed electrician repair it and an electrical permit be obtained from the electrical inspector. The fire inspector should also notify the electrical inspector in a referral notice or interoffice communication.

Other Electrical Hazards

Other common electrical hazards include machinery with missing electrical safety covers; open fuse or circuit breaker boxes; electrical motors with belt guards missing, greasy, or oily; and appliances that have had their cords or plugs altered. Fire inspectors should inspect all areas where electrical equipment is installed. Rooms containing high-voltage equipment must be properly marked and identified on the door so firefighters are made aware of the hazard within the room. All electrical panels and electrical equipment must have proper clearances of at least 36 inches to combustibles and must be accessible for servicing.

Elevator System

Elevator systems are regulated by the building's code and the ANSI A17.1 standard. All elevators that travel more than 25 feet are required to have emergency operation capability for firefighters. New elevators will have both phase I and phrase II recall and elevator controls for firefighter override and access. Phase I recall is when the elevator car is recalled to the ground floor or an alternate floor by a manual key or the activation of a smoke detector in an elevator lobby or the elevator shaft. Phase II recall is the ability to operate the car in firefighter override after the car is recalled in phase I. These smoke detectors are part of the elevator system and may or may not be connected to the automatic fire alarm system, depending on the building code requirements. It is important for the fire inspector to identify the proper activation method, as these systems must be tested for emergency operation. Another important aspect of phase II recall is that the elevator shaft must be protected by automatic sprinkler systems, and water may affect the braking systems on the elevator car. The elevator shaft system is usually protected by a preaction sprinkler system that will ground or secure power to the car prior to sprinkler head activation. Firefighters may operate the elevator car in phase II from the interior using the firefighter override control provided that the sprinklers have not activated in the shaft. The directions on firefighter control of the elevator car must be posted in the car over the control panel. Firefighters must also be aware that to use the car properly, the doors must fully open before they step out of the car; if they fail to do this, the doors will automatically close and lock them out of the elevator car. In most fire department high-rise firefighting procedures, a firefighter is assigned to operate the elevator controls to avoid this potential problem. In high-rises, the elevator car also has a two-way firefighter communications jack so that the elevator can speak to the fire command center (see Figure 7.15). This equipment should be annually tested for proper operation.

All elevator lobbies require signs warning people not to use the elevator in case of fire, but to use the exit stairs. In the current IBC and future building code editions, elevators are used as an element of means of egress if the elevator is properly protected. It has been demonstrated that the higher a person is in a building, the more likely the person is to use elevators as a means of escape. In the World Trade Center Tower 2 evacuation, almost 11 percent of the people who evacuated used the elevators even though the signs and fire alarm directed them to exit stairs.[7]

The elevator emergency and firefighter keys must be kept in a location that the fire department has access to. In some buildings, this may be the front desk. Keys may also be kept in a Knox Box or at the building fire command center. It is not uncommon for buildings to have several different elevators from different manufacturers within the same building. With the exception of New York City, all elevator cars are not keyed alike; therefore, the proper operating key should be kept in a secure Knox Box in the elevator

FIGURE 7.15 The instructions for firefighters on emergency car operation are missing in this elevator car. The car does have a connection to the firefighter communication system. *Courtesy of J. Foley*

lobby. This will ensure that firefighters have access to the proper elevator keys. Keys placed inside Knox Boxes should be properly tagged or labeled based on the fire department's coding system, and boxes should be inspected periodically to ensure that keys are in place and properly tagged, especially in buildings with frequent fire department responses.

UPS Battery Storage Systems

Uninterrupted power supply battery storage systems are used as standby power for many different types of electrical equipment including life safety and computer equipment. The batteries may be nickel-cadmium, lead–acid, valve-regulated lead–acid, or lithium ion–type batteries. The batteries may be stored in the same location as the equipment they serve provided that they are in cabinets and at least 10 feet from the equipment. If the storage system is in a separate ventilated room, the batteries may be in placed in storage racks. Liquid acid batteries require some form of spill control to neutralize any spill from the batteries. Sealed batteries do not require this. Rooms containing batteries must be identified on the door for firefighters and properly placarded as to the corrosive hazard inside. Battery storage rooms must also be protected by automatic smoke detectors.

Summary

Many of the common fire hazards that are encountered during fire inspections are common to all buildings. The fire inspector must be aware of the types of fires that occur in each type of building use they may inspect and key on those particular hazards during the fire inspection. Reviewing the annual fire statistics or national fire statistics can help identify common ignition sources and hazards and their impact on fire deaths, injuries, and property loss. These are the critical fire causes that require the fire inspector's attention, as they have the highest probability of causing a fire in a specific building use.

Fire inspectors must keep accurate records of fire inspections to ensure proper monitoring of building systems that they are required to inspect. The fire inspector should keep a record of compliance dates so that they can ensure the code requirements are being followed for each facility on testing and system maintenance. In most cases, a simple tickler file with index cards or a computer spreadsheet can be used to monitor system testing and maintenance schedules. The fire inspector can check each month's index cards or computer record and determine which property requires a follow-up call or inspection to ensure that the building and fire system have been properly cleaned or serviced.

Most fires occur because fundamental hazards like smoking, waste removal, improper chemical storage, or poor system maintenance are not addressed. If fire inspectors monitor these fundamental fire causes, they will reduce the number of fires in the community and lower the possibility of firefighter injuries and death. As technology changes and we move to a more green-conscious society, fire inspectors will continue to be challenged with new fire hazards. These hazards will include many home technologies, from solar panels to manufacturing of biofuels in the garage and refueling electric and compressed-gas vehicles in public parking facilities. The development and use of hydrogen cell electric generators will become more prevalent in both commercial and residential buildings. Green buildings will also begin to have living plants and parks on roof tops. How will these green areas be protected from lightning and potential brush fires, and what kind of challenges will they present if the building collapses? Fire inspectors will have to pay close attention to planting with mulch and foam plastic planting mediums. Energy conservation will increase the use of foam plastics on the exterior facades and rooftops of buildings, creating larger fire challenges to the fire department. The green movement not only will be a strategic and tactical challenge to firefighters, but also will require fire prevention inspectors to be vigilant regarding how to maintain these sustainable energy challenges in a firesafe manner.

Review Questions

1. Review the scope and purpose of your fire prevention code. What does the code allow you to inspect, and what types of fire hazards are addressed?
2. Fire inspectors must have a system to keep track of different compliance dates at properties. Describe how your fire prevention bureau would accomplish this task.
3. Most fire prevention codes contain broad-based violations like the ten general requirements. When should fire inspectors cite these as a violation of the fire prevention code?
4. Review your local fire prevention code and identify the types of records that must be provided for review and at what time interval they should be provided.
5. Combustible waste is a leading cause of fires. Devise a program to manage the removal of combustible waste in your community.
6. According to your fire prevention code, under what conditions are open burning or open flames permitted in and around buildings? Determine whether fire safety or other permits are required.
7. Outdoor cooking and heating appliances can be a leading source of structure fires. What regulations does your fire prevention code provide to control these hazards?
8. When are candles permitted to be used in places of public assembly? Are there criteria for these devices and are fire safety permits required?

9. Devise a system to reduce torch fires in your community. How would you institute programs or controls to address this problem? Explain the necessary fire safety precautions that must be addressed.

10. Describe how commercial kitchen cooking appliances should be inspected. What are the key elements of the inspection?

11. List the critical building services that would be encountered during a fire inspection. What are the key inspection points?

Suggested Readings

International Code Council. *Uniform Fire Prevention Code*. Washington, DC: International Code Council.

National Fire Protection Association. 2011. *NFPA 96: Standard for Ventilation Control and Fire Protection of Commercial Cooking Operations*. Quincy, MA: National Fire Protection Association.

International Code Council. 2012. *International Mechanical Code*. Washington, DC: International Code Council.

Endnotes

1. State of New Jersey, *Uniform Fire Code* (Washington DC: International Code Council).
2. John R. Hall, *Home Fires Involving Torches and Burning* (Quincy, MA: National Fire Protection Association, Fire Analysis and Research, 2009).
3. Ibid.
4. *1990 Morbidity/Mortality Weekly Report 1992* (Washington, DC: Center for Disease Control, 1992).
5. "Smoking Related Fires in Residential Buildings (2008–2010)" (Emmitsburg, MD: Federal Emergency Management Agency, 2012).
6. http://www.bloomberg.com/news/2011-01-13/u-s-foreclosure-filings-may-jump-20-this-year-as-crisis-peaks.html
7. Guylène Proulx, Amber Walker, and Rita F. Fahy, *Analysis of First-Person Accounts from Survivors of the World Trade Center Evacuation on September 11, 2001* (Institute for Research in Construction, National Research Council of Canada, 2004).

CHAPTER 8

Inspection of Hazardous Materials

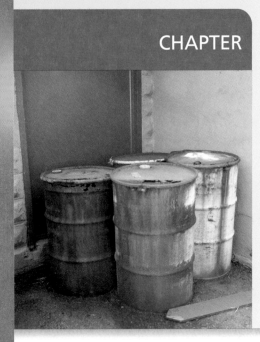

Courtesy of J. Foley

OBJECTIVES

After reading this chapter, the reader should be able to:

- Identify the different federal regulations that manage hazardous materials in the workplace.
- Identify the different hazardous materials classification systems used in facilities.
- Identify resources for determining the characteristics of hazardous materials.
- Review the design alternatives permitted in building codes for the storage, handling, and use of hazardous materials.
- Review the major categories of hazardous materials and the inspection points necessary to prevent fires.

Professional Levels of Job Performance for Fire Inspectors as Cited in NFPA 1031 and NFPA 1037

- NFPA 1031 Fire Inspector I *Obj. 4.1 Requirements of section 4.2 of NFPA 472*
- NFPA 1031 Fire Inspector I *Obj. 4.3.8 Recognize hazardous conditions involving equipment, process, and operations*
- NFPA 1031 Fire Inspector I *Obj. 4.2.12 Code compliance of incidental storage*
- NFPA 1031 Fire Inspector I *Obj. 4.3.13. Verify code compliance of incidental storage, handling, and use of hazardous materials*

- NFPA 1031 Fire Inspector II *Obj. 5.3.6 Evaluate hazardous conditions involving equipment, processes, and operations*
- NFPA 1031 Fire Inspector II *Obj. 5.3.8 Storage, handling, and use of flammable and combustible liquids*
- NFPA 1031 Fire Inspector II *Obj. 5.3.9 Evaluate code compliance of storage, handling, and use of hazardous materials*

Introduction

One of the most difficult inspections that a fire inspector must perform is assessing the storage, handling, and use of **hazardous materials** in a building. The inspection of hazardous materials in fire code enforcement is also where a change in use of the building can occur if certain amounts of hazardous materials are exceeded. Fire inspectors generally are out of their comfort zone when inspecting facilities that contain appreciable quantities of hazardous materials because the fire code requirements are complex and sometimes difficult to comprehend. In many cities, buildings with large amounts of hazardous materials may be relegated to a hazardous material team or hazardous materials administration unit for inspection. In either case, understanding of basic hazardous materials fire code requirements is vital to eliminating fire code violations and protecting properties, the surrounding community, and firefighters' health and safety.

Fire code provisions on hazardous materials have changed significantly since the 1980s as a reaction to changes in both federal and state environmental regulations. These changes were driven by a number of hazardous materials incidents that occurred. In 1978, residents of Love Canal in Niagara, New York, were affected by 21,000 tons of toxic chemicals that were buried by the Hooker Chemical Company in 1947. The Niagara school district purchased the land from Hooker Chemical Company knowing that toxic chemicals were buried on the site. By 1976, toxic chemicals began to show up in groundwater, and eventually cancer clusters began to emerge in the Love Canal area. There were congressional hearings, and the federal government eventually relocated over 800 families and demolished their homes, declaring Love Canal a toxic waste site. The site was later remediated by the Environmental Protection Agency (EPA). This incident caused the federal government to enact the Comprehensive Environmental Response, Compensation, and Liability Act, known as *CERCLA* or the *Superfund*. The Superfund established a national priority list of chemical sites requiring cleanup and provided the funding and oversight through the EPA to address chemical waste disposal. In 1982, Times Beach, Missouri, a small community with dirt roads, hired a contractor to spray the roads with oil to reduce the dust. The contractor bought waste oil containing dioxins from a chemical company that produced Agent Orange for the military. The waste oil was sprayed on Times Beach roads, causing cancer clusters and resulting in a $110 million cleanup by the federal government. In 1984, a major chemical incident occurred in Bhopal, India, at a Union Carbide chemical plant operated by a subsidiary company, Union Carbide of India Limited. The plant manufactured pesticides,

hazardous materials ■ Those chemicals or substances, which are physical hazards or health hazards, whether the materials are in useable or waste condition.

and the incident involved the release of 32 tons of methyl isocyanate gas, which is an intermediate material used in the production of the pesticide Sevin. The release occurred due to water being released into the reacting tank, which caused an explosion. The toxic cloud over Bhopal initially killed about 5,000 people, and the death toll quickly reached 18,000 people, making Bhopal the worst chemical accident in history. In the same year, a Union Carbide plant in Institute, West Virginia, released methylene chloride and aldicarb oxime, resulting in 135 injuries, which was deemed a near miss. The United States Congress was forced into action by these incidents, and a series of environmental protection laws were passed and signed by the president. The new laws included the Emergency Planning and Community Right to Know Act (Title III), which was part of the 1986 Superfund Amendment and Reauthorization Act or SARA. These federal law requirements would begin to appear in the model fire prevention codes within the next code adoption cycle. The enactment of these laws changed many of the fire code provisions for the storage, handling, and use of hazardous materials that we see in the fire codes today.

An examination of the International Fire Code will reveal that one-third of the code book deals with hazardous materials; therefore, a large part of a fire inspector's work is inspecting facilities that store, handle, or use hazardous materials. The fire inspector must be knowledgeable and proficient in identifying hazardous materials and applying these code requirements in the field.

Federal Regulations

Because there are many federal regulations that impact how the chemical industry operates, it is important that the fire inspector have a basic understanding of what the requirements are and how they fit inspections. Let's examine each federal regulation.

RESOURCE CONSERVATION AND RECOVERY ACT

The federal government has always been involved in regulating interstate commerce and hazardous materials. In 1976, the Federal Resource Conservation and Recovery Act, or RCRA, was passed to regulate solid and hazardous waste in the United States. The law contained a cradle-to-grave provision requiring that certain hazardous chemicals be tracked from the day they are created until the day they are disposed of. The law requires extensive booking on these hazardous wastes and relegates many functions of the law to state governments and their departments of environmental protection. The RCRA law also regulates underground storage tanks, or USTs, and imposes requirements for monitoring, double containment, release detection, spill control, and overfill of the tanks. The law also addresses the disposal of medical wastes.

COMPREHENSIVE ENVIRONMENTAL RESPONSE, COMPENSATION, AND LIABILITY ACT

CERCLA ■ (Comprehensive Environmental Response, Compensation, and Liability Act): The federal statute that empowers the EPA to identify and clean up sites at which hazardous substances were released into the environment (referred to as the superfund law).

In 1980, the Comprehensive Environmental Response, Compensation, and Liability Act (CERCLA) was passed by Congress and signed into law by the president. The **CERCLA** law is also referred to as the Superfund because it created a fund for the remediation of hazardous waste disposal sites in the United States. The purpose of the law was to identify and create a national priority list of hazardous materials cleanup sites. The regulation authorized the Environmental Protection Agency (EPA) to identify responsible parties for dump sites and compel them to clean them up. The regulation created a source of funding for the removal and remediation of toxic chemical sites. The Superfund sites can be identified by fire inspectors using TOXMAP from the United States National Library of Medicine. This resource can be found at the U.S. National Library of Medicine website.

SUPERFUND AMENDMENT REAUTHORIZATION ACT TITLE III

In 1986, President Ronald Reagan signed into law the Superfund Amendment and Reauthorization Act, or as it is commonly known, **SARA**. The new Superfund regulation imposed a greater commitment upon state and local governments to become more involved in the hazardous material problem within their jurisdictions. One of the most important parts of the regulation was a standalone regulation called Title III, the Emergency Planning and Community Right to Know Act, or EPCRA. This regulation mandated that each state form a State Emergency Response Commission and every municipality form a Local Emergency Planning Committee or **LEPC**. The role of the LEPC is to conduct the following functions:

- Identify hazardous materials facilities and transportation routes
- Identify emergency notification and response systems
- Designate emergency coordinators in the community
- Identify local response equipment
- Establish evacuation plans
- Produce emergency plans and coordinate and conduct training exercises on the plan

The LEPC should be composed of elected officials, emergency management, fire, EMS, and police representatives as well as public health and hospital officials, media, transportation, community groups, and owner/operators of facilities subject to hazardous material regulations. The fire inspector or fire official makes a perfect member for the LEPC because he or she is charged with the inspection of hazardous materials in storage, handling, and use at fixed facilities. The role of the LEPC has expanded since 9/11 and includes additional management and planning to thwart potential terrorism. You can locate the members of your local LEPC at the federal EPA website.

The SARA Title III right-to-know requirements are shared between two federal agencies, the Occupational Health and Safety Administration, or OSHA, and the Environmental Protection Agency, or EPA. The EPA regulates Section 312 of the law for extremely hazardous substances, of which there are 356 chemicals, and hazardous substances, of which there are 720 chemicals. EPA also regulates Section 313 of the law on extremely toxic chemicals, of which there are 650 chemicals. These chemical lists may be added to from time to time and can be found at the EPA website under the "List of Lists." All chemicals that are not on the EPA "List of Lists" are regulated by OSHA under Section 311 of the federal regulation. Over 500,000 chemicals are regulated by OSHA, but there is no specific list of these chemicals. All chemicals that require the issuance of a Materials Safety Data Sheet by the manufacturer fall under the enforcement authority of OSHA and the OSHA Hazardous Communication Standard.

OSHA HAZARDOUS COMMUNICATIONS STANDARD

Another important federal regulation in the workplace is the OSHA Hazardous Communications Standard, or 29 CFR 1910.1200, which is referred to as a worker "right to know" standard. The regulation requires the labeling of hazardous materials in the workplace as well as the training of employees in the safe handling of chemicals and the use of personal protective equipment (see Figure 8.1). The importance of this standard is that all hazardous materials in the workplace must be properly labeled in accordance with a nationally recognized hazard identification system. The common labeling systems used are NFPA 704, *Standard System for the Identification of the Hazards of Materials for Emergency Response*, The Department of Transportation labeling system, and the Hazardous Materials Identification System (HMIS). During the course of the fire inspection, the fire inspector should identify any improper labeling such as that seen in Figure 8.2.

The OSHA 29 CFR 1910.1200 standard also requires that MSDS be maintained on site for both employees and emergency responders to use. It also requires the employer to

SARA ■ (Superfund Amendment Reauthorization Act): The federal statute that requires EPA when selecting a remedial action to consider the standards and requirements in other environmental laws and regulations and to include citizen participation in the decision-making process.

LEPC ■ (Local Emergency Planning Committee): A committee of local stakeholders established by SARA to plan and review information on facilities having reportable quantities of hazardous materials within the community.

FIGURE 8.1 Hazardous materials response teams must drill to maintain their competency to exercise their responsibility under emergency operating plans. *Courtesy of J. Foley*

maintain a written hazardous materials communication plan for their program. The fire inspector should request a copy of the hazardous materials communication plan, as fire inspectors are required by the fire prevention code to maintain both a hazardous materials management plan and the hazardous materials inventory statements submitted under SARA on site. The hazardous materials management plan should include the type of storage and the maximum amounts of hazardous material present in each storage or use location. The hazardous materials management plan must also specify the location of emergency isolation valves or spill mitigation devices as well as the location of any other emergency equipment at the facility. The IFC requires that the hazardous materials management plan and inventory sheets be kept in a locked location that is approved by the fire official. The location must be in an area accessible to emergency responders such as a guard shack or remote information box.

FIGURE 8.2 Improperly marked acid drum. A proper labeling system must be used as seen on the original container. *Courtesy of J. Foley*

Reporting Requirements of SARA

In March of every year, fire marshals and fire officials will begin to receive SARA tier II reports from chemical facilities within their jurisdiction. The fire inspector should understand what these reports represent and how to use them in the preparation for a fire inspection at a hazardous materials facility. The chemical reporting requirements are the physical hazards and health hazards of each chemical. The EPA requires that facilities that store, handle, or use hazardous chemicals identified on the "List of Lists" report them to the local government, police, fire, and EMS responders if they exceed the law's threshold planning quantities under Section 312 or 313 of the federal regulation. The **threshold planning quantities,** or *TPQs*, have established quantity ranges under regulation Sections 311 and 312 from 500 pounds to 10,000 pounds of a hazardous material before they are required to be reported to either the EPA or OSHA. Under Section 313 for Toxic Release Inventories, because of the extreme toxic nature of these chemicals, the reporting thresholds are termed **reportable quantities**, or *RQs*, and can be very small amounts, such as one pound of the hazardous material. The point of the threshold planning quantities and reportable quantities is to establish a benchmark at which a chemical must be reported to local emergency responders and the public.

threshold planning quantity ▪ The threshold planning quantities (TPQs) establish the minimum quantity at which extremely hazardous or hazardous substances must be reported to the EPA. The threshold for extremely hazardous substances is the threshold planning quantity or TPQ of 500 pounds. Hazardous substances have a TPQ of 1,000 to 10,000 pounds.

reportable quantity ▪ Reportable quantities (RQs) are the amounts of hazardous chemicals that once exceeded must be reported to the EPA under the rules of CERCLA. RQs are adjusted to one of five levels: 1, 10, 100, 1,000, and 5,000 pounds of a hazardous chemical.

TIER I AND TIER II REPORTS

Federal regulations require facilities that must report hazardous chemicals under Sections 311, 312, and 313 to either provide the MSDS to the local LEPC and emergency services or to provide an inventory list on either a Tier I or Tier II report. A Tier I report must contain the following information: the maximum amount of hazardous chemicals at the facility during the preceding year, an estimate of the average daily amount of hazardous chemicals at the facility, and the general location. Federal regulations only require facilities to submit a Tier I report; however, each state or local LEPC can require that a Tier II report be submitted instead of a Tier I report. The Tier I or Tier II report must be submitted every March 1 to the SERC, the LEPC, and local emergency responders.

The report reflects the chemicals that were on site from the preceding year. The facility submitting the Tier I or Tier II report must use either the Federal EPA report form found on their website or the state form if required by the particular state. In either case, the report will have the following information:

- A list of all chemicals and hazardous substances required according to the threshold quantities, TPQs, or reporting quantities, RQs.
- Common chemical name
- Chemical Abstract Services (CAS) number
- Physical state (solid, liquid, gas)
- Physical and/or health hazards
  - Physical hazards are fire, sudden release of pressure, and reactivity
- Health hazards
  - Immediate (acute) effects or delayed effects (chronic)
- Inventory information
  - Maximum daily amount
  - Average daily amount
  - Number of days on site
- Storage information—this includes container type, pressure, temperature, specific information on storage amounts, and locations

Fire prevention officials who receive Tier II report information annually are required to maintain this information in their fire inspection property files; this information is also to be disseminated to the firefighters within the fire department, but not to the public. The public can usually obtain this right-to-know information through local health officials.

HAZARDOUS MATERIAL COMPUTER PLANNING RESOURCES

Another source of gathering information is the use of the Internet with programs such as E-Plan, which is a collaborative effort between the EPA, Department of Homeland Security, and the University of Texas to coordinate the EPA Tier II reporting nationwide in a database that is readily accessible to first responders and local, state, and federal officials. The E-Plan database currently has over 400,000 facilities reporting to it and over 24,000 unique hazardous chemical records. E-Plan also integrates other federal agencies' information from the Occupational Safety and Health Administration (OSHA), the Center for Disease Control (CDC), the Environmental Protection Agency (EPA), The National Oceanic and Atmospheric Administration (NOAA), and the United States Fire Administration (USFA).

A similar system is WISER, which stands for Wireless Information System for Emergency Responders, and is provided by the National Institute of Health. This program is also free to first responders and provides access to many resources for hazardous materials incidents, including the National Library of Medicines Hazardous Substance Data Bank. WISER has many useful tools to help identify hazardous chemicals, including NFPA 704 placarding requirements for each specific chemical. These tools may be found at the E-Plan and WISER websites.

Hazardous Material Identification Methods

One of the first considerations the fire inspector must have in dealing with hazardous materials is how to properly identify their presence. The fire prevention code defines a hazardous material in the same manner as OSHA:

> Those chemicals or substances which are physical hazards or health hazards as defined and classified in this chapter (of the fire code), whether the materials are in a usable or waste condition.[1]

The definition indicates four important points of evaluation for the fire inspector to contend with:

physical hazard
■ Hazardous materials that are explosive, flammable, or combustible; flammable solids or gases, oxidizers, organic peroxides, pyrophoric or unstable, water reactive solids or liquids, and cryogenic fluids.

health hazard ■ Hazardous materials that show a statistically significant health risk or which are toxic or highly toxic, corrosive, or radioactive.

- ■ **Physical Hazards:** This includes hazardous materials that support combustion, can detonate, are contained under extreme pressure, are caustics or acidic, cryogenic, radioactive, water reactive, or increase combustion, such as oxidizers and peroxides.
- ■ **Health Hazards:** These hazardous materials include those that have statistically significant acute or chronic health effects on exposed personnel.
- ■ **Useable condition:** These are the raw materials used in manufacturing of a product or process.
- ■ **Waste condition:** These are the hazardous material by-products or residual materials that must be properly disposed of after the manufacturing process.[2]

DEPARTMENT OF TRANSPORTATION LABELING SYSTEM

The OSHA standard requires that industry label hazardous materials with a nationally recognized identification system but does not identify a specific labeling system. The Department of Transportation (DOT) hazardous materials labeling requirements under 49 CFR 172-400 are the most common in industry for the transportation of hazardous materials. The DOT system is part of the North American Free Trade Agreement and is followed by both Canada and Mexico in regional transportation of hazardous materials. The DOT system identifies hazardous materials using nine hazard classes for placarding under the system. The classification placards must appear on each individual bulk container or packages as illustrated in Figure 8.3.

Proper DOT labels

FIGURE 8.3 Chemicals should have the proper DOT hazard class labels affixed to containers. *Courtesy of J. Foley*

The nine hazard classes and the subdivisions are as follows in Table 8.1.

The Department of Transportation classifications and labels are used for both bulk and nonbulk commodity transport. The OSHA Hazard Communication Standard requires that DOT labels affixed in transportation remain on the container until the container is cleaned. DOT labels are the most common in industry and are easily recognized by firefighters using the *North American Emergency Response Guide* that is provided free to responders by the Department of Transportation.

The DOT labels assist the fire inspector by identifying the general chemical family and hazards of the material for the purpose of segregation and separation of hazardous materials in storage. The separation of chemicals by hazard class solves many of the storage incompatibility issues, but not all of them. Fire inspectors must be aware that some chemicals will fit in multiple hazard classes or may have both physical and health hazards. The fire inspector must be aware that even within a single hazard class there may be some chemical incompatibility. As an example, Hazard Class 8 contains both acids, which have low pH, and caustics that have high pH. Acids and caustics can react violently when suddenly mixed together. Another example would be Class 2 compressed gases, which contain flammable gases like hydrogen, oxidizing gases like oxygen, and acidic gases like chlorine or ammonia. All of these materials must be segregated further by physical properties to avoid chemical reactions from a leaking container. Oxidizers in Class 5 also must be segregated into weak and strong oxidizers, as mixing the two can cause a reaction and fire. When examining storage, fire inspectors should take note of chemical groupings and check on their compatibility.

NFPA 704 PLACARDS: FIREFIGHTER WARNING SYSTEM

The NFPA 704 placarding system is required by the NFPA and IFC fire prevention codes on aboveground tank storage and storage rooms or warehouses containing hazardous materials. The IFC does have an exception to NFPA 704 placarding for storage areas: When the owner provides a list of hazardous chemicals and it is readily accessible to the fire department, the placards are not required. The system is designed to provide firefighters with basic information on hazards due to fire, health, or chemical reactivity

TABLE 8.1	49CFR 172-400 Code of Federal Regulations
HAZARD CLASS	**DIVISIONS**
Explosives 1.1	1.1 Mass explosion hazard
Explosives 1.2	1.2 Projectile hazard
Explosives 1.3	1.3 Fire hazard
Explosives 1.4	1.4 No significant blast hazard
Explosives 1.5	1.5 Insensitive explosive
Explosives 1.6	1.6 Extremely insensitive explosive
Nonflammable Gases 2.1	2.1 Flammable compressed gases
Nonflammable Gases 2.2	2.2 Nonflammable compressed gases
Nonflammable Gases 2.3	2.3 Poison compressed gases
Nonflammable Gases 2.4	2.4 Corrosive compressed gases
Flammable Liquids 3.1	3.1 Flash point below 0°F
Flammable Liquids 3.2	3.2 Flash point below 73°F
Flammable Liquids 3.3	3.3 Flashpoint below 141°F
Flammable Liquids 3.4	3.4 Flammable, combustible, gasoline, fuel oil
Dangerous when wet	Flammable solids, spontaneous combustion, dangerous when wet
Oxidizers 5.1	5.1 Oxidizers, chlorine, fluorine
Oxidizers 5.2	5.2 Organic peroxides
Poisons 6.1	6.1 Poisonous
Poisons 6.2	6.2 Infectious
Radioactive 7.1 (I)	7.1 (I) less than 0.5 millirems per hour at the surface of the package
Radioactive 7.2 (II)	7.2 (II) 0.5 to 50 millirems per hour at the surface of the package
Radioactive 7.3 (III)	7.3 (III) greater than 50 millirems per hour at the surface of the package
Corrosives	Acids or caustics (low or high pH)
Other Regulated Materials (ORM)	Cosmetics, commercial goods

Ref: 49CFR 172-400 Code of Federal Regulations

when these materials are exposed to a fire, spill, or are removed from their containers (see Figure 8.4).

The NFPA 704 system uses a diamond-shaped placard with four different colored quadrants. Each quadrant identifies a specific hazard of the material; flammability is red and is always at the top of the diamond, health risk is blue and is to the left side, and chemical instability is yellow and is to the right side of the diamond. The bottom quadrant is white and may be blank or may contain special information such as a W to indicate the material is water reactive. In each of the quadrants, the risk factor is identified by 0 to 4 for hazard assessment. The higher the hazard number, the greater the risk from fire, reactivity, or health hazard. The general risk categories are as follows in Table 8.2.

To sum the risk factors up, firefighters need to be taught that NFPA 704 symbols with threes and fours stand for chemicals that are dangerous when they are exposed to fire or

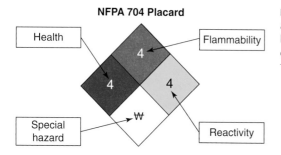

NFPA 704 Placard

Health

Flammability

Special hazard

Reactivity

FIGURE 8.4 The NFPA 704 placard provides information on flammability, health, and reactivity of a chemical with a 0–4 rating scale. The higher number indicates a higher risk factor. *Reprinted with permission from NFPA 704-2012, System for the Identification of the Hazards of Materials for Emergency Response*

out of their containers due to a rupture or accident. On large aboveground tanks, the fire inspector should make sure the 704 placards are of adequate size and visible on four sides for the fire department to read on their approach to identify the hazards.

HAZARDOUS MATERIALS IDENTIFICATION SYSTEM

The Hazardous Materials Identification System, or HMIS, is very similar to NFPA 704 with two exceptions. The first exception is that the HMIS system uses color bars instead of a diamond shape. The bars are blue for health, red for fire, and yellow for physical hazard.

TABLE 8.2	Hazardous Materials Identification System		
HAZARD RISK NUMBER	**HEALTH RISK**	**FIRE RISK**	**REACTIVITY RISK**
		COLOR	
	BLUE	RED	YELLOW
4	Very short exposure could cause death or major injury	Material will vaporize at atmospheric pressure and will burn readily	Materials themselves are readily capable of detonation or explosive decomposition
3	Short-term exposure could cause serious temporary injury or residual injury	Liquids or solids that can be ignited under ambient temperature	Materials that are capable of explosion or explosive decomposition but require a strong ignition source and must be confined
2	Continued or intense exposure could cause temporary or residual injury	Materials that must be moderately heated or exposed to high ambient temperatures	Material is normally stable but can undergo violent chemical reactions
1	Material upon exposure can cause irritation but only minor residual injury	Materials that must be preheated before ignition can occur	Materials normally stable but becomes unstable in elevated temperatures or may react with water
0	No hazard beyond that of ordinary materials	Will not burn	Normally stable under fire conditions, not water reactive

Source: Reprinted with permission from NFPA 704-2012, *System for the Identification of the Hazards of Materials for Emergency Response,* Copyright © 2011, National Fire Protection Association. This reprinted material is not the complete and official position of the NFPA on the referenced subject, which is represented only by the standard in its entirety. The classification of any particular material within this system is the sole responsibility of the user and not the NFPA. The NFPA bears no responsibility for any determinations of any values for any particular material classified or represented using this system.

The second exception to 704 is that the white bar in the HMIS system is used for personal protection. The HMIS system is geared more to laboratory technicians using chemicals than to firefighters responding to incidents. Major chemical manufacturers often use the HMIS symbols on bottled reagents. In the personal protection bar, there is typically a letter identifying the equipment necessary for protection. As an example, an "A" in the white bar indicates safety glasses, a "B" indicates safety glasses and gloves, while a "C" indicates safety glasses, gloves, and an apron. The HMIS has twelve levels of personal protection equipment, which should be posted in the laboratory if the system is in use.

MATERIAL SAFETY DATA SHEETS

Material Safety Data Sheet (MSDS) ■ A technical bulletin prepared by the manufacturer in accordance with the provisions of 29CFR1910-1200 to provide workers with detailed information about the properties of a commercial substance containing hazardous substances.

The **Material Safety Data Sheet**, or *MSDS*, is a very useful tool for the fire inspector to use in determining the properties of a specific hazardous chemical. Material Safety Data Sheets are required for all hazardous chemicals found in the workplace. The requirement for MSDS is part of the OSHA Hazard Communication Act and the right-to-know requirements under SARA Title III. MSDS provide better information than do DOT labels on a specific chemical. The MSDS provides the chemical name, common synonyms, the manufacturer's information, the physical and chemical properties, health risks, exposure factors, firefighting methods, environmental considerations, and safety concerns for handling the chemical. The MSDS is required by fire prevention codes to be maintained in a specific location on site as approved by the fire official. The MSDS location should be reviewed by the fire inspector during the inspection process and can be useful in determining the types of chemical storage present for compatibility and physical and health hazards. The key components of the MSDS for the fire inspector are the physical and health hazards of the chemical, which will assist them in identifying the correct hazardous material exempt amount table to use in the fire code. The fire code generally divides these tables into requirements for physical hazards and health hazards, and the table will provide an exempt quantity of hazardous material permitted to be used.

General Fire Prevention Code Requirements

control area ■ Location(s) within a facility where hazardous materials over the amounts requiring approval are stored, handled, and used. Control areas require a fire resistance rating of 1–2 hours and are reduced in number based on height above grade in the building.

Building and fire prevention codes take an integrated approach on the handling, use, and storage of hazardous materials in buildings. Code requirements are based on the physical and health hazards of the hazardous material and the quantities of each class of hazardous material that is present within the building. Because use group "H" is driven by the quantity of hazardous material in a building, a change of use can occur insidiously by either exceeding the permitted exempt amounts of hazardous material within a **control area** or not storing the materials in a control area at all (see Figure 8.5). The building and fire codes require that hazardous materials can only be stored, used, or handled in control areas or in H-use areas of a building. The purpose of the hazardous materials exempt amount tables is to provide a path for buildings to use limited quantities of hazardous materials and not be subject to the code requirements of the H-use group.

Building code officials depend on continuous fire inspections under the fire code to determine when and if a change of building use occurs at facilities using hazardous materials. After a building receives its certificate of occupancy, the building inspector has no further authority to conduct inspections unless the owner or operator obtains an additional building permit or a condition of the certificate of occupancy has been violated. The building code official may also have authority to inspect if the structure becomes unsafe or an imminent hazard but under normal conditions, the building inspector has no further authority to inspect. This is the main reason why the fire prevention code tables for hazardous materials mimic the building code's tables for exempt quantity of hazardous materials. Additionally, the fire prevention code is a maintenance code to ensure that the building is maintained to the original building code requirements. The fire code's

FIGURE 8.5 Hazardous materials must be segregated from other types of storage in properly fire-resistant control areas. *Courtesy of J. Foley*

maintenance requirements apply to the entire property including outdoor hazardous materials storage that would not be addressed in the building code. The fire inspector must be aware that the fire code will include additional tables for the storage, handling, and use of hazardous materials outdoors that are not building code related.

The International Building Code provides building design alternatives for the storage, handling, and use of hazardous materials. The IBC takes this approach to prevent every building from becoming use group H or high hazard when only small quantities of hazardous materials are present. Fire inspectors must be aware of these design alternatives so they can better determine whether a change of building use has occurred.

To make this difficult task simpler for the fire inspector to evaluate, let's divide the process into steps to gather the information necessary for a proper evaluation.

Step 1.   The fire inspector should determine the types of hazardous materials present and divide them into three materials groups: materials with physical hazards, materials with health hazards, and finally materials that have both physical and health hazards.

Step 2.   The fire inspector should take the three lists of materials and determine the quantity of each hazardous material that is present in the building and its location.

Step 3.   The fire inspector should now divide each list into the hazardous materials that are solids, those that are liquids, and those that are gases.

Step 4.   The fire inspector should determine on each list which hazardous materials are in storage, which are used in open systems, and which are used in closed systems.

Step 5.   Evaluate each area where the hazardous materials are stored, handled, or used and answer the following questions:

    1. Is the area enclosed in fire-resistant rate construction complying with the code requirements for a control area?

        1. Does the area have self-closing fire doors or fire shutters?

        2. Are there properly rated fire extinguishers in the area?

        3. Are the materials properly labeled and identified and is the area placarded?

TABLE 8.3	Permitted Number of Control Areas and Quantity Limits		
STORY	% OF EXEMPT TABLE PERMITTED	NUMBER OF CONTROL AREAS	FIRE RESISTANCE RATING
1	100%	4	1
2	75%	3	1
3	50%	2	1
4–6	12.5%	2	2
7–9 or higher	5%	1	2
Below grade 1	75%	3	1
Below grade 2	50%	2	1
Below 3	Not allowed	Not allowed	Not allowed

Source: Reproduced with permission. All rights reserved. www.iccsafe.org, Washington, DC, International Code Council

2. Is the use or storage area inside or outside the building or in a separate building?
   1. If outdoors, is the hazardous material stored under weather protection?
   2. How close is the storage to building and interior lot lines?
3. Is the indoor area protected by an automatic sprinkler system?
   1. What is the sprinkler system design density in gallons per minute per square foot?
4. On what floor is the control area located within the building?
5. Are the hazardous materials kept in code-compliant storage cabinets?
6. Does the area have proper ventilation and safeguards for spill control and spill containment?

Step 6. Take each list of hazardous materials and determine the permitted **exempt amounts** of each hazardous material based on the information you have gathered. Keep in mind the following key points:

exempt amounts
- The permitted amounts of hazardous materials requiring approval allowed for storage, handling, and use within a control area. Quantities that exceed the exempt amounts shall be placed in use group "H," high hazard.

- All hazardous materials must be stored, handled, or used in a control area.
- Approved listed storage cabinets will allow a 100 percent increase in the permitted exempt amount in the table.
- Automatic sprinkler protection of the appropriate design density will allow a 100 percent increase in the permitted exempt amount of materials.
- The permitted exempt amount of hazardous materials in the table applies only to the first floor of the building and will decrease by a specific percentage if the control area is on an upper floor (see Table 8.3).

CONTROL AREAS

Because of the increased fire and health risks presented by hazardous materials, building codes require that these materials be used only in fire-resistant rated areas deemed control areas. A *control area* may be the entire building or a portion of a building where hazardous materials are stored, handled, or dispensed. Control areas are required to be fire-resistant rated from 1 to 2 hours depending on where the area is located in the building. One- and two-hour fire separation walls must be constructed to the appropriate fire resistance rating design standards from the building code. Any openings for doors and windows must also meet the proper fire resistance requirements and must be self-closing. The control area may be above the first floor level in a building or below the first floor level; however, the percentage of permitted materials in the exempt amount tables will be reduced based on the location

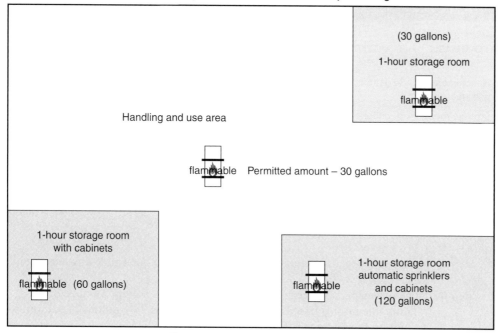

Factory F-1 use: Exempt amount of 1A flammable liquid is 30 gallons.

(30 gallons)

1-hour storage room

flammable

Handling and use area

flammable   Permitted amount – 30 gallons

1-hour storage room
with cabinets

flammable   (60 gallons)

1-hour storage room
automatic sprinklers
and cabinets
(120 gallons)

flammable

Flammable Liquids Control Areas

FIGURE 8.6 Each control area is permitted different increases in exempt amounts based on the type of protection provided. Sprinklers and storage cabinets allow 100 percent increase of exempt amounts.

(see Table 8.3). All control areas must also have proper rated fire doors, fire windows, and shutters, and dampers or fire stops must protect all openings. If a control area is located on an upper floor, all of the floors supporting the control area must also have at least a 2-hour fire-resistance rating on the structural supporting elements. The building code limits control areas and the amounts of permitted exempt quantities of hazardous materials within them based on their location in the building. The IBC limits control areas to four on the ground or first floor, three control areas on the second floor, and no more than two control areas up to the ninth floor, and only one per floor above the ninth floor. All control areas shall be protected by 1-hour fire barriers and floor ceiling assemblies up to the third floor, and 2-hour fire barriers and floor ceiling assemblies from the fourth floor or higher.[3]

How to Calculate the Permitted Exempt Amount of Hazardous Materials
The following example illustrates how the fire inspector can determine the maximum allowable quantity permitted in the building without altering its use group classification.

Example #1 (see Figure 8.6): The fire inspector inspects a one-story factory F-1 use that uses 1A flammable liquids in both production and storage. The building is 1-hour fire-resistant construction and has three storage areas, of which one is protected by automatic sprinklers. Two of the storage areas have listed flammable liquid cabinets and 1-hour fire barriers with ¾-hour fire doors protecting all three areas. The owner asks, how much 1A flammable liquid is permitted in the building under the fire code?

Step 1.   The fire inspector checks the building code and determines that it allows four control areas on the first floor of the building with 1-hour fire barriers and fire doors.

Step 2.   The fire inspector checks the fire code's permitted exempt quantity tables for physical hazards of 1A flammable liquids in storage and closed systems and determines that the permissible amount is 30 gallons in a control area.

TABLE 8.4	Exempt Amount of Flammable Liquids Allowed on First Floor				
MATERIAL 1A FLAMMABLE LIQUID	PERMITTED AMOUNT	PROPER CONTROL AREA	CABINET INCREASE 100%	SPRINKLER INCREASE 100%	TOTAL PERMITTED
Handing and use area	30 gallons	Yes	No	No	30 gallons
Storage room with cabinets	30 gallons	Yes	Yes	No	60 gallons
Storage room with cabinets and sprinklers	30 gallons	Yes	Yes	Yes	120 gallons
Storage room	30 gallons	Yes	No	No	30 gallons
Total Allowed					240 gallons

Step 3. The fire inspector evaluates each control area to determine any allowable increases as in Table 8.4.

Step 4. The fire inspector determines that the permitted exempt amount of 1A flammable liquids based on current conditions is 240 gallons maximum to maintain the Factory F-1 use group.

The factory owner could install additional flammable liquid cabinets in the storage room and handling area and increase the amount by another 60 gallons, or the owner could install the cabinets and sprinklers in the entire building and increase another 240 gallons to a maximum of 480 gallons of flammable liquids. If the 480 gallons were to be exceeded, the building would change use from Factory F-1 to High Hazard H-2 use.

Example #2 (see Figure 8.7): In this example, let's assume that the factory has now added two additional floors and is using 1A flammable liquids in control areas on the third floor.

FIGURE 8.7 Control areas on the third floor are reduced to two with a 50 percent reduction in permitted quantities; cabinets and sprinklers allow a 100 percent increase of the reduced table quantity.

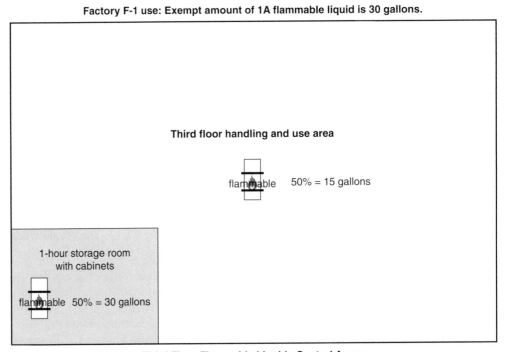

Factory F-1 use: Exempt amount of 1A flammable liquid is 30 gallons.

Third floor handling and use area

flammable   50% = 15 gallons

1-hour storage room with cabinets

flammable   50% = 30 gallons

Third Floor Flammable Liquids Control Areas

TABLE 8.5	Exempt Amount of Flammable Liquids Allowed on Third Floor				
MATERIAL 1A FLAMMABLE LIQUID	PERMITTED AMOUNT	PROPER CONTROL AREA	CABINET INCREASE 100%	SPRINKLER INCREASE 100%	TOTAL PERMITTED
Handling and use area	15 gallons	Yes	No	No	15 gallons
Storage room with cabinets	15 gallons	Yes	Yes	No	30 gallons
Total Allowed					45 gallons

Step 1. The fire inspector checks the building code and determines that it allows three control areas on the third floor of the building with 1-hour fire barriers and fire doors. However, the maximum allowable quantities are reduced by 50 percent.

Step 2. The fire inspector checks the fire code's permitted exempt quantity tables for physical hazards of 1A flammable liquids in storage and closed systems and determines that the permissible amount is 30 gallons in a control area. However, the inspector must reduce that permissible amount by 50 percent to 15 gallons.

Step 3. The fire inspector evaluates each control area to determine any allowable increases as in Table 8.5.

Step 4. The fire inspector determines that only 45 gallons of 1A flammable liquids are permitted on the third floor. Again, the owner could increase the permitted amount of flammable liquid by the use of automatic sprinklers, cabinets, and constructing a third control area on the floor. If the owner made these improvements, the maximum amount of 1A flammable liquids would be increased to 180 gallons total. If that amount were exceeded, the factory would become an H-2 High Hazard use building. The high hazard use has greater fire resistance ratings, automatic fire sprinklers and fire alarm systems, explosion venting, as well as means of egress with shorter travel distances.

STORAGE CABINETS

Hazardous materials storage cabinets are generally required to be listed under UL 1275, *Flammable Liquid Storage Cabinets,* or other UL listings for the specific products they must store. The cabinets may be for flammable liquids, aerosols, compressed gases, or corrosives; however, they must be chemically compatible with the products that are to be stored in them. Storage cabinets have limits as to the quantity of hazardous materials they can retain. Most flammable liquid cabinets are limited to 60 gallons of Class I and II flammable liquids and 120 gallons aggregate as long as Class I and II limits are not exceeded. Cabinets typically are unvented to improve fire resistance but are manufactured with plugs that may be removed to install vents. If vented, the vent must go to the exterior of the building or to an approved exhaust hood for vapor removal. The storage cabinet is constructed of double wall 18 gauge steel with a 1-1/2-inch air space for insulation. The cabinet's joints are riveted or welded and are generally liquid tight with a 2-inch sill at the bottom to contain any internal leaks or spills. Doors should have a three-point latching mechanism to ensure that the door remains properly closed when exposed to a fire.

NFPA 30, *Flammable and Combustible Liquids Code,* also allows the construction of nonlisted wooden flammable liquids storage cabinets. These cabinets can only be used for storage of flammable and combustible liquids and are not appropriate for corrosives or other hazardous materials. The nonlisted cabinets under NFPA 30 must be constructed of 1-inch marine grade plywood. All construction joints must be rabbited and fastened in two directions. The doors must be rabbited and overlap by 1 inch and must have a three-point latch and be tight fitting. The cabinet must be painted with an intumescent paint to reduce the combustibility of the plywood.

Hazardous materials storage cabinets are an important element in the protection of hazardous materials from fire exposure and accidental spillage within control areas. All cabinets must be properly identified as to the hazard of the material located inside the cabinet with labels such as "FLAMMABLE—KEEP FIRE AWAY" or the appropriate NFPA 704 markings.

Storage Arrangements

Storage arrangements of hazardous materials are an important aspect of fire prevention safety that must be observed so that accidental chemical reactions do not occur. Generally, chemicals should be stored based on their hazard classification and compatibility with surrounding materials. The IFC generally requires that incompatible hazardous materials be separated by 20 linear feet or noncombustible barriers that extend 18 inches above the highest container in the storage pile. The IFC regulates both indoor and outdoor storage of hazardous materials and requires that outdoor materials be protected with containment pallets and weather protection. If these conditions cannot be satisfied, then the hazardous materials must be placed in a storage facility that has secondary containment and spill control mechanisms. Hazardous materials storage must have adequate aisle space for firefighting purposes, and storage must also be secured from vandalism and motor vehicle impacts and must be maintained clear of smoking materials and combustible vegetation in outdoor settings.

Flammable and Combustible Liquids

Flammable and combustible liquids are the most common of all hazardous materials and will be found in the widest array of buildings. Flammable and combustible liquids have vapor pressures less than 40 psia. Any material above this vapor pressure would be either a liquefied gas or a compressed gas.

Flammable Liquids Flammable liquids are those liquids having a flash point below 73°F and boiling points below 100°F. Flammable liquids are further divided into three subclasses:

- Class IA—Liquids having a flashpoint below 73°F and a boiling point below 100°F
- Class IB—Flash point below 73°F and a boiling point at or above 100°F
- Class IC—Flash point at or above 73°F but below 100°F[4]

What makes the inspection of flammable liquids so important is the fact that they are always in a state ready to burn at normal ambient temperatures. If flammable vapors escape from a container, the potential for a fire is significantly increased if an ignition source is close by.

Combustible Liquids Combustible liquids are those that have flashpoints above 100°F. These are further divided into the following three subclasses:

- Class II—Flash point above 100°F but below 140°F
- Class IIIA—Flash point above 140°F but below 200°F
- Class IIIB—Flash point above 200°F[5]

Combustible liquids must be preheated to produce sufficient vapor to burn; however, some combustible liquids like K-1 clear kerosene are blended with a flash point around 106°F. While deemed a combustible liquid, K-1 kerosene is not that far from being a flammable liquid. The higher the classification of a combustible liquid, the higher the flash point. Fire prevention codes allow greater quantities of combustible liquids in a building because they do not produce sufficient ignitable vapors. The visual appearance of the liquid can also sometimes assist the fire inspector in determining the potential flammability hazard. The liquid's ability to flow can indicate the relative class of the flammable or combustible liquid. Thin, fast-flowing liquids are more likely to be flammable liquids as opposed to thicker liquids, which lean more toward the combustible liquid side. This is called *viscosity*, and the more viscous the liquid is, the less likely it is to release vapors.

Other Flammable Liquids Fire inspectors must keep in mind that hydrocarbons, while predominant, are not the only flammable liquids that exist. There are many chemical families that contain flammable liquids including the alcohols, ethers, ketones, organic acids, and others. Many of these materials are commonly used as solvents or constituents for other chemical reactions. Many of these flammable liquids are water-soluble, which can present fire protection problems when using firefighting foams. Today, because of the need for alternate fuels, a hydrocarbon-like gasoline is being blended with ethanol from corn. The percentage of ethanol may be 10 to 15 percent blended. Typically, flammable liquid storage facilities or tank vehicle loading racks and storage tanks may be protected with NFPA 11 aqueous film-forming foam systems, or AFFF. While AFFF works well on hydrocarbons, the ethanol will break down the foam bubbles. This is because ethanol is a polar compound, the same as water. These new blended fuels require specialized alcohol-resistant firefighting foams.

In the examination of flammable and combustible liquids, fire prevention codes generally establish safety requirements in three areas: the storage of the material, the dispensing of the material, and the general handling and use. The code requirements include labeling and signage utilizing NFPA 704 or another recognized labeling system. Some level of fire protection will also be required, including the placement of portable fire extinguishers or fire hose cabinets or fixed fire protection systems such as automatic foam appliances (see Figure 8.8).

All electrical equipment and devices within range of flammable liquid vapors require protection with explosion-proof fittings as prescribed in the National Electrical Code®, or NFPA 70. Control areas, depending on the quantities of material being used, may require spill control and secondary containment systems. Above-ground storage areas or tanks must be properly protected from vehicles if potential collisions exist. Vehicle impact protection requires the installation of protective bollards based on the size and weight of vehicles in the area. Tanks and piping also must be protected from galvanic actions and corrosion by the installation of dielectric fittings or protective coatings. If tanks are situated on permanent foundations, the foundation system must be at least 2-hour fire-resistant rated, and inspectors should make sure that combustibles and weeds are kept clear of the underside of the tank.

FIGURE 8.8 Remote activated firefighting foam systems may be used in protection of flammable and combustible liquids such as helipads. *Courtesy of J. Foley*

FIGURE 8.9 Aboveground storage tanks may require secondary containment systems and fire protection with firefighting foam. *Courtesy of J. Foley*

STORAGE

Storage requirements generally increase based on the quantity of flammable and combustible liquids being stored. Small quantities of hazardous liquids must be stored and used within control areas, and the appropriate storage cabinets complying with UL or NFPA 30 should be used. Storage can also be in drums, portable tanks, and underground or aboveground storage tanks. All storage of hazardous materials must be properly marked and labeled as to the hazard and content of the vessel. All ignition sources around storage areas must be controlled. This is generally within 15 to 30 feet of the storage pile or tank. Underground and aboveground storage tanks must have monitoring technologies and secondary containment systems that monitor the tank for potential leaks. Tanks must have overfill protection that stops the flow of product at 90 to 95 percent of the fill capacity. The vapor space of the storage tank must be properly vented and all vents must terminate at least 12 feet above grade level and at least 5 feet from any building opening. All storage tanks must be listed to the UL, ASTM, ANSI, or NFPA 30 standards for tank construction and use. Factory manufactured tanks may also comply with the requirements of the American Petroleum Institute (API) or Standard Tank Institute (STI) for testing and use. All storage tank installations must meet the requirements of the local building code for backfill depth, tank monitoring, and impact protection and vapor recovery.

Large, aboveground storage tanks are generally field fabricated and meet the requirements of NFPA 30 for normal and emergency venting as well as fire protection requirements for fixed fire protection systems (see Figure 8.9). Firefighting foam systems meeting NFPA 11 and 11A specifications are required to be installed under the following conditions:

1. Tanks' shell-to-shell spacing is less than 50 feet and the liquid surface exceeds 1,500 square feet and meets one of the following requirements:
   a. Contains Class I or II liquids
   b. Contains crude oil
   c. Located within 100 feet of processing equipment

d. Possesses an unusual hazard due to topography or proximity to adjoining properties.

e. Boilover liquids in storage tanks over 150 feet in diameter must also have gas inerting systems to protect them[6]

Abandonment of Storage Tanks   Often, the fire inspector will encounter abandoned aboveground or underground storage tanks. The fire prevention codes provide parameters for making these abandoned facilities safe by establishing time frames for repair or removal of the storage tank from the premises. Tanks that are abandoned for less than 90 days must be secured and have their fill lines isolated to avoid leaks. Tanks abandoned over 90 days must be emptied and have the fill lines secured, but the vent lines must remain open and intact to relieve internal vapors. Tanks that are out of service for more than 12 months are considered abandoned and must be removed from the premises. The removal process usually requires a demolition permit from the building officials and may also require additional permits and remediation by the appropriate environmental agency.

Building Storage of Flammable and Combustible Liquids   For other than tank storage, fire prevention and building codes have a hierarchy for how flammable and combustible liquids are stored. First, they must be used and stored within a control area and maintained in cabinets if the quantity is over 10 gallons capacity. If the quantity of a control area is exceeded, then the storage will be located in a liquid storeroom or cutoff room that is designated use group "H." Cutoff rooms have no direct access to the attached building, and to enter them, you must exit the building and enter from the outside. The fire separation wall between the cutoff room and building will meet the minimum fire-resistance ratings between the high hazard use and that of the facility. Flammable liquid storage rooms are also limited by quantity and storage pile height and must provide adequate aisle space for firefighting. Generally, at least 4-foot aisles will be provided for access by firefighters with hose lines. Automatic sprinklers or AFFF foam systems generally protect flammable liquid storage and cutoff rooms. If a flammable liquid storeroom exceeds the permissible quantity of flammable and combustible liquids, then the overflow must be stored in a flammable liquid warehouse. Flammable liquid warehouses are constructed to the requirements of the building codes and are completely separate structures. Flammable liquid warehouses are classified use group H. Flammable liquid warehouses must be sprinkler-protected, and storage is generally calculated by density of liquids in specific commodity classes and packaging configurations. These quantities determine the appropriate sprinkler design requirements in gpm/square foot over a designed operating area of 3,000 to 8,000 square feet. Water system demands must also include the minimum hose stream allowances for durations of 1 to 2 hours at the specified application rate based on the protected commodities. Flammable liquid warehouses may have storage rack sprinkler protection as well as fire hose stations for fire control purposes. Fire hose stations must be 1-1/2-inch lined hose or 1-inch solid rubber hose in flammable liquid warehouse applications. Spill control and secondary containment as well as weather protection must be provided if the control area is for outdoor storage. Outdoor control areas must be at least 20 feet from all public ways or lot lines, or 2-hour fire-resistant rated walls must be constructed and must extend 18 inches above the highest storage pile.

DISPENSING AND USE

Fire inspectors must also examine the locations within each control area where hazardous materials are used to ensure proper safety practices with dispensing and use of these materials. A key concern is the transferring of flammable and combustible liquids and the potential for static electricity. Flammable liquids should never be transferred by gravity or air pressure unless the transfer system is specifically engineered for that purpose and an inert gas is being used. Flammable liquids flowing through air may build and discharge static electricity, causing an ignition and fire. All flammable and combustible liquid containers and transfer equipment must be bonded and grounded to each other before any liquid transfer to eliminate differences in electrical potential.

On November 22, 2006, at 2:45 a.m., the small town of Danvers was rocked by an explosion of a small chemical manufacturer located in the community. The explosion injured ten residents and destroyed 24 homes and six businesses. A complete investigation of the incident was conducted by the U.S. Chemical Safety Board (CSB), and it was determined that the cause of the explosion was human error in the heating of an ink mixing tank by the use of manually operated valves by steam. The tank contained 2,000 gallons of ink. The tank contained a mixture of flammable liquids including heptane and propyl and isopropyl alcohol. The CSB determined that the heating of the tank occurred over several hours, vaporizing the liquids in the tank, and at 2:45 a.m., the vapors above their LEL found an ignition source, causing the explosion and extensive property damage. The CSB, through examination of regulatory requirements of the state of Massachusetts, made several recommendations including the following:

■ Better and more comprehensive fire inspection practices. The building had not been inspected since 2002 before the explosion

■ Better licensing and permitting of hazardous materials as the CAI/Arnel plant exceeded maximum permitted amounts of hazardous materials

■ Better training of fire department personnel in the inspection of hazardous materials

■ The development of an active local emergency planning committee, as Danvers did not have one at the time of the incident

■ Adoption of current NFPA standards and fire and building code requirements as well as better code enforcement

■ The use of automatic temperature and pressure interlocks on all processing equipment heating flammable or combustible liquids[7]

Additional information on this incident and videos are available at the Chemical Safety Board's website.

Combustible liquids, because of their higher flash points, can be transferred by gravity from a drum cradle; however, the drum faucet must be an automatic-closing and fail-safe device to avoid accidental spillage. Quarter-turn drum faucets (valves) should not be used when dispensing combustible liquids or other hazardous materials from 55-gallon drums in drum cradles. The fire inspector should examine the dispensing area for correct electrical equipment classification and make sure that other activities like cooking or heating appliances are not being performed in these areas. When processes require the heating of flammable or combustible materials above their flash points, the fire prevention codes require additional special considerations for safety distances to adjoining properties as well as increased levels of fire resistance of the structure. These structures must have explosion control systems to avoid catastrophic building failure by relieving internal pressure in a controlled fashion. Depending on whether the processing equipment is an open system to atmosphere or a closed system where the process is contained within a sealed vessel, the fire prevention code imposes different fire protection criteria for the vessel's protection. Vessels requiring heating must have temperature and pressure interlocks or limit switches to control the chemical reactions. The interlocks prevent the process from taking place unless all valves are in the proper position; more importantly, if power fails, the valves remain in a fail-safe condition to prevent an explosion. Limit switches operate on temperature and pressure and ensure that processes stay within specified temperature and pressure parameters. If a limit switch exceeds its pressure or temperature, the interlocks operate and the process is shut down.

Flammable Liquids and Cleaning   Flammable liquids are not to be used as cleaning agents unless they are in vessels specifically designed for such purposes. Cleaning machines such as parts washers must be UL listed and may not exceed the liquid limitation of the fire prevention code or the manufacturer. Machines with remote solvent reservoirs may not be used for parts soaking. The liquid should run over the part and back to the holding tank or reservoir. Parts washers also should have flash fire control systems such as fusible link surface covers that automatically close if the sump or sink catches fire. Areas where multiple machines are used must have proper natural or mechanical ventilation and must be separated by distance or barriers between machines. The mechanical ventilation rate in flammable liquid use areas is typically 1 cubic foot per minute per square foot of floor area. The fire inspector should check the mechanical code of the jurisdiction to determine the required ventilation rate. Machines that recycle solvents to remove impurities by distillation must have operation manuals available for the fire official's review and must be used within the manufacturer's specification and guidelines. Solvent distillation equipment also requires larger portable fire extinguishers (40BC) located within 30 feet travel distance in addition to the normal fire extinguisher distribution required under NFPA 10, *Standard for Portable Fire Extinguishers*.

Liquefied Petroleum Gases

Liquefied petroleum gases became popular in the 1920s as a residential and industrial fuel source. The first LP gas standard, known as *Standard 58,* was adopted by NFPA in 1932. The fire prevention codes and NFPA 58, *Standard for the Storage and Handling of Liquefied Petroleum Gases*, regulate LPG installations today nationwide. LP or LPG is a combination of propane, butane, isobutene, propylene, or butylene gases blended together as gas fuel. These materials are liquid within the container but immediately convert to gas once released. The expansion ratio is approximately 270 cubic feet of vapor to one liquid gallon. LP gases are heavier than air and should not be permitted to be used in basements, pits, or other locations where vapors could collect and be trapped at lower levels of a structure. LP gases are shipped and used in many different container sizes from fixed tanks of 30,000-gallon capacity to portable tanks of 2.5-pound capacity. The fire prevention code restricts the use of LPG in certain building use groups. Factories are limited to 735 pounds per manifold and must separate additional manifolds by 20 feet. Educational and institutional uses are limited to 50 pounds for research and experimentation; it may not be located within a classroom, however. Fifty pounds of LPG may be used for educational purposes, while institutions are limited to 12 pounds. Temporary use for public exhibition is limited to 12 pounds, while self-contained hand torches are limited to 2.5 pounds. LP gas warming stoves are permitted in restaurants, provided that the devices are listed and tested appliances.

LP gas installations must be located with respect to building lot lines and public ways at fixed facilities and must have adequate spacing between cylinders for safety and inspection. LP gas systems over a 400-gallon capacity are required to have a fire safety analysis and may require the installation of a NFPA 15–compliant fixed water spray fire protection system. LP gas installations exceeding 180,000 gallons must be protected with additional fire protection systems. Nonsmoking regulations must be enforced within 25 feet of LP gas installations and combustible materials and vegetation must be cleared to within 10 feet of the tanks (see Figure 8.10).

FIGURE 8.10 LPG gas tanks must be protected and marked and identified using NFPA 704 placards. *Courtesy of J. Foley*

In a casino hotel in Atlantic City, New Jersey, portable propane cooking appliances were being used during a special event. The porter could not get the butane cylinder to seat properly into the appliance when setting it up and forced the 12-ounce can into position. Once lit, the fuel can began to leak and erupted into a fire. A hotel chef hurried and grabbed a portable fire extinguisher and began to operate it when the metal butane can failed. The resulting explosion drove the 250-pound chef 20 feet through the air and into a wall. The chef suffered first and second degree burns to the face and hands as a result of the explosion. The ensuing fire was quickly extinguished with portable fire extinguishers. The cylinder contained only 12 ounces of butane, but the energy released from this explosion, which caused a small BLEVE, was significant. A BLEVE is a **b**oiling **l**iquid **e**vaporated **v**apor **e**xplosion that occurs when a flammable liquid is heated above its boiling point in a closed container, which increases the internal pressure within the container to the point of failure. Once the container fails, the liquid immediately converts to a gas and rapidly expands and explodes. Portable cooking appliances are often used in hotels and banquet halls and can be problematic if used incorrectly. Fire inspectors should examine portable equipment during the inspection and make sure that the equipment is properly listed and in good working order. They should also inspect the reserve cans of butane for proper storage in an aerosol cabinet.

Tanks must be protected from vehicle impacts, and cylinders weighing 100 pounds or more must be marked with proper warning signs as well as the gas name.

LP gas installations exceeding 250 gallons must have signs indicating the gas supplier and emergency contact telephone numbers. Fire protection systems are required to be installed under NFPA 58, *Liquefied Petroleum Gas Code,* for LP gas facilities over 4,000-gallon capacity. LP gas may not be stored in portable tanks on building roofs and will be limited for storage inside buildings. Generally, buildings frequented by the public are limited to 200 pounds, and those not frequented by the public are limited to 300 pounds of LP gas.

Compressed Gases

Compressed gases have the physical hazard of high pressure and are contained in pressure cylinders. Compressed gases exist in a vapor state at 68°F at 760 mm hg and may be liquefied or nonliquefied. Compressed gases, because they are in a normal vapor state, have boiling points below 68°F at 760 mm hg. Compressed gases represent a wide range of products that may be flammable, nonflammable, poisonous, corrosive, or toxic. Depending on the nature of the gas and its use, both the building and fire codes will require different countermeasures for fire safety. The hazards of compressed gases generally come from three types of occurrences. The first is exposure to fire or heat. The fire prevention code requires protection in the form of fire-resistant rated barriers, vaults, and clearance to combustible materials and the control of ignition sources around storage areas. The second hazard is a leak, and the codes address these concerns by the installation of leak detection monitoring equipment, use of cylinder safety caps, and mechanical ventilation of enclosures and storage areas. Facilities may also have containment equipment such as that seen in Figure 8.11. The third potential hazard is a cylinder being knocked over or impacted and becoming a missile or developing a leak, as happens when the valve is sheared off. The fire prevention code addresses this concern with requirements to secure the cylinders with chains or nest cylinders and by requiring valve hoods to be in place when the cylinder is not in use. Cylinders should always be stored in the upright position. When cylinders are in use, the valves should be equipped with the appropriate pressure regulator. Teflon® tape should not be used on straight thread connections where the seal is made by metal-to-metal contact. The Teflon® tape may weaken the threads and cause a

FIGURE 8.11 A cylinder recovery device may be used to address leaking cylinders. *Courtesy of J. Foley*

leak. Oxidizing agents such as oxygen, chlorine, fluorine, or nitrous oxide should not be used with any lubricating oils on the valve or regulator fittings, as these can accelerate a fire in the presence of an ignition source. Gasketing materials must be compatible with the gas in the cylinder and should follow the Compressed Gas Association (CGA) guidelines. Gas storage areas should have proper warning signs for the gas hazards that are present, such as "flammable gas," "oxidizer," or "toxic gas."

All high-pressure cylinders of compressed gases must comply with DOT and ICC regulations for pressure relief valves and hydrostatic testing of the cylinders. These hydrostatic tests are usually conducted every 5 to 12 years, depending on the cylinder's construction. Typical industrial compressed gases include acetylene, ammonia, oxygen, carbon dioxide, chlorine, fluorine, hydrogen, ethylene, nitrogen, propylene, silane, methane, and MAPP gas. Some of these gases are oxidizers like oxygen, chlorine, and fluorine, while others are flammable gases like hydrogen and acetylene, and some may be inert gases like helium argon and carbon dioxide. The main hazard of any compressed gas is pressure and exposure to heat. Fire inspectors should make sure that gas cylinders at facilities are not intermixed between hazard classes and are maintained in safe storage configuration inside fire-rated storage rooms or gas cabinets. If toxic gases are used, they must be in a gas cabinet, which has a negative pressure to keep any leaks within the cabinet. These cabinets also have monitoring alarms and vision windows to look inside. Gas cabinets are usually limited to three 100-pound cylinders. If compressed gas cylinders are stored in gas vaults, the vault must meet specific requirements of the building code for monitoring, fire protection, and mechanical ventilation. Ventilation rates are usually a minimum of 150 cubic feet per minute or 1 cfm per cubic foot of storage, whichever is greater. Cylinders stored outside should be properly nested in triangles and should be kept out of direct sunlight. Cylinders should be secured in cages such as those seen in Figure 8.12. Cylinders in use or storage should be secured to walls by chains or they should be located in stable racks to prevent toppling. All compressed gases, especially flammable gases and oxidizers, should be separated by at least 20 feet in distance.

FIGURE 8.12 Cylinders are kept secure in gas cages outside the construction site. *Courtesy of J. Foley*

Cryogenic Gases

Cryogenic gases are those gases that have boiling points below –130°F at 760 mm hg. These super cold temperatures present a risk of advanced hypothermia and skin damage on contact. Cryogenic liquids must be stored and used in compliance with NFPA 55, *Compressed Gases and Cryogenic Fluids Code*. Cryogenic liquids are kept in insulated vessels that have pressure relief systems to prevent overpressure and rupture of the vessel (see Figure 8.13). Cryogenic materials must be properly marked and identified by NFPA

FIGURE 8.13 Bulk storage of cryogenic gases is often found at hospital facilities. *Courtesy of J. Foley*

704 placards. Containers of cryogenics must be kept secure from accidental dislodgement and physical damage. Moving cryogenic materials within a facility should only be performed using carts or trucks compatible with the material. Permanent tanks should have limit controls to prevent overfilling of the vessel.

Oxidizers

Oxidizers are compounds that gain electrons from other elements, causing them to be reduced. The fire problem with oxidizers is that they increase the combustion rate, especially if they contain additional oxygen in the compound, such as hydrogen peroxide that is liberated in a chemical reaction. Oxidizing agents exist as gases, such as medical oxygen, chlorine, or fluorine gases. They also exist as liquids, such as nitric acid, perchloric acid, or sulfuric acid, and additionally, they can exist as solids such as chlorates, chromates, and nitrates. Combustion with the presence of oxidizers can be explosive. Fire prevention codes classify oxidizers into four distinct hazard categories:

- *Class 4 Oxidizers:* These can cause explosions by physical shock. They may undergo spontaneous combustion when in contact with other combustible materials. Some examples are perchloric acid, 91% hydrogen peroxide, and ammonium perchlorate.
- *Class 3 Oxidizers:* These increase the rate of combustion significantly and will undergo vigorous decomposition when exposed to heat. Some examples are fuming nitric acid, 60 percent perchloric acid solutions, and sodium chlorate.
- *Class 2 Oxidizers:* These oxidizers moderately increase the burning rate and may cause spontaneous ignition of combustibles. Some examples are 50 percent hydrogen peroxide, calcium chlorate, and magnesium bromate.
- *Class 1 Oxidizers:* These materials cause a slight increase in burning rate but will not cause spontaneous ignition of combustible materials. Some examples are hydrogen peroxide below 26 percent solution, chlorine, and nitric acid below 40 percent solution.[8]

The fire prevention codes allow minimal exempt amounts of oxidizers in all building use groups but usually require that these materials be maintained in hazardous materials storage cabinets. When the exempt amounts of oxidizers are exceeded, the material must be located in a separate detached storage facility. The storage facility must meet certain minimum distances to lot lines, building egress discharges, and the public ways based on the amount of oxidizers stored. Storage facilities must also have explosion control for Class 4 oxidizers and automatic sprinkler systems designed and installed to NFPA 430, *Code for the Storage of Liquid and Solid Oxidizers*. Spill control and secondary containment as well as automatic smoke detection systems are also required. Class 4 oxidizers must be separated from all other materials by at least 1-hour fire separation walls.[9]

Aerosols

Aerosols were responsible for the largest loss industrial fire in recent history. The fire occurred at the K-Mart warehouse in Falls Township, Pennsylvania, in 1982, resulting in a loss of $190,000,000. The warehouse was over 85,000 square feet and was divided into four fire areas by firewalls and opening protective 3-hour fire doors. The warehouse had automatic sprinkler systems supplied by an automatic electric and backup diesel fire pump.[10] The fire was the result of a forklift fire in an aisle containing aerosol carburetor cleaner storage. As the fire progressed, the cartons of aerosols became involved and rocketed around the warehouse and through open firewall doors. It wasn't long before the fire overtaxed the automatic sprinklers, and the building became a total loss. Building codes and fire codes were changed to address the separation of aerosol storage from other types of commodities after this incident. Traditionally, aerosols were treated as 1A flammable liquids under NFPA 30; however, after the K-Mart fire and full-scale fire testing in the 1980s, the NFPA developed NFPA 30B, *Code for the Manufacture and Storage of Aerosol Products*. The standard addresses the separation of aerosols from other types of commodities and also addresses the installation of Early Suppression Fast Response, or ESFR, sprinkler systems.

Aerosols are categorized into three levels of hazard based on their heat of combustion:

- Level I—aerosols with a total heat of combustion of 8,600 BTU/lb
- Level II—aerosols with a heat of combustion greater than 8,600 BTU/lb but less than 13,000 BTU/lb
- Level III—Aerosols with a heat of combustion of 13,000 BTU/lb or more[11]

Aerosols are permitted to contain a maximum of 33.8 fluid ounces per can. Cartons that contain aerosols must identify the level of aerosol within the carton or box. The fire prevention codes regulate Level II and III aerosols primarily, as Level I aerosols are considered Class III commodities under NFPA 13 for sprinkler protection, and NFPA 30B for aerosol storage. The fire prevention code regulates aerosols in building use groups A, B, E, F, I, and R to 1,000 pounds for Level II and 500 pounds for Level III, or a combined aggregate of 1,000 pounds total. If these exempt amounts are exceeded, the remaining storage must be in a flammable liquid storage room in proper aerosol storage cabinets. Aerosol storage may be segregated or nonsegregated and may or may not be protected by automatic sprinklers depending on the quantities being stored. Obviously, nonsprinklered storage areas have restricted quantities of aerosol storage, usually around 2,500 pounds of level II or 1,000 pounds of level III aerosols. Segregated storage may be accomplished by the installation of chain link fence or fire-resistant hard wall construction; NFPA 30B, *Code for the Manufacture and Storage of Aerosol Products*, requires that the sprinkler protection extend at least 20 feet beyond the aerosol storage area.

Warehouses specifically designed for aerosol storage and protected with proper ESFR sprinkler systems are not restricted on the total quantities of level II and III aerosols permitted. Warehouses are required to have adequate aisle spacing of 4 feet if commodities are palletized and 8 feet if they are in rack storage. Storage of aerosols outside must have proper clearances to the interior lot lines, building exit discharges, and public way for outdoor control areas.

The IFC also regulates retail display of aerosols by restricting the height and quantity of the materials on display. Cartons must be removed in retail display, and piles should not exceed 8 feet in fixed shelving. If the height is greater than 8 feet, then additional sprinkler protection and fire separation requirements will be applied.

Explosives

Explosives are used in mining operations, road and tunnel construction, and explosive demolition of buildings. Federal laws as well as the fire prevention codes regulate the possession and use of explosives. Fire prevention codes have requirements for explosives in manufacturing, storage, and handling and use. Additionally, all fire prevention codes require that a fire safety or explosives permit be secured from the local jurisdiction for activities involving explosives. The Department of Homeland Security through the Bureau of Alcohol, Tobacco and Firearms, or ATF, regulates explosives, and since September 11, 2001, keeps tight control over their distribution and ultimate use. The discharge and use of explosives requires licensing of the blasters at the federal and (usually) state levels. Fire prevention codes require these licenses as a prerequisite to obtaining fire safety permits from the fire prevention official.

Explosives are categorized as DOT Class 1 materials and are divided into six subcategories by the regulation:

- Division 1.1—Mass explosion hazards
- Division 1.2—Projectile hazards
- Division 1.3—Fire and minor blast hazards
- Division 1.4—Minor explosion hazards
- Division 1.5—Very insensitive explosives
- Division 1.6—Extremely insensitive explosives with no mass detonation hazard[12]

The DOT also provides 13 compatibility groups for explosives in transportation that are designated by the letters A to S. As an example, commercial fireworks are 1.3G (formerly Class B, special fireworks) and are compatible for shipment with other explosives in the Class G category such as 1.4g pyrotechnics (formerly Class C, common fireworks). Explosive or energetic materials must be stored in special magazines for safety. The type and construction of the explosive magazine are based on the type of explosive it must store. Magazines must comply with the requirements of NFPA 495 for explosives and NFPA 1124 for fireworks storage. Magazines may be permanent structures or mobile or portable containers. Magazines may include igloos, boxes, semi trailers, or other mobile-type containers. The magazines must be fire resistant and theftproof as well as weather resistant. Explosive storage areas must be clear of combustible vegetation and are required to be posted with signage to warn first responders of the explosive danger. Packing or unpacking of explosives may not take place within 50 feet of a magazine, and all tools must be made of nonferrous or nonsparking metals. Magazine placement must follow the American table of distances for explosives storage in relation to buildings, public streets and highways, and other explosive magazines. Barricades around the magazine allow certain reductions in the safety distances specified in the tables (see Figure 8.14).

These distance tables become important during the explosive demolition of buildings in cities where safety distances may require additional protections because of surrounding occupied structures.

FIGURE 8.14 A mobile explosive magazine, protected by concrete barriers to reduce safety distances. *Courtesy of J. Foley*

FIGURE 8.15 Nonelectric firing systems connected to explosives in columns. *Courtesy of J. Foley*

Fire inspectors must also be aware of the types of firing mechanisms used in explosive detonation. The safest systems use nonelectric detonators (see Figure 8.15). Nonelectrical systems must still be protected from lightning and static electricity but are safer from accidental discharge.

Electric explosive firing systems can be detonated by stray radar or radio transmissions and must be protected from these radio frequency sources. When explosives are being loaded for detonation, only authorized persons may be within the loading area. After the detonation, no one may enter the blast zone until the licensed blaster has inspected the entire site for unexploded materials. Building demolition using explosives can produce many challenges to fire inspectors, especially with surrounding occupied structures.

CASE STUDY | Fireworks and a Building Implosion

In Atlantic City, as in Las Vegas, the casino industry likes to combine entertainment with building demolition. These events are called "Vegas Shoots." The demolition involves a fireworks display prior to the explosive demolition of the building. The challenges for the fire official are that the event will attract a large audience and must be performed at night, which poses a problem in terms of site security. When the Sands casino, built in 1980, was demolished in Atlantic City, it was located within 100 feet of the adjoining Claridge casino. All of the emergency exits from the Claridge casino discharged on the Sands casino side of the building *(see Figure 8.16)*. The fire prevention official had to have the demolition contractors construct protected exit enclosures for the Claridge casino and had to secure all sleeping rooms from the twenty-fourth floor down to ensure that no occupants were on that side of the building before or during the demolition. The fire department assigned members to inspect each floor area and secure the sleeping rooms before the demolition could occur. The danger to occupants and spectators was potentially from flying rock that could escape the demolition site. *Figures 8.17 and 8.18* demonstrate the Las Vegas-style explosive demolition.

FIGURE 8.16 Casino exits had to be protected within 100 feet of the demolition site. *Courtesy of J. Foley*

FIGURE 8.17 Fireworks displays in conjunction with building demolition may attract large crowds; fire marshals must take great precautions to maintain the safety of the viewing audience. *Courtesy of J. Foley*

FIGURE 8.18 The 22-story high-rise Sands Casino is explosively demolished at the end of the fireworks display. *Courtesy of J. Foley*

Fireworks

Many states, like New Jersey, Delaware, Massachusetts, and New York, prohibit the use of common fireworks, except in public displays. The reason for this is the high injury rate related to fireworks accidents. In 2009, the NFPA reported that over 18,000 fires were started by use of common fireworks, and this resulted in $38 million dollars in direct property loss.[13] The U.S. Fire Administration reported 8,600 fireworks-related injuries in 2010.[14] Fireworks can present a real danger to a community, even in public displays; therefore, fire inspectors must take great care in inspecting and supervising the discharge of public fireworks displays. The fire prevention codes require that NFPA 1123, *Code for Fireworks Display,* be followed for outdoor fireworks and NFPA 1126, *Standard for the Use of Pyrotechnics Before a Proximate Audience,* be followed for indoor pyrotechnic displays. The discharge of pyrotechnics requires the fire official to issue a fire safety permit for the event. Most jurisdictions maintain a fire inspector on the site during and after the event as well as additional fire companies for fire protection. Outdoors fireworks displays usually involve 1.3g explosive products that are required to be fired at minimum safety distances to the audience viewing areas as determined by the shell size in NFPA 1123. This distance is 70 feet per inch of shell diameter being fired. Wind may also be a factor in establishing the viewing areas and debris fall-out zones, and fireworks should not be discharged in high winds above 20 mph. Fireworks may be manually or electronically discharged depending on the shell size. Firework shells less than 6 inches in diameter may be fired by hand with technicians using flares. Technicians, however, must have personal protective equipment including a fire-resistant shirt, hardhat, ear protection, and safety glasses during the igniting process. All fireworks displays using shells over 6 inches in diameter must be fired electronically from a remote location. All fireworks are discharged from mortar tubes, and they must be of the proper type and size for the shell being installed, the shell should slip in the mortar easily, and the mortar should be free of debris from past discharges. Mortars may be constructed of cardboard, high-density PVC, metal, or carbon fiber (see Figure 8.19). The mortar tubes may not be made of frangible materials that can fragment. All mortars must be properly placed in secure rack systems or buried so as not to be able to tip over.

Additionally, many fireworks companies use display pieces called *cakes,* which are small multimortar tube devices in a cardboard box. These devices must also be buried to

FIGURE 8.19 Fireworks shells and mortars during show preparation in Atlantic City, NJ. *Courtesy of J. Foley*

FIGURE 8.20 Display cake boxes must be buried before firing. *Courtesy of J. Foley*

WARNING: DANGEROUS EXPLOSIVE

a specific depth line on the box to prevent misfires from displacing the mortar tubes (see Figure 8.20). All combustible materials within the site must be controlled. Grass fields should be mowed and wetted before the fireworks event to prevent unnecessary grass fires. During the display, the fire inspector must have direct communication with the fireworks technician to ensure that safety can be maintained. The fire official has the authority to halt the firing for any safety concerns. At the conclusion of the display, all mortars should be allowed to cool for 15–20 minutes before any inspection is conducted for misfired shells (see Figure 8.21). Fire inspectors should never place their head over top of a mortar tube, but should inspect it from an angle or by using a mirror. Unfired shells should be drowned in water and removed and packaged for proper disposal. The fire

FIGURE 8.21 Fireworks must be allowed to cool before inspection. Here, several devices are still live and unfired. *Courtesy of J. Foley*

inspector and shooter should also conduct a first light sweep of the firing area to ensure that no fireworks devices have landed unexploded in the area.

Proximate pyrotechnics are used indoors in accordance with NFPA 1126, *Standard for the Use of Pyrotechnics before a Proximate Audience*. Indoor pyrotechnics are usually 1.4g explosives and involve theatrical productions such as concerts, sporting events, and ice shows. These devices are small and are fired electronically on music cues. They may include flame projectors using Lycopidium, gerbs, sparklers or concussion cannons, and airbursts (see Figure 8.22). Some devices may be propane fired, such as flame projectors. The safe display of indoor pyrotechnic devices depends on their location on the stage with relation to personnel when they are discharged. Adequate types and numbers of portable fire extinguishers must be placed in the area. The audience must be a safe distance from the devices on the stage, and a demonstration of each device should be witnessed by the fire official before approval for use. The fire inspector must also ensure that surrounding materials are properly fire-resistant treated and that fire protection systems are not disrupted or reduced as the result of these devices being used.

FIGURE 8.22 NFPA 1126 proximate audience pyrotechnics and firing system explosives are 1.4g. *Courtesy of J. Foley*

Summary

Fire inspection of hazardous materials is a significant challenge to fire inspectors in the field. Inspectors must be capable of identifying the particular hazards of the material and determining the permissible quantities of product that can be stored, handled, or used within a facility. Fire inspectors must have knowledge of the identification systems used to identify hazardous materials as well as how to gather information on chemicals through MSDS or information databases.

Fire inspectors will encounter hazardous materials in every building that they inspect, and they must ensure that they are properly stored, handled, and used within control areas and that quantities are under the exempt amounts that would alter the building use group. Hazardous materials must be protected by fire protection systems to prevent or reduce large fire losses. Fire protection controls include passive measures such as fire-resistant enclosures, walls, ceilings, and floors, and UL-listed storage cabinets, as well as active fire protection systems, including automatic fire sprinklers, fire alarms, explosion venting, mechanical ventilation, and static control and must be thoroughly inspected and tested or certified. Fire inspectors must study the requirements of the fire prevention codes and apply that knowledge to the storage, handling, and use of hazardous materials where these products exist.

Review Questions

1. Identify and review some of the major chemical incidents that have occurred in the United States. What were the causes of the incidents, and could proper fire code application have eliminated or reduced their consequences?
2. Inspect your local high school chemistry laboratory. Describe how the chemicals are stored and what labeling systems are used.
3. Devise a hazardous material inspection protocol based on the information received by the fire marshal under SARA Title III.
4. List and describe the types of information found on a material safety data sheet and explain how it can be applied to fire inspection activities.
5. Using your local fire prevention code, in the following diagram, identify the exempt amounts of flammable liquids that may be stored in each of the identified control areas.
6. Using the building code in your jurisdiction, list and describe the categories of High Hazard "H" use groups and the types of hazardous materials that would be found there.
7. You inspect a facility that has multiple compressed gas cylinders on the property. What code requirements would apply to the storage, handling, and use of these hazardous materials?
8. Go to a local home center and examine how aerosols are displayed and stored. Identify the aerosol classes and the code sections that would be applicable to the storage and display of these commodities.
9. Develop a list of general safety precautions using NFPA 1123 for the display of public fireworks. What safety measures must pyrotechnic technicians take during the firing of a manual fireworks display?
10. List and describe the fire prevention code requirements for mercantile display of nonflammable, noncombustible hazardous materials over the exempt quantities for storage.
11. Go to the Chemical Safety Board's website and review the Danvers Final Report. Compare and contrast how your jurisdiction inspects and licenses small chemical plants, and describe changes that should be made to make your community safer.

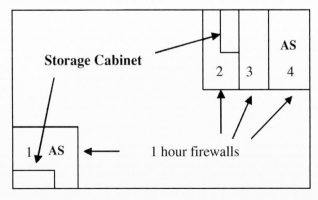

AS = Automatic Sprinklers

Suggested Readings

Burke, Robert. 2003. *Hazardous Materials Chemistry for Emergency Responders*. Washington, DC: Lewis Publishers, CRC Press.

Meyer, Eugene. 2009. *Chemistry of Hazardous Materials*. 5th edition. Upper Saddle River, NJ: Brady/Prentice Hall Health.

National Fire Protection Association. 2010. *Fire Protection Guide to Hazardous Materials*. Quincy, MA: NFPA.

Endnotes

1. *International Fire Prevention Code–New Jersey 2006* (Washington, DC: International Code Council, 2006).
2. Ibid.
3. *International Building Code–New Jersey edition 2009* (Washington, DC: International Code Council, 2009).
4. *International Fire Code–New Jersey 2006* (Washington, DC: International Code Council, 2006).
5. Ibid.
6. Ibid.
7. U.S. Chemical Safety and Hazard Investigation Board, *Investigation Report Confined Vapor Cloud Explosion, CAI Inc. and Arnel Company Inc.* (Danvers, Massachusetts, November 2006).
8. *2003IBC-Hazardous Materials and the I-Codes* (Washington, DC: International Code Council, 2003).
9. *International Fire Code–New Jersey 2006* (Washington, DC: International Code Council, 2006).
10. Jeff L. Harrington, "Lessons Learned From Warehouse Fires." *Fire Protection Engineering*, 1st quarter, 2006.
11. *International Fire Code–New Jersey 2006* (Washington, DC: International Code Council, 2006).
12. Eugene Meyer, *Chemistry of Hazardous Materials*, 5th edition. Pearson, Brady Publishing.
13. John R. Hall, *Fireworks–June 2010* (Quincy, MA: National Fire Protection Association).
14. Ti Yangling, *Fireworks-Related Deaths, Emergency Department Treated Injuries, and Enforcement Activities during 2010* (U.S. Consumer Product Safety Commission, June 2010).

Courtesy of J. Foley

KEY TERMS

aisle, *p. 218*

aisle accessway, *p. 218*

aisle convergence, *p. 219*

common path of travel, *p. 205*

exit, *p. 204*

exit access, *p. 204*

exit discharge, *p. 204*

floor area gross, *p. 202*

floor area net, *p. 202*

horizontal exit, *p. 211*

inches per occupant, *p. 199*

increased occupant load, *p. 203*

maximum permitted occupant load, *p. 214*

means of egress, *p. 196*

minimum design occupant load, *p. 202*

occupant load, *p. 213*

posted occupant load, *p. 214*

smokeproof enclosure, *p. 210*

stairwell pressurization, *p. 210*

travel distance, *p. 202*

unit of egress width, *p. 198*

OBJECTIVES

After reading this chapter, the reader should be able to:

■ Understand the history of means of egress requirements and how they were derived.

■ Identify the three components of an exit system and the relevant inspection points for each component.

■ Understand how occupant loads for buildings are determined by building codes.

■ Know how to inspect places of assembly for hazards in the means of egress and how to deal with issues of overcrowding.

Professional Levels of Job Performance for Fire Inspectors as Cited in NFPA 1031 and NFPA 1037

■ NFPA 1031 Fire Inspector I *Obj.4.3.2 Compute the allowable occupant load of a single use occupancy and portions thereof*

■ NFPA 1031 Fire Inspector I *Obj. 4.3.3 Inspect egress elements*

- NFPA 1031 Fire Inspector II *Obj. 5.3.5 Analyze the egress elements of a building or a portion of a building*
- NFPA 1031 Fire Inspector II *Obj. 5.4.2 Compute the occupant load*
- NFPA 1031 Fire Inspector II *Obj. 5.4.5 Verify the means of egress elements are provided*
- NFPA 1031 Plan Reviewer I *Obj. 7.3.4 Verify the occupant load*
- NFPA 1031 Plan Reviewer I *Obj. 7.3.5 Verify that adequate egress is provided*

Introduction

means of egress ■ A continuous and unobstructed path of travel from any point in a building or structure to a public way that consists of three separate and distinct parts: the exit access, the exit, and the exit discharge. A means of egress comprises the vertical and horizontal means of travel and shall include all intervening rooms and spaces, doors, corridors, hallways, passageways, balconies, ramps, stairs, enclosures, lobbies, escalators, horizontal exits, courts, aisles, and yard.

"This way to the **means of egress**" read the sign that P.T. Barnum, an American showman and owner of the Barnum & Bailey Circus placed in Scutter's American Museum in New York City in 1841. Barnum's plan was to keep people from lingering at the exhibit. Barnum believed that people would not understand that egress meant exit and that they would assume they were going to see another exhibit. Barnum knew that they would leave the building and would have to pay another quarter to reenter.[1] Means of egress from a fire code perspective is one, if not the most important, element of fire code enforcement. Key to maintaining safe exiting is a knowledgeable fire inspector who can analyze and determine the adequacy of the means of egress and correct any deficiencies that may exist. Fire prevention code is only as effective as the code enforcement program behind it, and good enforcement equates to knowledgeable fire inspectors.

Why Exits Are Important

In the examination of major U.S. fires, we will find that most of them have led to significant loss of life. Many of these fire incidents caused changes in the fire prevention codes. In the majority of these fires, means of egress violations played a significant role on the death tolls. Many of the fires where lives were lost had code violations such as obstructed exits, inadequate exit capacity, and failure of the exit's fire resistance that led to large death tolls. The challenge for fire inspectors in examining the means of egress falls into two distinct categories. First, the fire inspector must understand the technical requirements that apply from the building and fire code. Second, the fire inspector must understand human behavior and how people actually use the exit systems that are available in an emergency. From the technical code perspective, the components of the means of egress must comply with the jurisdictionally adopted building and fire code. Egress systems must meet the minimum required ceiling height and access widths for all of the egress components from the most remote part of the building to the public street. All exit access ways and exits must be well lit during normal and emergency conditions and they must be free of obstructions and available for use at all times. Exit doors must have proper opening protection and swing correctly in the direction of expected egress flow. Exits and exit access corridors must have the proper interior finishes to minimize potential flame spread and smoke development during occupant evacuations. Exits must be properly identified by illuminated exit signs and exit access ways, exits, and exit discharges must have

FIGURE 9.1 Exits behind public areas may be obstructed. This requires a behavior change for management and employees. *Courtesy of J. Foley*

emergency lighting. The building and fire codes require stairs to have proper handrails and guardrails. Exits must be open and accessible, as well as providing multiple escape routes. Most important is that the exit must be under control of the user.

As an inspection requirement of fire prevention code, the means of egress is relatively simple to inspect. Building and fire code requirements are well documented and easily identified by the fire inspector during fire inspection tours. Fire inspectors should be capable of readily identifying broken automatic door closers, unlit exit signs, and any obstructions within the path of exit travel. Fire inspectors will find these violations are very typical in all types of buildings. The more difficult challenge for the fire inspector is to determine how occupants will use the means of egress in an emergency. The fire inspector must realize that it is the occupants' behavior that causes most of the technical egress deficiencies. These occupant deficiencies include obstructing exits with items or storage, locking or obstructing exit doors, chocking fire exit stair doors open, or placing combustible materials inside of fire exit stairs (see Figure 9.1). The building occupants' interaction and interference with the egress system is by far the most difficult violation for fire inspectors to correct. Changing occupant behavior in most cases requires additional fire inspections and time. Fire prevention bureaus may assist by directing fire safety education programs for the building operators and occupants emphasizing the need for safe exit systems. Simply citing exit code violations often does not change the occupants' attitudes and behaviors that caused the violation in the first place. Changing the occupants' behavior takes time and education to personalize the fire risk to the occupants and ultimately change the behavior that causes the code violations.

The Human Element

When we examine elements of means of egress, we must understand that the majority of exit measurements and exit design requirements have existed for a very long period of time and are time tested. In 1903, the National Bureau of Fire Underwriters established

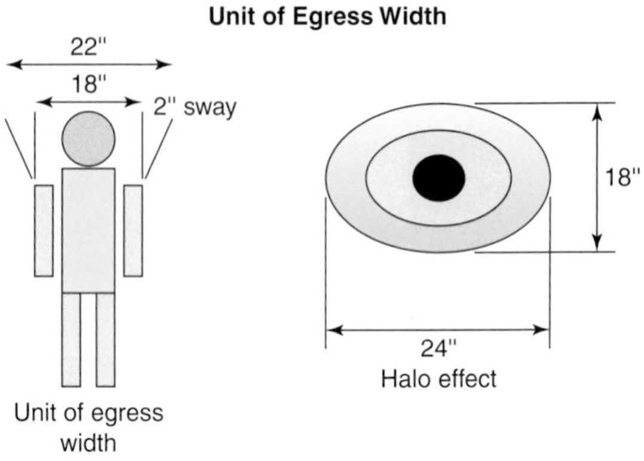

Unit of Egress Width

22"
18"
2" sway

Unit of egress width

24"
Halo effect

18"

FIGURE 9.2 The 22″ unit of egress width vs. the 24″ × 18″ halo effect.

unit of egress width ▪ The unit of egress width is a measurement of 22 inches based on the average size of a male at 18 inches and a sway of 2 inches in either direction when walking. The unit of egress width was the foundation of determining adequacy of exit elements for years until the 1980s when model building codes replaced it with inches per occupant

the first *National Building Code,* which began to establish uniform building requirements for means of egress. The NFPA Committee on Safety to Life, which was established in 1913, published the *Building Exit Code* in 1927.[2] This was the first code directed specifically at exit safety. The *Building Exit Code* eventually evolved after the great fire tragedies of the 1940s such as the Cocoanut Grove and the Winecoff Hotel fires and became NFPA 101, *Life Safety Code*®. The *Building Exit Code* and *Life Safety Code* are the foundation for the all of the model building code requirements of means of egress. These legacy documents are still reflected in the means of egress requirements of the International Building Code (IBC) and International Fire Code (IFC) that exist today. Many of the original exit requirements can be traced back to these early code documents. This is not to say, however, that all of the assumptions of the legacy code were totally correct. The codes are constantly evolving as we learn more about occupant behaviors through fire analysis and research. As an example, the 44-inch stairway was introduced into the building exit code in 1913[3] and was based on the **unit of egress width** model. The concept of the unit of egress width was based on the shoulder width of average males and was originally developed by the French. This width was determined to be 18 inches, and because a person who is walking sways up to 2 inches in either direction of his or her central axis, a unit of egress width was determined to be 22 inches (see Figure 9.2). The idea of the 44-inch stairway was that it could accommodate two persons per step side by side while ascending or descending the stairway in an orderly fashion. Today, based on scientific research, we know that people do not descend stairs in that fashion at all; in fact, they descend stairs in a crablike gait. At the time the original *Building Exit Code* was written, there were also two competing theories of how egress should be calculated. The first method was called the *capacity method* in which the building's stairway and exits would be required to hold all the occupants from each floor as an area of refuge. This method was employed in healthcare facilities, institutions, and high-rise buildings where population movement was difficult. The second method was called the *flow method*. The flow method assumed that people move through the exits and down the stairway at a fixed speed, therefore, stairways would only have to accommodate each floor's occupant load.[4] The average evacuation speed of the flow model was determined to be 45 persons per 22 inches of egress width on a stairway and 60 persons per 22 inches of egress width through a door over a period of 1 minute. These flow assumptions of 45 and 60 per unit of egress width were also used by the British in the London Transportation Study Report 95 when the underground rail system was redesigned. The London study determined that people move at about 250 feet per minute when spaced at about one person per 25 square feet. The speed decreases to 145 feet per minute or shuffling at about 7 to 8 square feet per person and all motion would cease below 2.25 square feet per person,[5] as this was considered the exit jam point. This study established the maximum flow capacity of an exit at one person per 3 net square feet. Three net square feet was used in building code up until 2003 IBC. Currently, the IBC establishes the maximum egress flow capacity at 5 net square feet per person based on newer information from evacuation studies.

Many of the early assumptions based on the concepts of the unit of egress width have proven to be no longer valid. Jake L. Pauls, a researcher in the field of human behavior and evacuation studies, conducted over 40 high-rise evacuation drills for the Canadian government that documented how people evacuate high-rise buildings. The study suggested that the assumed flow calculations of 45 and 60 persons per minute per unit of egress width were artificially high. Traditionally, building codes only considered occupant movement by whole and half units of egress width. A whole unit of egress width is

22 inches and half units of 12 inches. Any available egress width below 12 inches was simply disregarded in the flow calculation. Pauls demonstrated in the Canadian evacuation studies that a more linear function of inches of stairway per occupant would be a more appropriate measurement of occupant flow.[6] In the mid-1980s, most of the national model building codes adopted the concept of measuring egress components by **inches per occupant**. This change would take into account the total effective clear width of an element of egress instead of eliminating widths under 12 inches as was done using the unit of egress width method.

In the early 1970s, John J. Fruin and John Templer, who were researchers on stair design and pedestrian movement, developed the theory of the body ellipse, which was based on the earlier work of psychologist Albert Damon. The theory is that persons maintain an invisible territory or "halo" around themselves to avoid contact with other objects. Fruin determined this space to be 0.61 m (24 inches) by 0.41 m (18 inches). The implication from this was that the 22-inch unit of egress width is probably not sufficient in width for two persons to descend a 44-inch stairway.[7] Pauls further reinforced this concept of the halo effect during the occupant evacuation of high-rise buildings in Canada. The halo effect establishes that people have a territory surrounding them that extends about 6 inches in all directions making a halo 24 inches across the shoulders and 18 inches through the chest to back (see Figure 9.2) In observing people evacuating the high-rise buildings, Pauls noted that occupants had a psychological desire for space and interpersonal separation even during emergency evacuations. He also noted that this would include the wall surface area of the stairway. Occupants would move further from the wall depending on the roughness of the wall textures. The study identified distance changes by lighting, interpersonal relationships, and winter clothing as other environmental factors. All of these factors affected the speed and flow of evacuees. The most important discovery made in the Canada studies was that the maximum occupant flow from the 40 evacuations had a mean of 24 persons per unit of egress width and not the 45 persons that has always been assumed in the building codes based on unit of egress width (see Figure 9.3).[8]

The National Institute of Standards and Technology in their study report on the "World Trade Center Disaster" supported the Canada flow theory. The report identified that the maximum flow rate of World Trade Tower 1 with 7,900 occupants evacuating was 16 occupants per minute per unit of egress width.[9] This number was very much in line with Pauls' and other researchers' studies.

This information indicates three challenges for fire inspectors in the examination of means of egress systems:

1. The exit systems must be immediately available, well lit, unobstructed, and ready for use at all times.
2. Communications systems by fire alarms and exit signage must be effective to motivate people to start the evacuation process quickly. The problem of constant false fire alarms will negate occupants' response and slow evacuation.
3. Fire drills must be conducted on a regular basis to ensure occupant response and to assist in the development of effective emergency evacuation plans.

These challenges will require additional time and training on the part of fire inspectors in the course of performing their job, especially in high-occupancy buildings.

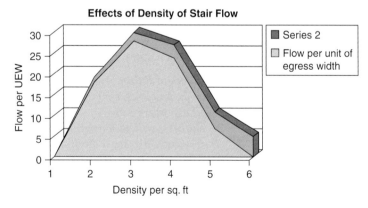

FIGURE 9.3 The Canada studies identified that the exit flow assumptions based on unit of egress width were artificially high as compared to actual flow in 40 high-rise buildings. *Based on: NFPA, Fire Protection Handbook, 15th edition, pages 6–9; Jake L. Pauls, National Research Council, Canada, "Movement of people in building evacuations," Human Response to Tall Buildings, pages 281–292, 1977.*

Safe Exit Design

Modern building codes predicate safe exit design on five basic principles:

1. All of the occupants of the building can be accommodated by adequate means of egress.
2. The available egress will provide for alternate paths of travel to safety.
3. The egress will shelter the occupants from the fire until escape is accomplished.
4. The means of egress will be a well-marked, well-lit, unobstructed path of travel to safety and will be under the control of the user.
5. The exit system will provide a reasonable, risk-free path of travel during normal everyday use.

The model building codes apply these principles to both new buildings and existing buildings under the fire prevention code. It is important for the fire inspector to keep in mind that these basic egress principles, especially #4, "under the control of the user," are critically important to occupant safety. Certain situations or circumstances may restrict user control by time delay or access controls or obstructed or locked exits. The building codes will require that special conditions be followed when egress control devices are installed. These conditions will require posted instruction for the occupant on how to operate the door or how long the delay time will be before the door opens. We will discuss these conditions for application of egress restricting and locking devices later in this chapter.

The building and fire codes generally divide means of egress components into eight general categories:

- Occupancy
- Number of exits
- Exit capacity
- Width of exits
- Travel distance
- Exit marking
- User restrictions
- Fire resistance

Let's examine each of these general requirements and how they are applied to the fire inspection.

Occupancy

The occupancy or use group of the building determines all of the applicable egress requirements in the building and fire prevention code. The building codes may divide occupancy and use into two specific categories: general occupancy and use, and special occupancy and use. General occupancy and use categorizes buildings into ten use groups:

1. *Places of Public Assembly:* These are buildings used for gathering of people for civic, religious, social, recreational, eating, and drinking purposes and awaiting transportation.
2. *Businesses:* These are buildings used for conducting business rendering professional services with small amounts of good and wares on site.
3. *Educational facilities:* These are structures of five or more students for the purpose of education up to the twelfth grade. They may also include day care facilities and vocational training centers.
4. *Factory and industrial uses:* These are facilities engaged in the manufacturing or fabrication of products from raw materials.
5. *High hazard occupancies:* These are facilities that manufacture, process or store materials that are hazardous by nature due to detonation, deflagration, flammability, or health hazard.

6. *Institutions:* These are facilities that house people with physical or mental disabilities due to age or medical conditions; they also include facilities for incarceration of persons who are criminals.
7. *Mercantile:* These typically are buildings that stock and display goods for resale to the public and are open to the general public.
8. *Residential:* These are structures where families or households sleep.
9. *Storage:* These are facilities used to store or transfer goods to other facilities.
10. *Utility:* These are buildings and structures that do not fit into any other category and would include outbuildings, fences, sheds, and other similar structures.

Each of the main building use categories may also have a number of subcategories to more clearly define the building's use. As an example, Places of Public Assembly has five subcategories of use:

1. *A-1 These occupancies are theaters with working stages and proscenium arches. The proscenium arch separates the stage and flyloft from the audience or auditorium.*
2. *A-2 These occupancies are generally nightclubs and entertainment centers with dancing and alcoholic beverages being served.*
3. *A-3 These occupancies are restaurants and diners or lecture halls and meeting rooms.*
4. *A-4 These occupancies are churches and religious buildings.*
5. *A-5. These occupancies are outdoor stadiums and grandstands.*

Residential use is another occupancy category with several subcategories that are based on the occupant's duration of living in the building. As an example, hotels are highly transient and are defined as use R-1. Apartments are less transient but not owner occupied, so they are defined as R-2, and single-family dwellings are categorized as R-3. The lower the category number, such as R-1, the more requirements that exist for the use in means of egress, fire protection, and other requirements of the building codes.

Special use and occupancy covers unique structures or building features that may produce additional fire safety challenges beyond that of occupancy and use. These structures include covered malls, high-rise buildings, atriums, public garages, airport control towers, and open parking structures. These special buildings will also be assigned a general use group but will have additional egress requirements. Understanding the proper occupancy use groups for a building is very important in determining the appropriate code requirements that are to be enforced.

NUMBER OF EXITS

Every building code establishes a minimum number of exits from a building. The minimum number of exits is based on the number of intended occupants. As an example, the International Building Code, or IBC, establishes the minimum number of exits from a building as shown in Table 9.1.

A fire inspector who looks at this table may think, "Well, how come some buildings only have one exit?" The building codes address single-exit conditions by creating exceptions to the two-exit requirement. Single-exit exceptions are usually very limited and affect very small occupancies. As an example, most building codes allow single-exit conditions in a building when the building is under 3,000 square feet and the maximum travel distance to the exit door does not exceed 75 feet at grade. This may vary based on the building use group and occupant load. If any of the conditions established by the exception cannot be met, then two exits will be required for the building. Several state fire prevention codes may also include retroactive provisions for existing buildings related to egress. The key factor for the fire

TABLE 9.1	Minimum Number of Exits
OCCUPANTS	MINIMUM NUMBER OF EXITS PER STORY
0–500	2
501–1,000	3
1,000 OR MORE	4

TABLE 9.2	Table Egress Width per Occupant (inches per occupant)			
USE	**WITHOUT SPRINKLERS**		**WITH AUTOMATIC SPRINKLERS**	
	STAIRWAYS	**DOORS, RAMPS, CORRIDORS**	**STAIRWAYS**	**DOORS, RAMPS, CORRIDORS**
A, B, E, F, M, R, S	0.3	0.2	0.2	0.15
High Hazard	0.7	0.4	0.3	0.2
Institution-1	0.4	0.2	0.2	0.2
Institution -2	1.0	0.7	0.3	0.2
Institution -3	0.3	0.2	0.3	0.2

travel distance ■ Travel distance is the permitted length of travel allowed by building codes in exit access until the exit is reached. Travel distance is calculated in a rectilinear fashion along the natural path of travel in a room or space.

minimum design occupant load ■ The occupant load that determines the minimum number of exits that a building, room, or space requires under a building code.

floor area gross ■ Gross floor is the length times the width of a building, room, or space without deduction of any nonoccupied space. Gross floor area is then divided by the occupant load factor for the building's use group to determine the minimum occupant load for design.

floor area net ■ Net floor area is the determination of occupant loads in buildings, rooms, and spaces based on what can actually be occupied. Net floor area calculations deduct the space of walls, chairs, tables, and spaces that people cannot occupy. The area is then divided by a factor from the building code based on use to determine occupant load. Net floor area is generally used in buildings with large occupant loads like places of public assembly.

inspector to consider is that the building code establishes the minimum number of exits that must be provided in a building. This minimum number of exits generally increases because of **travel distance** restrictions that are also in the codes.

EXIT CAPACITY

The exit capacity generally defines the maximum occupant load permitted within a building. The occupant load calculation also establishes the minimum number of occupants that a room, space, or building must be designed to accommodate. Under building and fire codes, occupant loads serve completely different purposes. In the application of the building code, the occupant load establishes the minimum number of occupants that must be considered in the building's design. The established minimum occupant load not only drives means of egress requirements in the building, but establishes the minimum ventilation system, numbers of bathrooms and parking spaces, and general floor space requirements. Architects must design to these minimum requirements, but may also increase the occupant loads above the minimums provided that all the occupants have adequate egress and the occupancy density does not exceed a maximum number of occupants permitted per square foot of floor space.

The building codes have tables for determining the maximum floor area per occupant that is used in the calculation for the **minimum design occupant load**. This table identifies occupancy based on how the particular space is going to be used and not the use group of the building. As an example, a typical business space requires 100 gross square feet of floor space per occupant, while a restaurant with tables and chairs calculates the number of occupants at 15 net square feet. See Table 9.2.

Key important features of the table are the terms **floor area gross** and **floor area net**. The gross floor area is calculated by multiplying the length of the building with the width of the building or room or space to determine the area in square feet. The area is then divided by the appropriate floor area factor from the table.

> Business Building 100 feet by 150 feet
>
> $L \times W$ = Gross Area
>
> $100' \times 150'$ = 15,000 square feet
>
> 15,000 square feet/100 gross = 150 occupants

EXAMPLE

In this example, the designer must plan for a minimum of 150 occupants in the building design, even if it is only going to have no more than 20 occupants. The designer may also increase above the 150 occupants provided that all other egress and building code requirements are addressed. This is called **increased occupant load**. The occupant load calculation for floor area net is slightly more complicated. The term *net* means actual usable space after all nonusable spaces have been deducted from the gross floor area. These non-usable spaces would include the width of walls, furniture, chairs, or other equipment that takes physical space away from the floor. The net square footage is the area of the floor that can physically be occupied by a person. To simplify the calculation, the code floor area tables offer factors already taking into account used space by tables and chairs, or chairs only. Standing space may be calculated at the maximum permissible number of occupant per 5 net square feet. If the person is sitting in a chair, the factor is 7 net square feet, and if the room has tables and chairs, it is calculated at one occupant per 15 net square feet. The table eliminates the need for measuring these moveable objects and having to deduct them from the available square footage.

Example:

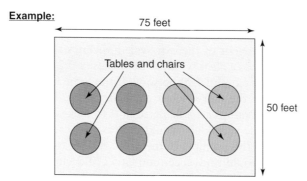

Area: 75' × 50' = 3,750 sq. ft/15 net = 250 occupants (tables and chairs)

The occupant load of a building room or space may also be calculated by a combination method of both the actual count, such as fixed seating in a auditorium and area count based on net square footage. Typically, fixed bench-style seating is calculated at one occupant per 18 inches of seating. Booths are calculated at one occupant per 24 inches of bench. Once the occupant load is established by the building code, it must be maintained under the fire prevention code. The fire inspector must ensure that the established occupant load is maintained over the useful life of the building.

WIDTH OF EGRESS COMPONENTS

The capacity of each component of an exit system is determined based on the occupants that are intended to use them and certain required minimum widths specified in the building codes. As an example, building codes establish a minimum of 36 inches for any corridor, ramp, and stairway when it serves a total occupant load less than 50 occupants. Corridors that serve more than 50 occupants must be a minimum of 44 inches wide or more based on the occupant load they are intended to serve. The building codes establish certain minimum corridor widths for other use groups as well, such as hospitals (I-2) at 96 inches for the movement of beds between fire barriers or educational buildings with 100 or more occupants must have corridors of 72 inches to allow for lockers and standing room. Fire inspectors must be familiar with these minimum width requirements to evaluate the egress system in the inspection of all occupancies. These are established in the building codes in order to maintain a minimum level of safety.

The capacity of egress width per occupant may also be modified in the building codes based on the building having a complete automatic fire suppression system (see Table 9.2). As an example, in the International Building Code (IBC), a nonsuppressed assembly building has an occupant flow rate calculated at 0.3 inches per occupant for stairways and

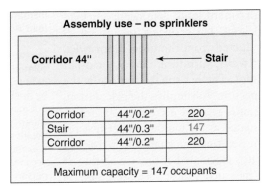

Assembly use – no sprinklers

Corridor 44"   ← Stair

Corridor	44"/0.2"	220
Stair	44"/0.3"	147
Corridor	44"/0.2"	220

Maximum capacity = 147 occupants

FIGURE 9.4 Calculating egress capacity in a nonsuppressed assembly.

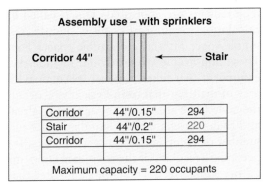

Assembly use – with sprinklers

Corridor 44"   ← Stair

Corridor	44"/0.15"	294
Stair	44"/0.2"	220
Corridor	44"/0.15"	294

Maximum capacity = 220 occupants

FIGURE 9.5 Calculating egress capacity in a suppressed assembly.

exit access ■ This is any portion of a building room or space that a person can travel on the path to an exit.

exit ■ The exit is the fire-protected element of the means of egress or the door leading directly outside the building. Exits usually require a fire-resistance rating and opening protection on interior or exterior stairs because travel distance is unlimited within the exit.

exit discharge ■ This is the outside portion of the exit systems that leads to the public way or a deeded, dedicated area away from the building for occupant safety.

0.2 inches per occupant for all other components such as doors, corridors, and ramps. The most restrictive component of the exit system dictates the exit's occupant capacity (see Figure 9.4). If a door has a clear width of 36 inches at 0.2 inches per occupant, it will accommodate an occupant flow of 180 persons. If the same assembly is protected by an automatic sprinkler system, the stairways are calculated at 0.2 inches per occupant, and all other doors, corridors, and ramps are calculated at 0.15 inches per occupant. In the example in Figure 9.5, automatic sprinklers would increase the capacity of the corridor from 147 to 220 occupants and the example of the 36-inch door calculated at 0.15 inches per occupant, thus the door would now accommodate 240 occupants, or 60 more than the same door in a nonsuppressed occupancy.

The reason the exit capacity calculation allows additional occupants in a suppressed building is because the sprinklers will control the fire, and the occupants will have greater time to escape. This may or may not be a perfect assumption based on the fact that significant smoke generation can still occur even when the building is protected by automatic sprinklers.

EXIT ACCESS TRAVEL DISTANCE

All exit systems are comprised of three distinct and separate components: **exit access**, **exit**, and **exit discharge** (see Figure 9.6). Exit access is all walkable space within a room or building that leads to an exit. The exit access is restricted in the code by travel distance because it is unprotected from smoke and fire. The exit is the fire-rated protected element that takes the occupants from the exit access to the exit discharge. Exits are fire protected and the travel distance within the exit is unlimited. Interior fire-resistant stairways may travel hundreds of stories because they are protected. The exit discharge is the final egress component that leads from the exit to an established area of safety or a public street. These spaces must be deeded and dedicated for that purpose so they may not be built upon. For example, if a building's exit emptied onto a neighboring vacant parking lot that was not part of the original building property, the building owner would require a deeded easement for the exit discharge onto the neighbor's property until it reached the public street. If there were no deed restrictions, the neighbor could build up to their property lot and block the exit.

Travel distance is limited in exit access by the building codes. Building codes sometimes divide exit access travel distance into segments of travel within a room or space and within a corridor. Generally, room and space exit access travel distance is limited to 50 or

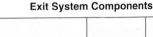

Exit System Components

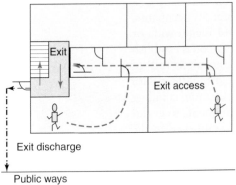

FIGURE 9.6 Main components of means of egress.

TABLE 9.3	Exit Access Travel Distance	
USE	**TRAVEL DIST NON-SPRINKLER**	**TRAVEL DIST AUTOMATIC SPRINKLER**
A, F-1, M, R, S-1	200	250
I-1	NOT PERMITTED	250
B	200	300
F-2, S-2, U	300	400
H-1	NOT PERMITTED	75
H-2	NOT PERMITTED	100
H-3	NOT PERMITTED	150
H-4	NOT PERMITTED	175
H-5	NOT PERMITTED	200
I-2, I-3, I-4	NOT PERMITTED	200

75 feet before two or more exits are required. In the International Building Code, travel distance is measured from the most remote point in exit access to the closest available exit (see Table 9.3). This distance incorporates the permitted travel distance within rooms or space.

The IBC again allows additional travel distance for exit access in buildings protected by complete automatic fire sprinkler systems because fires will be smaller and controlled. The presence of combustible materials is permitted in exit access, so travel distance is limited to 200 feet before an exit must be reached. If the building has an automatic sprinkler system, then an additional 50 to 100 feet is added to the nonsprinklered travel distance.

The fire inspector must also understand how to measure travel distance correctly within a room and space. Measuring the travel distance is not necessarily the shortest distance between two points. The IBC requires a rectilinear measurement from the room's most remote point along the interior walls to the entrance to a corridor. In the example given here, this distance is 20 feet longer than the diagonal distance measured across the room. The concept is that people in an emergency smoke condition may have to use the walls as guidance to the exit, therefore the distance along the walls must be considered in the measurement.

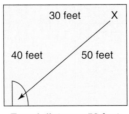
Travel distance 50 feet

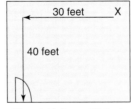
Travel distance 70 feet

COMMON PATH OF TRAVEL

Building codes use the concept of common path of travel in the limitation of travel distance within exit access. **Common path of travel** is defined as follows: "That portion of exit access which the occupants are required to traverse before two separate and distinct paths of egress travel to two exits are available."[10] The concept refines travel distance similar to that of dead-end corridors in that travel distance is limited when the occupant does not

common path of travel ■ The common path of travel is any portion of the exit access that must be traversed where the occupant has only a single choice of direction. Common paths of travel are usually limited by building codes.

have two choices of direction to go to the exit. The idea is to assure divergence before the occupants have traveled too far along the path of egress. Typically, common paths of travel are limited to 30 feet in assembly aisles and 75 feet in business uses. The concept of common path of travel usually establishes multiple exit door requirements in rooms or spaces before occupants enter the corridor where they have two choices of direction to an exit.

THE EXIT

Travel distance becomes unlimited once the occupant has entered the exit. The exit is the protected component that has a fire-resistance rating and proper opening protection. The exit door must be self-closing and positive latching. The fire-resistance rating on the exit door must be no more than 30 minutes less than that of the enclosure walls; typically these are "B" labeled 1–1½ hour fire-rated doors. Generally, interior exit stairs are required to have at least 1-hour fire resistance rated up to three stories and a 2-hour fire-resistance rating for four stories and above. The exit can be used for no purpose other than means of egress. Fire inspectors should make sure they are free of hazards such as mops, buckets, ice machines, chairs, ashtrays, ladders, and trash receptacles. There should be no combustible materials in the exit.

THE EXIT DISCHARGE

The exit leads to the portion of the system called the exit discharge. The exit discharge is usually outside the building but still on the property. The exit discharge will require emergency lighting in almost all building use groups. The exit discharge, like the other egress elements, must be an accessible and unobstructed path of travel. The fire inspector should inspect the exit discharge for locked gates, trash cans, or other obstructions in the path of travel as well as even walking surfaces. The exit discharge terminates at the public street or another deeded and dedicated public space away from the building. Fire inspectors must make sure that they inspect the entire exiting system and that the egress is clear and accessible all the way to the public way.

REMOTENESS OF EXITS

When more than one exit is needed, the building code requires that the exits be remote from one another to maximize the occupants' safety. The general test for remoteness of exits is to measure the diagonal distance of the building, room, or space and divide that by 2. The exits must be at least half the diagonal distance apart to be remote. In buildings protected with automatic sprinklers, the diagonal distance for remoteness is usually reduced to one-third the distance.

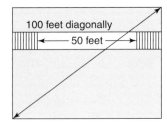

EXIT MARKINGS

All exit systems must be properly marked when more than one exit is required. Exit signs and directional exit signs must be located along the path of travel in order to ensure occupant direction toward the exit (see Figure 9.7). Exit signs must be listed by Underwriters Laboratories and may be either internally illuminated or self-luminous if they meet the requirements of the UL 924 *Emergency Lighting and Power Equipment* standard. Exit signs are required to be illuminated at all times and must provide a minimum of

FIGURE 9.7 Enlarged
exit signs used in facilities
with large floor areas.
These assist in identifying
the means of egress.
Courtesy of J. Foley

5 foot-candles (54 lux) of light. A foot-candle is the measurement of light from a candle on a surface 1 foot away and 1 foot square in area. This is also referred to as a lumen. A lux is the metric equivalent of the foot-candle. The conversion from foot-candles to lux is 1 foot-candle equals 10.764 lux. The light measurement can be taken with a light meter on the illuminated walking surface. Exit signs usually require 6-inch lettering with ¾-inch strokes and may be either red or green in color. The placement of directional exit signs is usually at any change in corridor direction or not more than 100 feet apart or no further than the listed viewing distance of the exit sign. In 1968, Underwriters Laboratories published *Bulletin of Research 56*, which evaluated exit sign visibility 12 feet away from the sign under smoke conditions. The results of the report concluded that most exit signs would be obscured within 5 to 12 minutes once smoke entered the corridor area.[11] Current building codes are starting to require the installation of photo-luminescent egress path marking in high-rise building stairways (see Figure 9.8). Photo-luminescent marking

FIGURE 9.8 Photo-
luminescent exit markings
installed in exit stairways.
Courtesy of J. M. Foley

FIGURE 9.9 Exit corridors must be well lit at all times and must have emergency lighting. *Courtesy of J. M. Foley*

systems were installed in the World Trade Center after the 1993 bombing and were credited in assisting people out of the building on September 11 after other life safety systems were critically damaged by the aircraft hitting the building.

The current 2009 International Building Code requires the installation of these photo-luminescent low-level lighting systems in new high-rise buildings in the exit stairways. The photo-luminescent markings must be placed on the stair nosing, handrails, stair edges, any obstacles like standpipes, and around door openings so that people can evacuate rapidly even if the emergency lighting systems in the building fail. These systems must provide a minimum of 1 foot-candle on the walking surface for the first 60 minutes in accordance with the UL 1994 standard.

In addition to the installation of exit signs, the means of egress must be normally illuminated and must be provided with emergency lighting from a secondary power source (see Figure 9.9). The secondary power may be an emergency generator or batteries within the individual emergency lighting units. Emergency lighting secondary power supplies are required to provide 60 minutes of power at 1 foot-candle (11 lux) at the walking surface. Places of public assembly are permitted to reduce the normal egress lighting during a show to 0.2 foot-candle, provided that the lighting returns to normal levels during emergencies or upon activation of the fire alarm or sprinkler system. The testing of emergency power supplies by fire inspectors is an important aspect of occupant safety.

Emergency generators must be tested under load annually and in accordance with NFPA 111, *Standard on Stored Electrical Energy Emergency and Standby Power Systems*. Fire prevention codes require the testing of emergency generators to ensure the operation of both the generator and the automatic transfer switches, or ATS. Generally, these tests must be under load and power transfer should occur within 10 seconds or less of primary power failure. If the ATS switches are not physically tested on a regular basis, they may malfunction or hang up during a real emergency.

Another important aspect of emergency lighting is to ensure that the exit discharge is properly covered by emergency lighting on the path of travel to the public way. Exit discharges are usually required to have emergency lighting in all use groups but the utility use group.

FIRE RESISTANCE AND USE RESTRICTIONS

Fire resistance requirements generally increase along the path of egress travel. In most cases when occupants leave a room or space, they enter a corridor. Because choices in a corridor become limited, they must go either right or left. Corridor walls require a fire-resistance rating, and doors along the corridor have to be self-closing and also carry a fire-resistance rating. There are exceptions to this in buildings protected by automatic sprinkler systems where the corridor may not require a fire rating at all.

Corridors are still considered part of exit access path and therefore will have limited travel distance. These limitations along with the remote exit requirements will create a redundancy in the exits within the building. In some buildings, dead-end corridors may exist and are usually limited to the maximum travel distance of 20 feet unless fire alarms or automatic sprinkler systems are installed. Exit access corridors are permitted to have furnishing and other combustible materials present, provided that they do not encroach on the required minimum effective egress width of the corridor. Interior finishes will also be restricted in the corridors based on the flame spread rating and smoke generation of the interior finish material. Flame spread is usually tested in accordance with the Steiner tunnel test or ASTM E-84, *Standard Test Method for Burning Characteristics of Building Materials*. The test requires a 19.5-foot sample that is placed in a horizontal position at the top of a 25-foot-long tunnel. The flame spread is measured as it progresses down the sample away from the ignition source. The sample is then compared to red oak, which has a flame spread of 100 feet. The exit access corridor terminates at the entrance to the exit. As stated earlier, exits may not be used for any other purpose than means of egress. Fire department standpipe and hose connections are permitted in the exit but must be outside the effective travel width of the stairs. The fire-resistance rating of the exit is based on the type of construction and the building height. The exit will always have a higher fire-resistance rating than the exit access corridors.

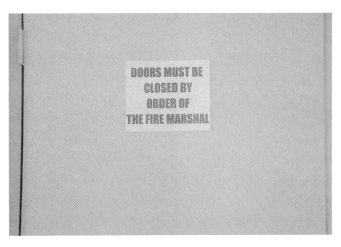

FIGURE 9.10 Fire doors must be kept closed at all times to protect the exit. *Courtesy of Damian Moosang*

The fire doors leading to the exit must be listed and labeled to meet the requirements of NFPA 80, *Standard for Fire Doors and Other Opening Protectives*. Fire exit doors must remain closed at all times to prevent smoke passage into the exit system (see Figure 9.10). The fire-resistance rating of fire exit doors should be equal to, or 30 minutes less than, that of the wall they are installed in. As an example, a 2-hour fire-resistant stair enclosure would require "B" labeled 1–1-½-hour fire doors. All of the door hardware, including the automatic self-closer, the handles, and the crash or panic hardware, must be listed and labeled for fire door application. Fire doors traditionally have been labeled based on the types of opening they protect in a firewall or a fire separation wall. Typically, for a firewall separating buildings or fire areas within a building, this is a class A opening that requires a class A fire door of 3-hour fire-resistance rating. Openings in vertical shafts, stairways, and elevators rated at 1–2 hours fire resistance are class B openings. Class B openings require "B" labeled 1–1-½-hour fire doors. Openings in fire separation walls of moderate fire exposure are considered class C openings. Examples would be storage, mechanical, or heater rooms. Class C openings require 45-minute fire doors or fire windows.[12] Corridor doors in fire separation walls with low fire exposure risk are considered class D openings and require only 20-minute "D" labeled fire doors. Fire door labels should be legible and should never be painted over.

All labels on fire doors are installed by the door manufacturer and are not to be field installed by the door supplier (see Figure 9.11).

Fire inspectors also need to inspect the exit enclosure for any penetrating items such as pipes, electrical conduits, ventilation ducts, smoke removal ducts or other building systems. All building systems that enter an exit must be maintained by the proper fire stop meeting the required fire resistance rating of the enclosing wall.

Exit stairways in high-rise buildings must also have systems to prevent the intrusion of smoke into the stair. This may be accomplished by either a smokeproof enclosure or pressurizing the stairway with a higher pressure than that of the corridor.

A smokeproof enclosure requires either a ventilated vestibule that separates the exit stair from the corridor or an outside balcony that takes the occupant outside before entering the exit stair. Smokeproof enclosures were the predominant method of stairway protection before modern ventilating and fire alarm system technology. Because the

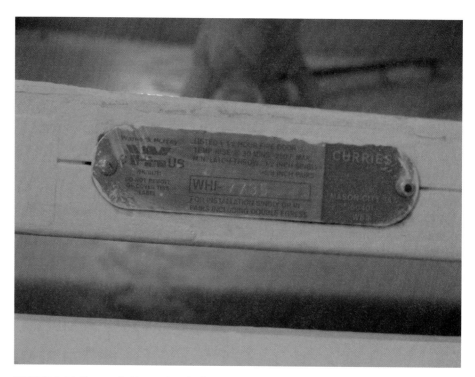

FIGURE 9.11 Fire marshal inspecting the fire door label in exit stairway. *Courtesy of J. Foley*

smokeproof enclosure is physically separate from the building, it is expensive to construct and generally has been replaced by internal stairs with mechanical **stairwell pressurization**. Modern high-rise buildings rely upon stairway pressurization to keep any smoke out of the exit. Pressurization requires the installation of pressurization fans and barometric dampers to control internal stair pressure. The pressurization system is connected to the building's automatic fire alarm system as an auxiliary function that operates in fire alarm mode. The design of these systems is very difficult because of stack effect. Stack effect is the normal movement of air in a tall structure across a neutral pressure plane. Stack effect changes with temperature differentials between the inside air temperature of the building and the outside air temperature. The stack effect is what creates strong air movement or winds around tall buildings. Typically, air moves into the structure below the neutral pressure plane and out of the structure above the neutral pressure plane. This air movement may force smoke into exit stairs via small openings even when doors are closed. The system design engineer must achieve a delicate balance that maintains a higher pressure in the stairway than the corridor but still allows the exit doors to operate without excessive force by the occupants. Most stairway pressurization systems use a single fan and ducted supply openings every three to five floors. The installation of stairway pressurization systems is covered in NFPA 92A, *Standard for Smoke-Control Systems Utilizing Barriers and Pressure Differences*. Generally, the building code requires a minimum stairway pressure of 0.5 to 0.15 inch of water column measured by a manometer. The maximum pressures should not exceed the minimum door force permitted by the building code to open an exit door, which is usually between 8 and 15 pounds. As an example, a stairway pressure of 0.4 inch of water column would require 30 pounds of force to open the stairway door. This would be unacceptable pressure. The maintenance and testing of stairway pressurization systems by the building owner is critical. If exit doors do not close properly, pressure is lost, or if barometric dampers fail to open properly, the stair may overpressurize, making doors difficult to open or causing them to slam shut on the occupants. The fire inspector should require the recertification of these systems by a certified testing company as part of the annual fire inspection.

Other Egress Elements

The model building codes may also allow other types of egress components such as horizontal exits, fire escapes, revolving doors, sliding doors, and access control doors to be used as means of egress. Let's examine each of these components and their use in an egress system.

HORIZONTAL EXITS

Horizontal exits are exits that pass through a fire-resistant wall to an area of refuge on the opposite side. The horizontal exit must have a fire-resistance rating of least 2 hours with the opening protection of fire-rated doors and fire dampers where the wall is penetrated. The area of refuge on either side of the horizontal exit must be capable of safely accommodating the normal occupant load plus the evacuating occupant load. Horizontal exits are usually restricted to 50 percent of the required exits off a building floor although there may be exceptions in the institutional use groups.

horizontal exit ■ The horizontal exit is an exit that leads to an area of refuge through a fire-resistant wall. Most building codes allow only 50 percent of the floors exits to be horizontal exits.

FIRE ESCAPES

Fire escapes have existed since 1887 when the first U.S. patent was issued to Anna Connelly. Anna Connelly made a rudimentary metal exterior fire escape that was low cost and could be used on both new and existing buildings. The use of fire escapes is no longer allowed on new buildings by the building codes. All new building exits must be either interior or exterior exit stairways. Fire escapes have been eliminated because of safety issues, which include corrosion, weather protection, size, and the fact that people used them for outdoor gardens and places to hang laundry. Fire escapes may exist on many older buildings, and are permitted to continue to be used provided that they are safe and properly maintained. Fire escapes were originally designed to allow escape of occupants in buildings that traditionally had open interior stairways. The fire escape also provided the fire department with access as well as providing a path of escape to the apartment occupants. Most fire escapes are 22 inches wide and may have a counterbalanced stair or drop ladder at the last landing before grade. The fire prevention code identifies the maintenance requirements of fire escapes, which include painting, corrosion protection, proper structural connection to the building and operable and accessible fire-rated windows along the path of travel. Generally, most early fire escapes did not require fire-rated windows because they usually served only one apartment on each floor level. Fire escapes that serve multiple apartments on a floor may require opening protection of the windows. Fire escapes were no longer considered a viable primary means of egress after the Triangle Shirtwaist fire in New York City that killed hundreds of garment workers. Fire escapes did continue to be used as secondary means of egress, and can be found all around the country. Fire inspectors must examine the fire escape for rust, weakness, and failing connections and ensure they are safe and usable. An unsafe fire escape should be evaluated by a structural engineer, and any improvements should be made as the engineer specifies (see Figure 9.12).

Under today's codes, exterior stairways have replaced fire escapes as a means of egress.

FIGURE 9.12 Fire escapes may deteriorate in weather and require constant maintenance. *Courtesy of J. M. Foley*

Exterior stairways are required to be protected by fire-resistant exterior walls, and all openings must have fire-rated windows and doors. The fire-resistance rating must extend out 10 feet vertically and horizontally from the exterior stairway.

REVOLVING DOORS

After the Cocoanut Grove fire in Boston in which hundreds of occupants died as the result of a revolving door, building codes have limited use of revolving doors as a means of egress element. The IBC restricts the use of revolving doors to 50 percent of the exit capacity and restricts the capacity to 50 occupants per door. The revolving door must also collapse in the direction of egress travel when subject to 130 pounds of force being applied 3-1/2 inches in on the door wing. Revolving doors must also have an out-swinging exit door of equal exit capacity located within 10 feet of the revolving door.

SLIDING DOORS

A sliding door may be placed in the path of egress provided that it is fire rated and has both automatic and manual controls. Sliding doors must be able to be opened without any special knowledge on the part of the occupant. The sliding door must open with no more than 15 pounds of force at the handle or a force of 250 pounds force applied perpendicular to the door's surface. The sliding door must be provided with emergency power and must be electronically supervised by the fire alarm system. Automatic sliding doors must reopen to the minimum egress width within 10 seconds of activation by the occupant. These doors often are used in entrances to buildings or to separate areas like atriums for smoke control purposes. Sliding fire rate doors may also be used to separate mixed-use groups, as fire barriers, or for vertical shaft enclosures.

ACCESS-CONTROLLED EGRESS DOORS

Access control is permissible under most building codes provided that a multitude of safety requirements are met. As you may remember, one of the most important aspects of safe exiting is that the egress is under the control of the user. Access control removes some of the occupant control for the purpose of security or hands-free ingress and egress. These doors may be encountered in a mercantile use, like a supermarket, or in exit stairways in multistory business buildings. The IBC requires six safety conditions to be met for access-controlled doors:

1. The door must have a sensor that opens it on approach from the egress side of the door. The door must unlock on a signal from the fire alarm, or loss of power to the sensor.
2. The door must be fail-safe in that loss of power to any component will unlock the door.
3. The door must have a manual unlocking device located no more than 4 feet above the floor and within 5 feet of the door. The manual pull station must be clearly identified by a sign and instructions on how to operate it, and once it operates, the door shall not relock for at least 30 seconds.
4. The activation of the building fire alarm shall unlock the door and it will remain unlocked until the fire alarm system is reset.
5. Activation of the automatic sprinkler system shall unlock the door and it will remain unlocked unit the fire protective signaling system is reset.
6. Assemblies, mercantile, business, and educational use groups may not be secured from the egress side during periods that the building is open to the general public.[13]

The fire inspector should identify the type of access control doors in buildings and test them for proper operation. This is especially important on exit stairway doors. The fire inspector should make sure the doors do not relock immediately, as occupants could be trapped in a smoke-filled stairway.

Occupant Loads

As discussed earlier, there are several different definitions of the term *occupant load* and how and why it is calculated. Building codes use occupant loads to determine minimum occupancy requirements in a building. The building codes also allow the increasing of occupants within the building provided that sufficient egress is provided and that the maximum permitted occupant load is not exceeded. The fire prevention code takes a different perspective in that the established occupant loads and all egress system components determined by the building code must be maintained. The fire prevention code also requires the posting of occupant loads, which is a number that cannot be exceeded without the fire marshal's or fire official's permission. Let's examine each occupant load and how it relates to exit safety.

DESIGN OCCUPANT LOAD

The design **occupant load** is calculated using the building code for the express purpose of establishing a minimum number of occupants that the building and the associated building systems must accommodate and support. Even if the designer did not intend to accommodate that minimum number of occupants, the building must still be designed to that minimum occupant load number. Design occupant load is basically a starting point for minimum building design. The design occupant load is calculated by dividing the square footage of the building by the maximum floor area allowance table in the building code (see Table 9.4). The table will express that floor area in either gross square feet or net square feet based on the building use.

As an example, for a 30,000-square-foot office building and with a maximum floor area table for business use of 100 gross square feet, we would simply divide 30,000/100 to get a design occupant load of 300 persons. This number becomes the minimum design occupant load for the building. In the case of net square footage, the designer would divide only occupiable space by the net square footage factor from the allowable area table. The fire inspector should be able to get these occupant load numbers from the local building department records on the property or by reviewing the building's approved egress plans (see Figure 9.13).

INCREASED OCCUPANT LOAD

Increased occupant load allows the building designer to provide more occupants above the minimum design number in the building based on actual count or a combination of the design occupant load and additional occupants. Increased occupant loads, once established, must be provided with properly sized means of egress and other building systems necessary to support the increased occupants. The occupant load may be increased up to the maximum permissible occupant load as long as all other conditions are satisfied in the building code.

TABLE 9.4	Maximum Floor Area Allowance per Occupant
OCCUPANCY	**FLOOR AREAS**
Assembly – fixed seats	Number of seats
Assembly – chairs only	7 net
Assembly – standing room	3 net
Assembly – tables and chairs	15 net
Business area	100 gross
Industrial area	100 gross
Education – classrooms	20 net
Education – shops	50 net
Mercantile basement and grade	30 gross
Mercantile – areas on other floors	60 gross
Mercantile – storage areas	300 gross
Institution – inpatient treatment	240 gross
Institution – outpatient	100 gross
Institution – sleeping areas	120 gross
Parking garages	200 gross
Residential	200 gross
Storage areas, mechanical rooms, equipment rooms	300 gross

occupant load ■ The occupant load of a building is the permitted number of persons that can occupy a building, room, or space.

FIGURE 9.13 Fire marshal reviewing occupant egress drawings for public assembly. *Courtesy of J. M. Foley*

MAXIMUM OCCUPANT LOAD

The **maximum permitted occupant load** is established as the maximum number of people that can be accommodated effectively by the means of egress and does not exceed one occupant per 5 net square feet of occupiable space. Again, all elements of the means of egress must accommodate that maximum occupant load. The model building codes are designed to provide performance objectives for determining occupant loads, but not to restrict the ability of the building designer to provide a large number of occupants as long as sufficient exits are provided in the design. The maximum permitted occupant load can be affected by several different component factors. These limiting factors may include door and corridor capacity, travel distance to exits, main entrance capacity, lack of automatic sprinklers, and other building code requirements that are driven by occupant load, such as plumbing and ventilation systems.

POSTED OCCUPANT LOAD

Under the fire prevention codes, occupant loads take on a somewhat different meaning. Fire prevention codes deal with the means of egress based on the requirements established by the building code used at the time of first construction or occupancy. Many states also have retroactive provisions dealing with older buildings and requiring them to be brought up to a more compliant safety standard but not necessarily the current levels of the adopted building code. Fire inspectors and fire marshals are required to establish a **posted occupant load** for certain buildings, rooms, or spaces within their jurisdiction. The posted occupant load may be the occupant load established by the building code at the time of occupancy or may be a lower number based on actual use of the building. All fire prevention codes require the posting of assembly room occupant loads. The posting may require the inspector to determine the maximum number of occupants for different room configurations such as standing room, chairs only, and tables and chairs. Whatever number of occupants is finally posted, that number becomes the permitted occupant load for that space and cannot be exceeded without permission from the authority having jurisdiction. Posted occupant loads may not be increased above the maximum permitted occupant load for a room or space that was established by the building code.

SPECIAL PUBLIC ASSEMBLY EXIT REQUIREMENTS

Generally, assembly buildings with occupant loads greater than 300 persons require that the main entrance or exit to the building accommodate 50 percent of the total occupant load. As an example, in an assembly building if the main exit can accommodate only 200 persons, the maximum permitted occupant load for the building is 400 persons based on the 50 percent rule. The main exit would be a limiting factor to increasing the occupant load regardless of any other available exits. In addition to the 50 percent main exit requirements, all other exits must accommodate 66⅔ percent of the total occupant load in the space. The general concept is that in public assembly uses, the exit door capacity always exceeds the permitted occupant load of the space by 17 percent once 300 occupants is exceeded. It should be noted that the NFPA 101, *Life Safety Code*® in 2006 and the IBC in response to the Station Fire in Rhode Island altered the main exit requirements. The *Life Safety Code* now requires the main exit to accommodate 66⅔ percent of the occupants, and the remaining exits must accommodate 50 percent of the total occupant load. This change provides 17 percent greater capacity at the main exit. The IBC applies the 50 percent main exit rule to A-2 nightclubs with 100 occupants and the 66⅔ percent rule to A-2 nightclubs of 300 or more occupants.

Determining the Adequacy of Exits

Fire inspectors are required to understand how occupant loads are calculated in order to determine when a building's room or spaces become overoccupied. When inspecting the exit system, the fire inspector must review the entire system, looking at all components of exit access, exit, and exit discharge. To conduct an effective inspection, the fire inspector should do the following:

1. Examine the room and all connecting doors, corridors, stairs, and ramps as a system. Remember, exits must go somewhere and end at the public way. Look for the restricting or minimum required element widths that may reduce the occupant flow, such as 36-inch corridors that can serve only 50 occupants.
2. Measure from the most remote areas of the space to the exits and ensure they are within the code-permitted travel distance. Again, look for restrictions such as stair or door widths or fixtures obstructing the exit access, keeping in mind common path of travel restrictions and exit convergence by additional occupants.
3. When measuring exit component width, remember that only clear width counts. Doors are measured from doorstop to doorstop, or the face of the door if it opens 90 degrees to the doorstop. Effective clear width is what will be divided by the appropriate inches per occupant factor, not the actual door size.

The inspector must also use the correct component factor depending on whether the building has a complete automatic sprinkler system or not. Make sure all of the egress components collectively can handle the total occupant load within the space. In public assemblies remember that the 50 percent and 66⅔ percent rules for the main entrance and exit must be considered in the occupant load.

Finally, inspect the safety aspects of the exit system, exit signs, emergency lights, panic hardware, automatic door closers, obstructions, locking devices, natural lighting, and surface transitions. The inspector should also examine the interior finishes and any draperies or decorations that may be required to be fire retardant or fire treated. Make sure that all access–delaying devices are properly tested and work.

OVERCROWDING

One of the most difficult situations fire inspectors are expected to deal with is overcrowding of public events. Often, fire inspectors are contacted by police officers or other government agencies to respond to a situation of overcrowding. Fire prevention inspectors should have a plan in place for such emergency occurrences. The first step in developing a plan for overcrowding should be preemptive by first identifying the types of buildings in the jurisdiction and the types of events that occur that have a potential for overcrowding conditions. These may be concerts, sporting events, public meetings, visiting dignitaries, or social gatherings. The fire inspector should provide these venues with information on how your agency will deal with overcrowding and how they can manage their facilities to avoid any unpleasant experiences. Providing information for special event planning has two important functions. First, it notifies the building owner of the enforcement agency's expectations to control crowds and identifies the consequence of failure to manage their establishment safely. Second, it provides the enforcing agency with an opportunity to assist the building owner in compliance before any overcrowding incident can occur. In large cities, the fire prevention bureau may have to deal with hundreds of special events annually, but smaller fire departments also have events that occur in their jurisdictions that can have equally devastating outcomes because of overcrowding. One way to effectively manage these events is through the fire safety permit process. Requiring permits for large public events by either the fire department or the municipality provides advance information to the fire official and allows municipal agencies to work with venue operators before difficult situations such as overcrowding can arise. Requiring permits provides critical time to plan for special events, identifies potential problems, and assigns fire marshals or additional security personnel to the event to maintain safety.

Fire prevention codes provide the authority to fire inspectors to deal with overcrowding situations. The fire inspector must use discretion, however, to avoid risking adversarial actions that can escalate into unruly crowd behavior and injuries. Fire inspectors generally have the authority to request that the room or building be emptied and a recount of the occupants be conducted upon reentry if they suspect the building is overcrowded. This approach works, provided that everyone is cooperative, which often isn't the case. The fire inspector can use this approach in situations such as open public meetings or hearings where people generally are in a cooperative mindset. In overcrowding situations involving alcohol consumption or politically charged issues, the approach of the fire inspector requires more discretion and patience. The fire inspector should first request assistance from the local police department; these situations can get very difficult and require some restraint to resolve. The fire inspector should require the owner or operator of the facility to first institute crowd control measures by not permitting any further occupants to enter the facility. The second step is to ensure that as occupants leave, they are not permitted reentry until the posted occupant load is reestablished. The fire inspector needs to remember that the fire code places the burden of controlling occupancy on the owner of the establishment and not the fire inspector. The fire inspector is required to use due diligence under his or her authority and impose whatever remedies are available to him or her by code regulations. In some states, for instance, substantial fines can be imposed as well as punitive closing of the establishment for up to 60 days for multiple incidents of overcrowding. Generally, these fire code penalties escalate based on repeated occurrences where the owner has not lived up to their responsibility to maintain a safe business establishment.

The fire inspector's duty in overcrowding situations is to remedy the situations to maintain a safe environment for the occupants; discretion must be exercised, however, so that your actions do not increase the risk in certain volatile situations. Advance planning and education of facility operators in crowd control can decrease the likelihood of these potentially dangerous situations.

Special Locking Arrangements

As a general rule, locks and latches on egress doors are prohibited by the fire prevention code. The driving concept is that doors must be operable without the use of a key or any special knowledge on how to operate the door. Locking devices such as deadbolts, chains, and bars are prohibited on exit doors. Locks may be installed with approval of the fire marshal of fire official on exit doors, but each use group will have certain exceptions or conditions of the lock's use, and in all cases, they must be installed with building code approvals. The fire prevention code will require in all cases that the exit doors always be unlocked when the building is occupied.

All model building and fire codes have specific requirements for the installation and use of locking or delay-restricting egress controls. In many buildings, such as correction facilities, juvenile detention centers, mental care facilities, and pediatric hospitals, these devices are necessary for the protection of society or the patients. Building codes allow locking and delayed egress devices provided that they comply with all of the code rules and provide additional safeguards. As an example, a bank closes at 3:00 p.m. to the public, and the front door then is locked. Usually, a security guard is posted until the last of the public leaves, then the door is locked. The bank employees are still in the building finishing the day's work; to the fire inspector, will this be a violation of the fire code? Under the IBC, at the discretion of the building code official, a key may be left in the lock on the main bank door as long as the public has left the building. The lock must be readily distinguishable as being locked and a sign must be in place that the door must be unlocked when the public occupies the building. Many building officials may also allow a single-action thumb latch on the main door so that in an emergency, the door can be readily opened. NFPA 101, *Life Safety Code®*, allows the locking of the main entrance door provided that the occupant load is less than ten persons in certain occupancies. A good example of this is when companies have outside janitorial services or a watchman in the building during closed business hours or weekends; the door to the building will be locked from the outside.

Special locking devices are also common in hospitals, especially on pediatric floors. These electric security locks may be on the stairway doors or the fire barrier doors to restrict access to the floor. These locks are necessary so that children are not taken from the facility. Building codes require that special locking devices be connected to, and unlock upon a signal from the automatic fire alarm and automatic sprinkler systems in the building. The locks must fail to the safest possible position on loss of primary power and must be capable of being remotely operated from a fire command center, the nurse's station, or another location that is continually occupied. Special locking evacuation procedures must also be addressed in the emergency operations plan that is required by the code to be submitted to the fire marshal or fire official for approval. All staff in the secured area must be provided with keys or access codes to unlock the doors. Building codes also require emergency lighting at these secured doors.

In commercial applications, special locking devices may also be employed to prevent employee theft or unwanted access from other floors via the fire stairwells. The IBC allow egress delay devices in all use groups with the exception of public assemblies and high hazard use groups. To properly install an egress delay device, it must meet the following six code compliance conditions:

1. The door unlocks on actuation of the automatic fire alarm or automatic sprinkler system.
2. The door unlocks on loss of power to the controlling mechanism.
3. The door can be unlocked upon a signal from a fire command center.
4. When a force of 15 pounds for 1 second is applied to the door, an irreversible process begins that will open the door within 15 seconds (30 seconds maximum permitted delay). The door may not relock and shall remain open.

5. The process shall activate an audible alarm in the immediate door area.

6. A sign must be placed on the door 12 inches above the hardware that says, "Push Until Alarm Sounds, Door Will Open in 15 Seconds."

Time delay for the door releasing device is not permitted to be field adjustable. Fire inspectors need to pay close attention to the installation of any special locking or delay restricting devices and ensure that they are properly installed with permits and approvals from the local building department. Often, locksmiths install these devices without the knowledge of the building code officials, and those locks do not meet the required fire code standards. Fire inspectors need to emphasize that because these special locking devices take away user control, they must meet building code safety requirements. Fire inspectors should test and witness the operation of these special locking devices during their annual fire inspections to ensure proper operation.

Places of Public Assembly

Every building that is inspected has egress elements to be examined; however, places of public assembly are by far the most challenging for the fire inspector. Public assemblies have large occupant loads and involve activities that attract collective populations. A collective population is one that has low cohesion because they do not know the other people around them in the room or space. Collective populations have no formal leadership, and in fact during emergencies will depend on emergent leaders to provide direction on what to do. Collective populations are quite different from formal populations or groups like those found in school settings where the teacher or authority figure directs the students to the appropriate exit and meeting place during a fire drill. Collective populations are by far the most difficult groups to get to evacuate in emergencies. The reasons for this difficulty include leadership gaps, poor ability to communicate, and lack of interpersonal relationships among the group. Each individual must determine the risk to themselves. Collective populations also exhibit mimicking behavior following the actions from others in the crowd. This is why we see crushing incidents at large or crowded events like the E-2 nightclub incident in Chicago, where 21 people were killed and 50 were injured. This is why fire inspectors must pay attention to detail in reviewing exiting plans and evacuation procedures for places of public assembly. In large crowd venues, the security staff must be adequately trained and supervised in how to react to crowd incidents. Security must use proper demeanor not to incite the crowd and cause further problems. Exit paths must be redundant and direct to disperse the crowd and not overload the egress elements. The fire inspector should beware of any events that have no ticketing, as these events can easily become overcrowded such as indoor craft shows, flea markets, or free concerts.

Fire inspectors must regularly inspect public assembly venues to ensure that exit code violations are addressed and proper crowd management is being performed.

Assembly uses may have fixed seating, such as a movie theater or showroom, or may have tables and chairs, such as a ballroom or lecture hall. People sitting in these seats must have access to the exits by direct routes. These direct routes are provided by adequate **aisle accessways** and exit **aisles**. The fire inspector must be versed in the applicable code rules that apply to these particular portions of the exit system. Often, the fire inspector has to inspect a large-venue floor layout with chairs, or tables and chairs, in hotels, social halls, fire halls, and banquet facilities. The questions of aisle accessway, cross aisles, and exit aisles must be properly addressed.

The International Building Code uses a concept of catchment areas to establish a balance in aisles and aisle accessways. A catchment is defined as the portion of a room or space that is naturally served by an aisle access way or aisle to the closest exit. The catchment area assumes a balanced occupant flow to all of the available means of egress. Typically, a room or space is divided into quadrants or sections based on the number of

aisle accessway ▪ The aisle accessway is a component within the exit access of a room or space that a person must traverse between seats or tables and chairs to reach an aisle.

aisle ▪ The aisles are a component of exit access within a room or space that leads to the exit access corridor or exits within the room or space.

room exits available. The tables and chairs would then be spaced so that occupants have aisle access-ways, which lead to cross aisles, or exits, which lead to the exit doors in the catchment area. The means of egress elements increase in size as more people enter the exit pathway. An aisle accessway or aisle is mea-sured at a point 19 inches perpendicular from the table. This 19-inch space allows seating room for a chair without impacting the aisle or aisle accessway. As an example, for an aisle accessway serving one table of four occupants that has a travel distance from the furthest seat of less than 6 feet, there is no mini-mum aisle accessway width; however, 19 inches is still required from the edge of the table. If another table were across from it, the minimum space between the tables is (19 + 19) inches or 38 inches. If the table is now 8 feet long, adding two additional occupants, the minimum aisle accessway width of 12 inches must be added between the tables. This makes the width between tables now (19 + 12 + 19) inches or 50 inches (see Figure 9.14). The aisle width must be uniform to

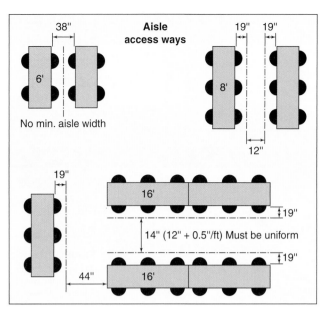

FIGURE 9.14 Diagram of minimum aisle accessway widths.

the farthest point in the aisle accessway. The 12-inch aisle accessway minimum width is maintained until the travel distance reaches a point of 12 feet. At the 12-foot point, an additional 0.5 inch must be added for each additional foot traveled until a maximum of 22 inches is reached. Aisle accessways may not exceed 30 feet in a common path of travel. At that distance, they must connect to a cross aisle or exit aisle where the occu-pant has a choice of two directions.

The aisle accessway terminates at the cross or exit aisle. The number of aisle access-ways and occupants within the catchment area determines the required cross or exit aisle width. All aisles, however, must comply with the minimum width requirements for aisles of 36 inches for those serving less than 50 occupants and 44 inches for those serving more than 50 occupants.

The cross or exit aisles are required to be increased in size if they exceed the carrying capacity specified for a 44-inch aisle. The fire inspector can determine the required aisle width by taking the total occupant load of the catchment area using the aisle and multi-plying it by the inches per occupant factor from the building code. The inspector must use the correct factor from the table based on whether the building is completely suppressed with automatic sprin-klers or not.

aisle convergence
■ Aisle convergence is the adding of occupant loads together that converge on a single path of travel to an exit within the room or space.

AISLE CONVERGENCE

Aisle convergence must also be accounted for within each of the assembly catchment areas. Convergence occurs if two or more aisles lead to the same exit along a common path of travel where the occupants must all go in the same direction. The paths of the converging aisles must then be added together (see Figure 9.15). The converging aisles must lead to exit doors or exit access corridors comparably sized to accommodate the total converging occupant load. If a cross aisle provides a choice of two or more exits, than the aisles are not converging.

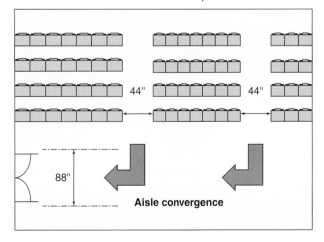

FIGURE 9.15 Diagram of aisle convergence, one-way travel.

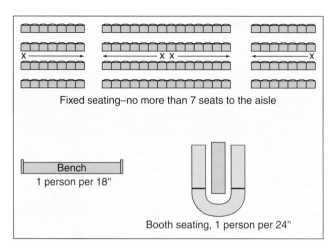

Fixed seating—no more than 7 seats to the aisle

Bench
1 person per 18"

Booth seating, 1 person per 24"

FIGURE 9.16 Fixed seating is calculated as one person per seat or 18 inches on benches and one person per 24 inches in booths. In theater seating, persons should not cross more than six seats to the aisle.

ASSEMBLY WITH FIXED SEATING

In occupancies such as gymnasiums or movie theaters, it is easy to determine the occupant load by the number of physical seats or benches present. Bench seats are calculated at one person for each 18 inches of bench. Rows of seating will typically have at least 12 inches between them leading to aisles or stairs. When floor seating is also provided in an arena or ballroom for concerts or other special events, care must be taken to ensure that adequate space and aisles are provided. Typically, rows of chairs should be no longer than 14 consecutive seats when served by two aisles and seven consecutive seats if located on a dead-end aisle (see Figure 9.16). An occupant should not have to cross more that six seats to reach the closest aisle.

Chairs should be secured in some fashion so that the occupants cannot indiscriminately move them and lose the aisle accessway space. Securing the chairs can be achieved by mechanical clips on the chair or electrical zip ties on the chair legs. Experience has shown that if the first three chairs at the ends of the aisles are secured, everything else usually stays in place providing the 12-inch aisle accessways in the row. Standing room should not be permitted in auditoriums unless it is part of a designed plan and the occupant load is calculated for standing room separate from the chairs. Allowing patrons to exit their seats and move up to the front stage will overload the front catchment areas and underutilize the rear exits. Overcrowding of exit catchment areas during an emergency can lead to crowd movement and possible crush injuries. Fire inspectors must be aware of the type of venue being presented and the potential type of crowd that may attend. Certain types of entertainment today can cause crowds to act unruly. Venues may have moshing pits where patrons are pushed or thrown from the stage, causing crowd surges. Also, concerts where the entertainers throw items from the stage can lead to trampling injuries. These venues need to be carefully managed by both the owner's security force and the public code enforcing agencies. Fire marshals and fire officials need to review the owner's crowd management practices and ensure cooperation with local fire and police agencies.

Summary

The inspection of the means of egress is one of the most critical components of a fire inspection and life safety program. Exits must be available at all times and must be under the control of the user. Fire inspectors must understand the application of building and fire code requirements for means of egress in order to maintain safety in public assemblies and other buildings. Advance planning must be part of the fire prevention program for public assembly to avoid overcrowding situations. The fire inspector must pay attention to detail and make sure that all of the elements of exit access, exit, and exit discharge are thoroughly inspected and any violations cited are corrected before the event begins. The fire inspector needs to review the building owner's evacuation plan and make sure it contains adequate crowd control measures.

Review Questions

1. Identify and discuss the three components of the means of egress and how they may be used by the building occupant.
2. Contrast the difference in measurements between the unit of egress width and inches per occupant.
3. Calculate the occupant load of a building within your jurisdiction using your local building fire code.
4. Review an egress plan from your local building department and discuss how the occupant load was calculated.
5. Conduct a practice fire inspection of your school and identify all of the components within the means of egress and any conditions, which may be code violations.
6. Develop a plan for dealing with overcrowding in public assemblies. Note who should participate, what safety elements should be included, and how the plan should be implemented and reviewed.

Suggested Readings

International Code Council. 2009. *International Building Code*. Washington, DC: International Code Council.

International Code Council. 2009. *International Fire Code*. Washington, DC: International Code Council.

NFPA 1, *Fire Code*. Quincy, MA: National Fire Protection Association, 2006.

NFPA 101, *Life Safety Code*®. Quincy, MA: National Fire Protection Association, 2006.

NFPA 5000, *Building Construction and Safety Code*®. Quincy, MA: National Fire Protection Association, 2009.

Endnotes

1. Bob Brooke, "History of the Circus in America," History Magazine.com, Oct/Nov 2001, (Apr 6, 2010) http://www.history-magazine.com/circuses.html
2. Richard Bukowski, *Emergency Egress from Buildings* (Gaithersburg, MD: NIST), 171.
3. Ibid., 168.
4. John L. Bryan, "Concepts of Egress Design," *Fire Protection Handbook*, 15th edition (Quincy, MA: National Fire Protection Association, 2009), 6–9.
5. Ibid., 6–8.
6. J. L. Pauls, *Movement of People in Building Evacuations* (National Research Council of Canada, 1977), p. 286.

7. Alyson J. Blair, *The Effects of Stair Width on Occupant Speed and Flow of High Rise Buildings* (University of Maryland: Department of Fire Protection Engineering, 2010), 10.

8. John L. Bryan, "Concepts of Egress Design," *Fire Protection Handbook,* 15th edition (Quincy, MA: National Fire Protection Association), 6–9.

9. Richard Bukowski, *Emergency Egress from Buildings* (Gaithersburg, MD: NIST), 172.

10. *International Building Code 2009*, 205.

11. Underwriters Laboratories, "Study of Smoke Rating Development in Standard Fire Tests in Relation to Visual Observation," *Bulletin of Research* no. 56 (Northbrook, Illinois, 1963).

12. *NFPA Inspection Manual,* 4th Edition (Quincy, MA: National Fire Protection Association), 253.

13. *International Building Code 2009* (Washington, DC: International Code Council), 214.

Courtesy of J. M. Foley

KEY TERMS

fire barrier, *p. 229*

fire endurance, *p. 224*

fire load, *p. 225*

fire modeling, *p. 225*

fire resistance, *p. 224*

fire resistance rating, *p. 226*

fire separation walls, *p. 229*

fire severity, *p. 225*

firestopping, *p. 233*

firewall, *p. 228*

intumescent fire retardant, *p. 235*

opening protective device, *p. 230*

OBJECTIVES

After reading this chapter, the reader should be able to:

■ Know the five different types of building construction.

■ Understand the function of different types of fire-resistant walls.

■ Understand the fire testing aspects for determining fire resistance ratings.

■ Understand the fire inspection points and critical features of fire-resistant construction.

Professional Levels of Job Performance for Fire Inspectors as Cited in NFPA 1031 and NFPA 1037

■ NFPA 1031 Fire Inspector I *Obj. 4.3.4 Verify the type of construction for an addition or remodeling*

■ NFPA 1031 Fire Inspector II *Obj. 5.3.2 Occupancy classification*

■ NFPA 1031 Fire Inspector II *Obj. 5.3.3 Determine the building area, height occupancy class, and type of construction*

■ NFPA 1031 Fire Inspector II *Obj. 5.4.6 Verify the construction type of a building or portion thereof*

■ NFPA 1031 Plan Reviewer I *Obj. 7.3.2 Verify the classification of the occupancy type*

■ NFPA 1031 Plan Reviewer I *Obj. 7.3.3 Verify the construction type*

Introduction

An important aspect of fire code enforcement is to ensure that if a fire should occur, its potential to grow, propagate, and attack the structural integrity of the building is reduced or controlled by the proper maintenance of passive fire protection features. Traditionally, building codes have employed a concept of fire protection called *compartmentation*. This is the segmenting of the building into smaller fire-resistive compartments to reduce the ability of the fire to spread. Compartmentation provides the fire department a better chance of controlling the fire when automatic sprinklers are not present. These passive fire protection requirements were primarily driven by specification codes on how the compartment would have to be constructed and the type of noncombustible materials that would have to be used in the construction. An excellent example of these specification codes is the Empire State Building in New York City. The Empire State Building was completed in 1931 and was called the seventh wonder of the world. The 1,250-foot, 102-story building was constructed in 402 days. The reinforced concrete and steel building weighed over 23 pounds per cubic foot, which is three times the weight of current high-rise construction. This construction method was so strong that the building withstood the impact of a B-25 bomber in 1945 when the plane accidentally collided with the building on the 78th floor. While 14 people were killed in the accident, the ensuing fire was controlled within 40 minutes by the FDNY.[1] The main reasons for such effective suppression were the inherent fire-resistive nature of the construction materials that were used, which limited the fire's ability to spread from compartment to compartment. It is important to note that in that period of time, plastics were not common materials in furniture and office equipment that would have increased the fire severity had they been present as they were in the World Trade Center disaster. Fire-resistive compartmentation was still a major component of the Building Official's Code Administration (BOCA) building code until 1978. It was in the late 1970s and 1980s that the building code began to change dramatically from passive construction specification to performance-oriented requirements and the installation of active fire protection systems. Passive fire resistance requirements were lowered or eliminated when water-based fire suppression systems and fire alarm systems were installed in buildings. As fire and building codes develop in the future, they will continue to become more performance oriented in their requirements. This is especially true with the advances in computer fire-modeling technology that allows the building designer to subject each design to different fire perils before the building ever is constructed. It is important for the fire inspector to make sure that passive fire protection elements are properly maintained to support the active fire protection systems or restrict the fire growth should the systems fail to operate properly.

A material's **fire resistance** is measured through the performance of the material in specific fire tests. Fire testing generally is based on estimated fire load and the amount of heat it would generate versus the time it would take to consume the fuel. The curve that develops between the temperature increase of the burning fuel and the time it takes to consume the fuel is called a standard time/temperature curve. The traditional test method used to determine the fire resistance of building assemblies is the ASTM E-119 fire resistance test. The purpose of the test is to establish **fire endurance** requirements

fire resistance ■ The ability of a building assembly to withstand an exposure to heat based on the standard time/temperature curve.

fire endurance ■ The ability of a building assembly to withstand a specific fire exposure based on a fire load and the time or severity of the heat produced before total fuel consumption occurs.

TABLE 10.1	Standard Time/Temperature Curve Points	
TEMP F	**TEMP C**	**TIME ENDURANCE**
1,000°F	538°C	5 minutes
1,300°F	704°C	10 minutes
1,550°F	843°C	30 minutes
1,700°F	926°C	1 hour
1,850°F	1,010°C	2 hours
2,000°F	1,093°C	4 hours
2,300°F	1,260°C	8 hours

Based on *Fire Protection Structure and Systems Design Reader* Ginn Publishing Lexington Mass p. 31

of materials based on the standard time/temperature curve for the expected fire severity and duration in hours. The concept is simply how long will it take the fire to consume the available fuel and whether the assembly will be intact at the end of the test. **Fire loads** were measured in the 1920s through the 1940s for basic types of residential and commercial occupancies, and tests were conducted to establish the amount of time it would take for a fire to consume different fuel packages. These tests would determine the fire endurance rating required in hours for different building elements. The time element established the expected **fire severity** that the building material or element would have to endure. The standardized time/temperature curve (see Table 10.1) became the basis of the ASTM E-119 fire resistance test for building assemblies. The curve would identify specific points of fire endurance based on fire severity exposure in the test apparatus.

The standard time/temperature curve test method still applies in building codes today to establish the fire resistance rating of different types of wall, floor, and ceiling assemblies, as well as fire-rated doors and windows. The ASTM E-119 test, like all fire tests, is one-dimensional and examines only one element of fire exposure. The test is conducted in a controlled environment and does not take into account changes in the current living environment such as plastic materials or the fuel source's distance to the assembly. As we know, fire can be affected by hundreds of variables from the geometry and type of fuel load, to the available ventilation within the room. As fire protection engineers improve **fire modeling** through computer application, more of these interrelated variables may be examined together to assess the impact they will have on a fire's growth and development and may eventually begin to replace these older fire-testing methods with more multidimensional variable models.

fire load ■ The amount of combustible materials in a typical occupancy expressed in pounds per square foot or BTUs per square foot to determine the amount of heat release during a fire.

fire severity ■ The time it takes to consume a specific fire load and the amount of heat that will be produced over that time period.

fire modeling ■ Computer applications simulating the conditions that would be found in an actual structure fire. Fire modeling provides analysis of building designs and systems before their construction to help identify potential fire safety issues.

ASTM E-119 TEST FOR WALL, CEILING, AND FLOOR ASSEMBLIES

The ASTM E-119 test itself is performed on a 100-square-foot wall, ceiling, or floor assembly sample that is loaded in a specially designed oven and subjected to temperature increases based on the standard time/temperature curve. Depending on the final use of the assembly, the test sample may be restrained or unrestrained and may be subject to an imposed building load as well. The test measures heat transmission on the unexposed side of the wall and may require a hose stream application test at the end to ensure the assembly still has structural integrity. If a wall assembly is tested for 2-hour fire endurance, it will see temperatures of 1,850°F for that duration. At the conclusion of the test, the assembly is hit with a hose stream of 30 psi for 1 to 5 minutes depending on the sample's hourly fire resistance rating. Wall assemblies of 1 hour or less are not subject to the hose stream test.

In an actual fire, the same wall may be exposed to higher temperatures for a shorter period of time, which is different from the test conditions. The test also only examines new assemblies and does not measure changes that may occur over time to the construction materials or the finishes that may be applied to them. The standard also has very specific construction methods for the assembly to be tested and does not take into account any field deviations in the actual construction of the assembly within a building.

The results of this fire test produce a detailed specification-driven construction method for the wall, ceiling, or floor design for the specific hourly **fire resistance rating** for the tested sample. Underwriters Laboratories, Factory Mutual Research, and the U.S. Gypsum Association then publish these approved designs for use by professional architects and engineers. Each tested assembly is given a specific design number, which the architect will specify on buildings' architectural drawings. It should be noted that the fire-resistant design must match the test specification for the assembly to comply with the building code's performance requirements for the hourly fire resistance rating. This specification includes the type of screw or nail patterns to be used, the staggering of wallboard joints, and the taping methods as well as thickness of wallboards and structural stud spacing. Any deviations in the method of construction may not yield the same fire endurance rating as the designed tested assembly.

ASTM E-152 TEST FOR FIRE DOOR ASSEMBLIES

Fire doors are tested to the ASTM E-152 and NFPA 252, *Standard Methods of Fire Tests of Door Assemblies*, which uses the same basic test methods of evaluating fire endurance as the ASTM E-119 test. The test for a 1-1/2 hour "B" labeled fire door requires the assembly of a 100-square-foot, 2-hour firewall sample with the fire door and frame mounted in the assembly. The assembly would then be loaded into the oven and exposed to heat based on the standard time/temperature curve to determine the hourly fire rating. At the end of the test, the assembly is hit with a fire hose stream to check glass and door integrity. If the door passes, it is issued a label for that level of fire endurance (see Figure 10.1). Fire doors are tested with all the associated hardware installed, and this hardware is also labeled for fire door application. Any deviations in hardware on the door once it is installed will void the door listing; this is especially true when new hardware nonlisted devices are installed on existing fire doors. Fire inspectors should examine the labels on the door, doorframe, and associated door hardware to ensure proper code compliance.

In this chapter, we will explore the relationship between the fire resistance requirements of the building codes and the long-term maintenance issues that fire inspectors encounter in conducting a fire prevention inspection as they relate to fire-resistant rated assemblies and associated building materials.

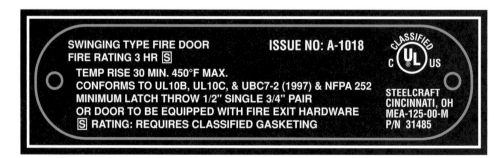

FIGURE 10.1 All fire-rated doors will be labeled. Note the maximum temperature rise of 450°F in 30 minutes; this is why the fire doors cannot be modified. The application of combustible materials would spread the fire from one side of the door to the other. Also note that the label requires fire-rated hardware and proper gasketing.

Types of Building Construction

The first question that every fire inspector must answer in the examination of a building is "What type of construction is this building?" In today's building environment, this question may not always be easy to answer without careful examination of the building's components. Sometimes, buildings may not be what they appear to be from the outside. A building may look like it is masonry construction, but you may find it is wood frame with brick veneer.

The basic point of reference for the fire inspector in determining a building's construction must be a thorough understanding of the five basic types of building construction specified in all building codes. The type of construction determines the fire endurance requirements for each specific structural element of the building including the following:

- Exterior walls
- Firewalls
- Fire separation walls
- Fire and smoke barriers
- Vertical shaft enclosures
- Exits and exit access corridors
- All structural columns supporting the building

These requirements are spelled out in tabular form in both NFPA 5000 and ICC building codes. NFPA also describes the types of construction in NFPA 220, *Standard on Types of Building Construction.*

The five main types of building construction are as follows:

- *Type I Noncombustible Construction:* These buildings are composed of noncombustible building elements such as concrete and steel and contain very limited amounts of combustible materials. High-rise buildings and larger structures are typically type I construction. The category is further divided into two subclasses based on the fire endurance ratings of structural elements:
  - *Type IA:* Generally 4 hours fire resistance supporting more than one floor and 3 hours fire resistance supporting a single floor or roof.
  - *Type IB:* Generally 3 hours fire resistance supporting more than one floor and 2 hours fire resistance supporting one floor or a roof.
- *Type II Noncombustible Construction:* Type II construction uses noncombustible materials on all structural elements and may have masonry or metal exterior walls and roofs supported by metal trusses or joists. Type II buildings are usually strip malls and stores or warehouses. Type II structures exist with either protected or nonprotected structural elements and general fire resistance ratings as specified:
  - *Type IIA:* 2 hours fire resistance supporting more than one floor and 1-1/2 hours fire resistance supporting one floor or a roof
  - *Type IIB:* 1 hour fire resistance supporting more than one floor and 1 hour fire resistance supporting one floor or a roof
  - *Type IIC:* 0 hours fire resistance supporting more than one floor and 0 hours fire resistance supporting one floor or a roof
- *Type III Ordinary Construction:* Type III construction is also referred to as ordinary construction and consists of both combustible and noncombustible building elements. The exterior walls are usually masonry or cement block units, and the roofs and floors are combustible frame or lumber. Ordinary construction presents unique fire challenges to firefighters because of fire cuts on the floor and roof joist systems. The fire cuts are designed to allow the collapse of the floor or roof systems in a fire without damaging the exterior masonry firewalls (see Figure 10.2). Ordinary construction may be either protected or unprotected.
  - *Type IIIA:* 1 hour fire resistance supporting more than one floor and 1 hour fire resistance supporting one floor or a roof

- *Type IIIB:* 0 hours fire resistance supporting more than one floor and 0 hours fire resistance supporting one floor or a roof
- *Fire inspectors should make firefighters aware of the collapse potential of buildings of ordinary construction in their jurisdiction to improve firefighter safety.*

- *Type IV Heavy Timber Construction:* Heavy timber construction is combustible and may have either combustible or noncombustible exterior walls. The structural frame of the building must have at least 8-inch nominal wood columns and 6-inch by 8-inch nominal wood beams. These may be solid lumber or laminated beams. Type IV structures may also be called post and beam or timber frame construction. Generally, heavy timber buildings offer relatively good fire resistance because of the minimum dimensions of the lumber supporting the structure.

- *Type V Combustible Frame Construction:* Type V is basic frame construction and may be one of three types: balloon frame built before the 1940s; platform frame built from 1940 to the 1970s; or lightweight frame, which is the predominant construction method used today utilizing lightweight trusses and truss joist wooden I beams for floor and roof assemblies.
  - *Type VA:* 1 hour fire resistance supporting more than one floor and 1 hour fire resistance supporting one floor or a roof
  - *Type VB:* 0 hours fire resistance supporting more than one floor and 0 hours fire resistance supporting one floor or a roof[2]
    - Most single-family homes today are type VB construction and may have a 1-hour fire-resistant wall and ceiling assembly if they have an attached garage.

All buildings fall into one of the five types of construction, and sometimes a building has multiple types of construction present depending on the building's square footage and whether additions have been added to it. The type of construction will establish the fire resistance ratings of building structural elements. The fire resistance ratings may be further modified in the building codes by the installation of active fire protection systems in the building. As an example, the IBC building code allows the lowering of one class of construction if the building is fully suppressed; therefore, a building required to be 1A construction can be reduced to IB if automatic sprinklers are installed. Inspectors must be aware that this trade-off applies only when the building is 100 percent protected by automatic sprinklers. Partial systems are not credited under the building codes.

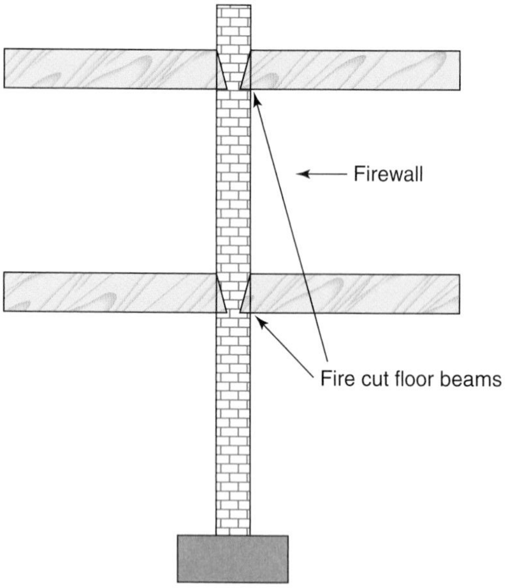

firewall ▪ A fire-resistant wall having protected openings, which restricts the spread of fire, is independently supported, and extends from the foundation to or through the building roof.

FIGURE 10.2 Firewalls are supported independent of the building and must remain intact should the building on either side collapse; this is accomplished by fire cuts that allow the release of floor joists from the wall pockets.

Firewall

Fire cut floor beams

Types of Fire-Resistant Walls

A challenge for the fire inspector is to determine which walls are fire resistant and what the fire resistance rating is for a wall. Walls damaged by holes and penetrations must be repaired with the proper fire-rated gypsum boards if they are fire resistant. Understanding the functions of different types of walls can assist the fire inspector in identifying them during an inspection.

FIREWALL

A **firewall** is sometimes also referred to as a *party wall*. A firewall is a fire-resistant barrier that physically separates two buildings. Firewalls must comply with the fire resistance rating for the particular type of construction and have limited openings permitted in the wall itself. Any openings that do exist in a firewall must be protected by a fire-rated protective device such as a fire door or window. Another important function of the firewall is that it must be independently supported from other structural connections and it must go from the building's footing to or through the roof. In older construction, the firewall typically extends 2 feet 8 inches above the roofline, identifying where the firewall is located from the outside of the building. In new construction, firewalls are located between buildings and generally go to the underside of the roof if it is properly firestopped and is of noncombustible construction. If the roof is combustible construction, then the firewall must have fire retardant treated plywood out to 4 feet on both sides of the top of the firewall. Firewalls are designed so that a fire occurring in the building on either side of the wall will not cause a structural collapse or affect the integrity of the firewall. Roof beams and floor joists attached to firewalls are inserted into pockets within the wall. The joists are fire cut to allow the beam to release from the wall without toppling it (see Figure 10.2). Firewalls may also be used as dividing walls in a building in cases where the building has exceeded the area limitations for the type of construction under the building code. Firewalls have the highest fire resistance rating of 2 to 4 hours fire resistance. Door openings in firewalls between buildings are considered Class A openings and must be protected with fire doors of equal fire resistance rating to the firewall. Class A fire doors are rated to 3 hours fire resistance and may be large sliding doors on fusible link pulley systems in industrial buildings to allow for passage on forklifts and goods. The fusible link is a simple mechanical system that uses two metal tabs held together by a eutectic metal that melts at a specific temperature such as 165 degrees Fahrenheit. When the link melts, the counterweights are released, causing the fire door to slide closed.

FIRE SEPARATION WALL

The **fire separation wall** is a fire-resistant wall that divides spaces, such as the separation of mixed-use groups within a building or the separation of an exit enclosure from the rest of the building. These walls are the most common type of fire-resistant rated wall encountered and must extend from one rated fire assembly to the next one above. All fire separation walls must have opening protection for doors, ducts, or any other item that may penetrate the wall enclosure. Fire separation walls are rated from 1 to 2 hours fire endurance.

fire separation walls ■ Fire-resistant walls that are used to divide or separate areas of a building but are not continuous like firewalls from the foundation to the roof.

FIRE AND SMOKE BARRIERS

Fire barriers and smoke barriers are typically found in institutional use groups where the movement of occupants due to restraint or sickness is difficult. The fire or smoke barrier generally divides a floor into smaller sections and runs from the exterior wall on one side of a building to the exterior wall on the other side. In hospitals, the fire inspector can identify these by the fire- and smoke-rated doors in the corridor. These walls segment the floor into smoke compartments. Fire barriers may also be installed in other types of occupancies for the separation of exit stairs around open interior stairways. If the fire barriers must also function as smoke barriers, any opening protection devices must be rated for both smoke and fire. Fusible link fire dampers that are used in fire barriers may not be used in smoke barrier walls.

fire barrier ■ A fire-rated assembly used to divide a floor area and prohibit the passage of fire, heat, and smoke across the barrier. Fire barriers extend from the outside wall on one side to the opposite outside wall.

FIRE EXIT ENCLOSURES

Building codes also require that access to exits be protected by fire-resistant enclosures. Corridors may also be required to be fire rated and usually have a fire resistance rating slightly lower than the exit enclosure fire rating. Generally, exits stairs are rated at 1 hour fire resistance up to four stories and 2 hours fire resistance above four stories, exit access corridors are generally rated at 1 hour fire resistance, but this may vary based on use group and the presence of automatic fire sprinklers. The fire inspector can identify the fire-resistant walls by examining the UL labels on the corridor doors and stairway doors. Typically, Class D, 20-minute fire doors and Class C, 45-minute doors are placed in 1-hour corridor wall construction. Fire exit stairway doors typically have a Class B fire resistance rating of 1 to 1½ hours depending on the height of the stairway enclosure. All of the doors installed in fire-resistant rate walls must be self-closing and tight fitting and have proper UL labels on both the door and frames as well as the door hardware.

Opening Protection

We have discussed the different types of fire-resistant walls commonly found in most buildings. It is important for the fire inspector to properly identify the rating of the wall. The fire inspector can sometimes tell the wall's fire resistance rating by the UL listings of devices installed within the wall. All of the **opening protective devices** installed in a wall must be fire resistant rated and will usually be slightly lower than the rating of the wall. The fire inspector should keep in mind that there is added cost to the installation of fire-rated assemblies; therefore, they are not installed haphazardly within buildings. In other words, if you see a fire-rated door, shutter, window, or damper in a wall assembly, the wall is a fire-resistant wall.

FIRE DOORS

The most important function of a fire door is to prevent the fire from extending from one compartment to another or from the compartment to the means of egress. To perform this function, the door must be tested using actual heat exposure based on the expected severity of a fire in the building. Building codes determine the minimum fire resistance rating of the door based on the fire resistance rating of the wall assembly it will be installed in. Fire inspectors must understand how a fire door is tested in order to properly evaluate fire code compliance during the fire inspection. Fire doors are tested to NFPA 252, *Standard Methods of Fire Tests of Door Assemblies*, and UL 10B, *Fire Tests of Door Assemblies*, or UL 10A, *Tin-Clad Fire Doors*. The fire door is loaded in a fire-rated wall assembly and subject to the standard time/temperature curve as in the E-119 test for the duration of 3 hours or less depending on the rating the manufacturer is trying to attain. The transmission of heat through the door is not a failure consideration during the test. The door is allowed to buckle up to 1½ times its thickness but must stay in the opening during the entire test. The door need not function at the end of the test, and in fact, most metal fire doors that are exposed to heat will expand and be nonoperational in a short period of time. At the conclusion of the fire test, the door is subject to a fire hose stream and must remain in place. If the fire door has a vision panel or window installed, 70 percent of the glass window must stay in the window during the hose stream application. Fire doors may not be modified after installation; this is an important fire inspection detail. Often, building owners add architectural accents to a fire door or install carpet or wall coverings on the door. These conditions violate the listing of the door unless they were tested on the assembly. The door also must have UL-listed fire-door-rated hardware that was tested with the door or it is in violation. If the fire door is a two-leaf configuration, the door must have an astragal and coordinator to close the gap between the fire door leafs. The astragal is the long flat bar that runs down the door edge and covers the space between the door leafs. The coordinator is a mechanical arm device at the top of the door that allows the door with the astragal to close last. Frequently, astragals are removed because the door coordinator is broken. Astragals and coordinators are vital components that must be maintained on the fire door. There are UL-listed astragal door sweep brush kits that can be installed on fire doors to replace metal astragals, but the fire inspector should verify that they are the correct type for the installation.

FIRE DOOR OPENINGS

Fire doors are specified by the openings that they are installed in (see Figure 10.3). Table 10.2 shows the

Types of Fire Door Openings

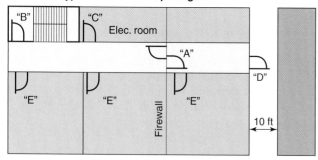

FIGURE 10.3 The type of fire door to be installed is determined by the type of wall it is installed in. Class A openings are in firewalls, Class B openings are in 2-hour fire separation walls, Class C openings are in 1-hour walls protecting special hazards, Class D openings are in exterior walls, and Class E openings are in corridor walls.

TABLE 10.2 | Fire Door Opening Specifications

OPENING	LOCATION	HOURLY FIRE RATING	WINDOW SIZE
A	Interior building separations	3 hours	None
B	Interior fire separations	1–1½ hours	100 square inches
C	Interior	¾ hour	1,296 square inches
D	Exterior	1–1½ hours	None
E	Interior	30 minutes	1,296 square inches
E	Interior	20 minutes	1,296 square inches

Based on *NFPA Inspection Manual,* 4th edition, Quincy, MA: NFPA, 1976, p. 253.

typical opening locations as well as the fire resistance rating and the permitted size of any vision panels or windows installed in the fire door.

FIRE DOOR INSPECTION

The fire prevention code requires that all fire doors be inspected in accordance with NFPA 80, *Standard for Fire Doors and Other Opening Protectives*. Fire inspectors may encounter all sorts of fire doors, especially in older buildings. Fire doors may be composite, hollow metal, metal-clad wood doors, tin-clad wood doors, sheet metal doors, rolling steel doors, and steel curtain-type doors. The key element for the fire inspector to examine is the UL label on the door and the door hardware. Door labels may not be painted over to the point where they cannot be read to identify the fire rating of the door. Fire inspectors must consider the following critical inspection points:

- The door must be in good condition, level, and tight fitting in the doorframe.
- The fire door must be self-closing and positive latching so that it cannot be forced open by pressure.
- The door may not be undercut more than ¾ inch at the door bottom.
- If the door has panic hardware, it should take no more than 15 pounds of force to release the door latching mechanism.
- The fire door should have the proper window glazing, wire glass, or glass that is compliant with the UL fire-testing standard.
- The fire door should not be modified in any fashion such as architectural trim or interior finishes.
- NFPA 80, *Standard for Fire Doors and Other Opening Protectives*, requires that all roll or curtain-type fire doors be tested for proper operation annually after a physical inspection of the door is performed (see Figure 10.4).
- The door shall be tested for proper operation by cutting the fusible link.
- The door must close no slower than 6 inches per second and no faster than 24 inches per second for personnel safety.

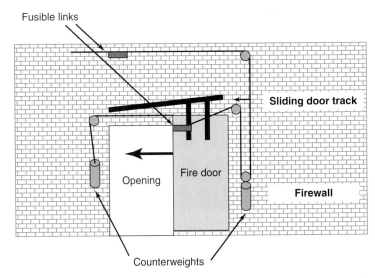

FIGURE 10.4 Sliding or roll-type fire doors must be inspected and operationally tested annually. They should be tested by cutting the fusible link to ensure that all cables and pulleys move freely and the door closes properly.

- Fire-rated doors on automatic hold-open devices or individual smoke detector checks must also be tested for proper operation.
- All hold-open devices must be fail-safe in that if electricity to the device is lost, the door automatically closes. Additionally, the doors may require automatic closing upon a signal from the building's fire alarm system.

FIRE WINDOWS AND SHUTTERS

Fire windows and shutters may be used on the exterior portion of a building to protect it from fires from adjacent buildings. Fire windows and shutters may also be used on the interior of the building to protect corridors from rooms with pass-through openings such as coatrooms or cafeterias. In either case, the fire window or shutter must be of the proper fire resistance rating for the wall in which it is installed. Fire windows and shutters must meet either ASTM E163, *Methods for Fire Tests of Window Assemblies*; UL 9, *Fire Tests of Window Assemblies*; or NFPA 257, *Standard on Fire Test for Window and Glass Block Assemblies,* fire testing standards. These shutter devices, like sliding fire doors, may operate on a fusible link system to automatically close and protect the opening. Fire shutters are required to be drop tested annually the same as sliding fire doors. Fire windows must have rated glazing capable of withstanding a fire exposure based on the hourly fire resistance rating of the opening they are protecting. Fire windows are required in exterior walls generally when exposures are within 30 feet of a building or less depending on the use group of the building. As an example, high hazard use group "H" requires exterior fire resistance ratings at less than 30 feet, whereas R-1 residential use groups such as hotels require fire resistance ratings at less than 10 feet. An exterior wall opening also is required to be fire protected if it is located within 15 feet vertically of an exposed roof structure. Exterior stairways and fire escapes also require opening protection along the path of travel and within 10 feet vertically and horizontally of the stairway by fire windows or fire shutters. The building code also requires that on long walls with many fire shutters, at least every third window shutter must be able to be opened from the outside by the fire department for access. These shutters must be identifiable from the ground by the fire department.

FIRE DAMPERS

Fire dampers may be either mechanically or electrically driven and are used inside ducts and mechanical ventilating systems that penetrate fire-resistant walls and floor ceiling assemblies. Fire dampers must comply with either UL 555, *Fire Dampers,* or UL 555S, *Smoke Dampers,* listings depending on their intended use. Fire dampers must be installed in line with the fire-resistant wall assembly. Fire dampers may also be rated for smoke penetration in smoke barrier walls. The smoke damper is tested by measuring the leakage rate across the damper in cubic feet per minute at an air movement pressure of 2,000 feet per minute or 4 inches of water column pressure. Typical leakage rates may go from 8 to 80 cubic feet per minute (cfm) depending on the manufacturer's desired rating for the smoke damper. Important fire inspection points for fire and smoke dampers are, first, to make sure that the damper is in line with the rated fire-resistant wall. Second, the duct will be connected to the damper by a breakaway collar. The inspector should make sure that the duct retaining bolts are between the duct and the fire/smoke damper assembly (see Figure 10.5). These bolts are used to align the assemblies during installation, then clips hold the assembly together and the bolts are removed. It is also important to make sure that the fire damper assembly's wall retaining angles are not caulked with nonapproved firestopping sealants. These fire sealants are intumescent and expand under heat, causing penetration or separation of the fire damper from the wall. Only sealants approved by the manufacturer may be used as firestopping around fire and smoke dampers. The fire or smoke dampers must also have an inspection access opening in the duct for inspection and maintenance of the damper. All fire dampers should be inspected at least every 4 years

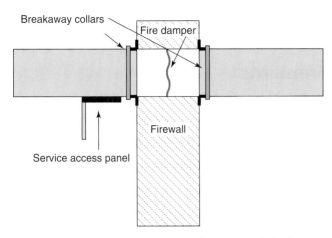

FIGURE 10.5 Fire dampers must be installed in line with the fire-rated wall. Ducts connected to the damper will have breakaway collars with alignment bolts that need to be removed. There should also be an access hatch to inspect and service the fire damper.

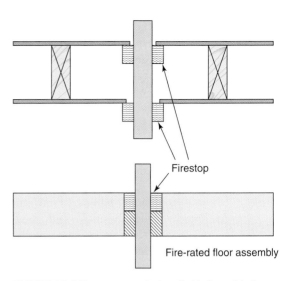

FIGURE 10.6 Firestops must be installed in line with the fire-rated wall. There are many types of UL-listed firestop materials, and manufacturers' information may be required to determine proper installation. Firestops should be tested to ASTM 814.

(6 years in hospitals) for proper mechanical operation. Fire and smoke dampers may collect dirt and debris in the duct over time, becoming clogged and failing to operate properly. Inspection certification should be provided to the fire inspector that an outside testing agency has inspected the fire or smoke dampers at the proper inspection intervals. Fire and smoke dampers that are electromechanical must be tested annually if they are part of a smoke control or stairway pressurization system. Fire dampers must be rated to 75 percent of the fire resistance rating of the wall assembly. This is helpful in determining the fire rating of the wall. If the fire damper is rated at 45 minutes, the wall is rated at 1 hour. In smoke control and stair pressurization systems, the dampers on the exterior of the building should be examined for any signs of rust or corrosion from weather.

FIRESTOPPING

The building codes require that openings that connect vertical and horizontal spaces be *firestopped* to prevent fire spread from floor to floor or compartment to compartment. **Firestopping** materials must be noncombustible and properly installed to seal or close any opening under fire conditions. The firestop must also meet the required fire resistance rating of the wall or floor/ceiling assembly upon which it is installed. If the penetration is in a 2-hour fire-resistant rated floor assembly, the firestop must have a 2-hour fire resistance rating. Firestopping materials come in many forms from mineral wool fiber to intumescent materials and mechanical firestops. The materials may be endothermic and elastomeric ones that absorb heat and still remain flexible. Through floor and ceiling firestops are tested to ASTM E814, *Standard Test Method for Fire Tests of Penetration Firestop Systems* (UL 1479, *Tests of Through-Penetration Firestops*) and are assigned an "F" rating for fire resistance, and a "T" rating for the time it takes the material to exceed 335°F. They will also have an "L" rating for air leakage through the firestop specified in cubic feet/minute. All approved firestopping methods must be tested and listed in the UL Fire Resistance Directory, or the Sheet Metal and Air Conditioning Contractors National Association (SMACNA), or Factory Mutual's fire resistance guides. The key inspection point for the fire inspector is to make sure the openings are firestopped using a proper listed design technique and an approved fire-stopping system (see Figure 10.6). This must

firestopping ▪ A method of sealing openings in walls, ceilings, or floors to prohibit the passage of fire into the internal components of a structure.

be verified by certification and documentation on the products used and the manufacturer's specifications for the listed installation. Most major manufacturers of firestopping materials provide proper firestopping systems and instructions on their installation based on the type of penetration needing repair. Most manufacturers post this information on their websites for customer use.

Interior Finish Requirements

The interior finish materials within a building must be properly installed and must meet the minimum building and fire code requirements of the jurisdiction. Improper or non-approved interior finish materials like those found in the Station nightclub fire in Warwick, Rhode Island, can lead to disaster by increasing flame propagation and releasing deadly smoke and gases. While most media outlets relate the Station fire disaster to pyrotechnics, it was actually the interior finish materials, in this case, foam rubber materials affixed to the wall, that were the culprit spreading the fire rapidly and causing the deaths of 94 victims. Interior finishes played a major role in many fire disasters throughout history from the Cocoanut Grove in Boston, to the Beverly Hills Supper Club in Kentucky, and the MGM Grand Hotel fires in Las Vegas, as all of these fires spread and grew rapidly due to interior finish materials.

The main fire test for analyzing interior wall finishes is the ASTM E84 Steiner tunnel test. The Steiner tunnel is a 25-foot tunnel in which the interior finish material sample of 19 feet 5 inches is affixed to the top of the tunnel (see Figure 10.7). A gas piloted flame is used to ignite the sample, and measurements are taken through observation windows on the flame's propagation along the material's surface over a 10-minute time frame. The test also measures the smoke created by the burning interior finish. The material is then compared to two reference materials. The reference materials are a cement board that has a flame spread rating of zero and red oak planking that has a flame spread of 100. In calibrating the tunnel, it takes approximately 5 minutes and 30 seconds for the flames to propagate to the end of the red oak sample as a point of reference. The flame spread rating that is achieved gives an indication of the material's

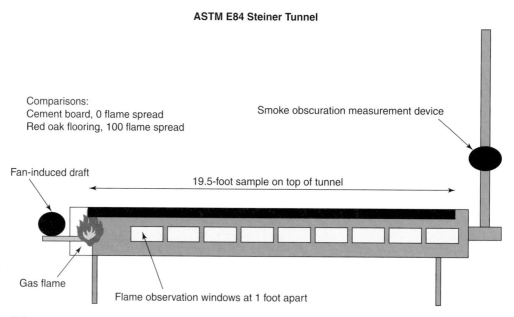

ASTM E84 Steiner Tunnel

Comparisons:
Cement board, 0 flame spread
Red oak flooring, 100 flame spread

Smoke obscuration measurement device

Fan-induced draft

19.5-foot sample on top of tunnel

Gas flame

Flame observation windows at 1 foot apart

FIGURE 10.7 ASTM E-84 Steiner tunnel is used to determine flame spread and smoke generation of interior finish materials.

burning characteristics as compared to those of red oak. As an example, a material with a flame spread of 200 would have a surface burn rate of 2.5 minutes or twice as fast as red oak. A key concern with the tunnel test is that it examines only a single aspect of flame spread on materials in the horizontal position. Many interior finish materials like plastics may perform better in a horizontal position than a vertical position. As an example, plastics that retreat from heat sources have a low flame spread rating in an ASTM E84 tunnel test. If you perform a vertical fire test on the same plastic material, it will rapidly burn. Fire inspectors must be aware that plastic interior finish materials require additional fire tests and evaluations. The Steiner tunnel also develops a smoke obscuration factor for the fire gases produced during the fire test. Building codes restrict the smoke development rating to a maximum of 450 for all interior finish materials. The E84 tunnel test produces three categories of flame-spread ratings for interior finish materials:

- Class I, or class A flame spread ratings are 0–25 (meaning that it burns at 25 percent of the rate for red oak).
- Class II, or class B flame spread ratings are 26–75.
- Class III, or class C flame spread ratings are 76–200.

The IBC and NFPA building codes do not allow flame spreads above 200 to be used in any interior finish application. The building and fire codes also require that the flame spread ratings be restricted in specific areas within a building such as vertical exit stairways and exit passageways, which may only have Class I flame spread rating. Exit access corridors are permitted to have a Class II flame spread rating, and rooms and spaces may have a Class II or III interior finishes flame spread rating based on specific use of the room. The building codes also allow interior finishes to be reduced by one category when automatic sprinkler systems are installed in the building. Fire inspectors often encounter older buildings where interior finish materials have no known flame spread ratings. In these cases, there are products available such as **intumescent fire retardant** paints that, when applied in accordance with the manufacturer's specifications, can provide a reduced flame spread rating. Fire inspectors should request the owner to provide the manufacturer's specifications and testing requirements for the paint and an affidavit that it was properly applied.

intumescent fire retardant ■ A fire-resistant material that expands and forms a char layer when exposed to high temperatures, reducing the fuel's ability to undergo pyrolysis and produce vapors to burn.

FOAM PLASTIC MATERIALS

Foam plastics have different burning characteristics than conventional wood and natural fibers and tend to burn with greater thermal output and produce more toxic and acrid smoke in large quantities. Building codes prohibit the use of plastic materials in interior finish if they generate smoke in excess of 450 or have a flame spread above 75 in the E-84 tunnel test. The use of foam plastics requires that they be protected in some fashion by a thermal barrier of drywall or some other noncombustible material. Foam plastics may be used without thermal barriers if they comply with a specific battery of fire tests like UL 1040, *Fire Test of Insulated Wall Construction*; UL1715, *Fire Test of Interior Finish Material*; and FM standard 4880, *Evaluating Plastic Interior and Exterior Finish Materials*, which are full-scale vertical corner burn tests.

INTERIOR FINISH MATERIAL ATTACHMENT

All interior finish materials must be properly attached to the structure to withstand a fire exposure of 200°F for a 30-minute period. This prevents the materials from becoming detached and contributing to the fuel load or becoming an entanglement hazard to firefighters. Trim and interior finish materials attached to fire-rated structural components must be affixed directly to the component with no void space behind the material unless the voids are properly firestopped every 10 linear feet in all directions.

NONCOMBUSTIBLE MATERIALS

The term *noncombustible* has a very specific meaning under the building and fire codes when these materials are used in interior finish and decoration. The test for elementary materials such as cement block, brick, or drywall to be categorized as noncombustible is ASTM E136, *Standard Test Method for Behavior of Materials in a Vertical Tube Furnace at 750°C*. The test requires the heating of the sample material to 1,382°F or 750°C, and three failure points are measured:

1. The material fails if the internal temperature increases by 54°F above the oven temperature at the test's beginning.
2. The material flames for longer than 30 seconds.
3. The material loses more than 50 percent of its weight during the test period.

Any material that fails the test is considered to be combustible and is limited as a combustible interior finish material by the code. The importance of this test to the fire inspector is to recognize that materials that are inherently flame resistant or flame retardant treated are not noncombustible materials. They are combustible materials that are ignition resistant and, therefore, are limited in use by the building and fire codes. Typically, combustible finish and decorations are restricted under the 10 percent rule. The building codes limit combustible interior trim such as wainscoting, chair rail, and baseboard to 10 percent of the aggregate wall and ceiling area of a room or space. This is considered a negligible amount of combustible materials. Decorative materials under the fire codes are also limited to 10 percent of the aggregate wall and ceiling areas for hanging draperies, banners, curtains, streamers, and the like. Fire inspectors must be aware that just because a material is treated to be ignition resistant it is not therefore noncombustible. Materials that are classified as noncombustible are not limited by the codes; however, they must pass the ASTM E136 fire test.

FLOOR FINISHES

The flooring within a building is also regulated by the building codes, depending on the space and location of the material within the building. Flooring must meet either the DOC FF-1 pill test from the Department of Commerce or the ASTM E648, *Standard Test Method for Critical Radiant Flux of Floor-Covering Systems Using a Radiant Heat Energy Source*, flooring test. Carpeting flammability is not as severe as wall and ceiling trim. All commercially produced carpet in the United States will meet the Department of Commerce pill test (DOC-FF1). Carpets are permitted in all enclosed rooms and spaces within a building no matter what the use group. The critical radiant flux test (E648) measures the amount of radiant energy necessary to drive a flame across the surface of a flooring material. The test qualifies two separate classes of flame spread:

- *Class I* flame spread requires 0.45 watt/cubic centimeter.
- *Class II* flame spread requires 0.22 watt/cubic centimeter.

A minimum of Class I or II floor covering must be provided in vertical exits and passageways and exit access corridors in use groups A, B, E, I, M, R-1, and R-2.

Fire inspectors should be aware that some imported flooring materials might not meet the code specifications. Any nondomestically produced rugs or carpets should be properly documented as to their flame spread ratings. Using the Internet, today, building owners can purchase carpets and other materials from all over the world and have them shipped to the United States. If the fire inspector suspects that the flooring materials are not compliant, he or she should request documentation or certification that the flooring meets U.S. building and fire code standards.

Decorations

Decorative materials are items that are added to a room or space for aesthetic purposes and are not considered furniture or part of the structure. Decorations may be permanent, such as drapes or wall hangings, or they may be temporary based on a function such as a party or a special event. In either case, the challenge for the fire inspector is to determine the degree of hazard that the decorative material presents and to determine the correct method to treat the material if it is non–code compliant. The fire danger of decorative materials is that they are often the first things that are ignited in a building fire or can become a contributing factor to fire spread. Decorations often are lightweight and have large surface-to-mass ratios, making them easy to ignite unless they are properly treated to make them ignition resistant. The fire prevention code requires that decorative materials be compliant with NFPA 701, *Standard Methods of Fire Tests for Flame Propagation of Textiles and Films*, using either a small- or large-scale fire test, depending on the type of material to be tested. The NFPA 701 small-scale test generally is used for textiles that are not being used in a folded arrangement or in multiple layers. Stage curtains or hanging drapes require a large-scale NFPA 701 test. The small-scale test uses samples 3.5 inches by 10 inches long that are cut in both directions of the fabric weave. Each sample is subject to a 12-second flame exposure, and the char of the fabric may not exceed 6.5 inches. When the flame is removed, the materials must self-extinguish within 2 seconds, and any dripping of burning material must self-extinguish. The NFPA large-scale test requires that 10 samples of proper size (7 feet in length) be tested with a flame exposure at the bottom of the sample for 2 minutes. The char may not exceed 10 inches, and it must self-extinguish in 2 seconds with no burning continuing on the test apparatus floor. All char measurements and after flaming are recorded on the test report for the material.

Fire inspectors should also use the NFPA 705, *Recommended Practice for a Field Flame Test for Textiles and Films*, match test to evaluate the fire retardant effect on the sample materials. The NFPA 705 flame test is an exclusionary test and does not reflect compliance or noncompliance with NFPA 701. The flame test is a method for fire inspectors to use their discretion and professional judgment as to whether they believe the material is properly treated or is noncompliant and requires additional fire retardant treatment or removal from the building. The NFPA 705 field test is sometimes referred to as the *kitchen match test* and should be conducted in a safe location or outdoors. The test requires the fire inspector to obtain a 1 inch by 4 inch sample of the fabric. This usually can be taken from a hem or fold in the material. The sample will be ignited by flame for 12 seconds and must self-extinguish in 2 seconds when the flame is removed. The sample may not be completely consumed, and if it drips, the drips must self-extinguish and not continue to burn. The fire inspector may use this test along with the manufacturer's certification on the textile to verify its fire resistance. Manufacturers' certifications identify the material as "FR" for "flame resistant" or "IFR" for "inherently flame resistant" and also identify the chemicals used to treat the fabric. Typically, fire retardant treatments are hydroscopic and give the fabric a damp feel to the touch. Treating fabrics is a temporary solution that will degrade over time and must be re-applied annually or as specified by the manufacturer for the fabric to retain its fire retardant qualities. Fabrics identified as "IFR," meaning they are inherently fire retardant, do not lose the fire retardant effect over time as it is chemically bonded in the fabric itself and not a topical treatment. Inherent fabrics typically have a permanent label affixed to the goods for inspection purposes, such as outdoor tents.

Fire inspectors must be aware that not all fabrics can be treated with fire retardant applications. Many plastic materials, like polyester, resist topical fire retardant treatments as they are nonabsorbent and do not perform very well. Most plastic fibers must be treated inherently to become fire retardant. Generally, fire retardant treatments work in one of two ways: they either retreat from the heat source and self-extinguish, or they form a char that resists

FIGURE 10.8 Fire retardant treated materials are not noncombustible and are limited to 10 percent of the aggregate wall and ceiling areas. Codes also prohibit overhead decorations that could become involved in a fire and fall and block occupants' exits. Ceilings like the one in the photo should be questioned by the fire inspector. If the room is protected by automatic sprinklers, they must not be obstructed. *Courtesy of J. M. Foley*

the propagation of the flame up the material. These two types of flame-resistant materials should not be used together in multiple layers, as one will negate the fire retardant effect of the other even if both materials are properly treated. The fire inspector should field test and examine all installations of decorative materials to ensure proper fire code compliance.

Other props and decorations constructed of foam plastic materials may also be used in buildings and require evaluation for fire retardancy by the fire inspector. The IFC provides that any decoration containing foam plastics be subject to evaluation under UL 1975, *Fire Tests for Foamed Plastics Used for Decorative Purposes.* The UL 1975 test examines the heat release rate of the foam plastic to determine the level of fuel contribution that the foam makes to a fire. The test permits heat release rates between 250 to 500 kilowatts of energy depending on whether automatic fire sprinklers are installed in the building. In any case, the decorator or material supplier must provide the fire retardant certification and documentation to the fire inspector that the materials meet and comply with these safety standards. Decorative materials installed overhead are prohibited unless the fire marshal or fire official approves them. Overhead decorations were a major contributing factor in the 1942 Cocoanut Grove fire in Boston (see Figure 10.8). If the fire inspector approves swags or some form of overhead drapery, he or she must ensure that it remains 70 percent open for sprinkler activation and that it does not impair the sprinkler water from reaching the floor. Fire inspectors should not permit rooms to be 100 percent draped in fire-treated fabrics. Remember the 10 percent rule, limiting these materials to 10 percent of the aggregate wall and ceiling area.

EXAMPLE

Suppose you have a ballroom 100 feet by 75 feet by 10 feet in height. To calculate the permitted amount of fire-treated materials, you would perform the following calculations:

$$\text{Ceiling Area} = 100 \times 75 \text{ feet} = 7,500 \text{ square feet}$$
$$\text{Wall Area } 2 \times 100 \times 10 = 2,000 \text{ square feet}$$
$$2 \times 75 \times 10 = 1,500 \text{ square feet}$$

Total: 11,000 square feet $\times$ 10% = 1,100 square feet of fire-resistant fabric is permitted.

This would equate to eleven 10 foot by 10 foot fabric swags.

CERTIFICATIONS OF FLAME RETARDANT APPLICATION

The fire prevention codes require that decorative materials and fabrics be supplied with certifications as to how the manufacturer treated the materials. The certification must state the fire tests it is in compliance with, must identify the chemical treatment that was used, and must specify the need for reapplication of the fire retardant treatment if it lasts longer than 12 months. Fire inspectors should also examine the conditions that the manufacturer specifies for retaining the fire retardant effect. These conditions include the permitted number of launderings, exposure to direct sunlight or high humidity, and whether the materials are approved for outdoor use. Fire treatment certification should be provided with a sample of the fabric so that samples may be properly matched during field inspection by the fire inspector.

OTHER FIRE RETARDANT TREATED MATERIALS

The fire inspector must also be aware that the fire prevention code requires certain occupancies to have fire retardant treated draperies and furnishings. Any curtains and drapes installed in use groups A, E, I-2, I-3, and R-1 must be fire retardant, treated in accordance with NFPA 701, and meet both small- and large-scale fire test requirements or be noncombustible. These flame retardant certifications should be reviewed during the annual fire inspection. If the material is not inherently treated, it must be retreated annually or as the manufacturer specifies. The fire inspector should note this in the building file record.

The fire prevention code also regulates furniture and mattresses in hospital I-2 uses and requires that they meet certain fire safety standards unless they were purchased before the adoption of the building code or the building is protected by automatic sprinkler systems throughout. NFPA 261, *Standard Method of Test for Determining Resistance of Mock-Up Upholstered Furniture Material Assemblies to Ignition by Smoldering Cigarettes*, is the fire test standard that applies to furniture, and it permits an energy release of 500 kilowatts or less when subject to fire. The furniture must also be cigarette ignition resistant. Cigarette ignition resistance is measured by NFPA 260, *Standard Methods of Tests and Classification System for Cigarette Ignition Resistance of Components of Upholstered Furniture*. To be deemed cigarette ignition resistant, the material may char no more than 1½ inches with no ignition of the material. All bedding and mattresses must also be fire retardant in accordance with federal standards under DOC 16CFR part 1632, which permits a char length of 2 inches with no ignition of the mattress. All mattresses must have a label from the testing agency approving compliance with the federal mattress safety standard.

DECORATIVE VEGETATION AND PLANTS

The fire inspector encounters different types of decorations in buildings on a seasonal basis. The addition of hay bales in the fall and cut Christmas trees in the winter must be addressed through proper fire code enforcement, especially in heavily occupied structures such as schools, hospitals, day care centers, and day nurseries. The fire prevention codes restrict the use of cut trees in buildings unless they are protected by automatic sprinkler systems. In protected facilities, the cut trees must be properly managed and not be permitted to dry out. Additionally, they must be upright and stable in their base, and decorative items such as cardboard packages may not be stored below them. Candles are prohibited in the immediate area of the tree, and all electrical equipment and tree lights must be UL listed and used in accordance with the manufacturer's specifications. Trees must be checked daily for dryness and must be removed once needles fall off when a branch is shaken or if the branches become brittle when bent between your fingers. Live-balled trees are permitted in all structures but must be maintained moist. Decorations should not impede or encroach upon exits or required exit access space.

Interior decorations are difficult to control, and the fire inspector must stay on top of seasonal activities, especially in elementary or high school buildings. Decorations that are placed in areas that obstruct exit access must be prohibited. Fire prevention codes limit the amount of paper artwork that can be placed in school hallways to 20 percent of the wall area. On a typical 10-foot wall, this allows 24 inches of art to be placed in line along the wall area. Storage of items in the corridors such as clothing, book bags, or boots also is prohibited unless automatic sprinklers, smoke detectors, or metal lockers are provided.

Fire inspectors must maintain vigilance in their community and be proactive in letting the public and building owners know what types of decorative materials are permitted under the fire prevention code as well as the fire retardant documentation and treatment certifications required for inspection of those decorations.

Common Fire Code Violations

Fire inspectors must ensure proper inspection of the passive fire protection construction features in the building. Identifying the common fire code violations begins with properly identifying the building's type of construction. The fire inspector must be capable of identifying the firewalls, fire separation walls, exit enclosures, and fire and smoke barriers within the building. This identification issue has long been discussed, and in the 2009 IBC identification signs are now required by section 703.6.[3] The signs are placed in the attic, basement, or void space above the drop ceiling at a 30-foot interval so that contractors working in these spaces do not penetrate the fire-resistant wall without proper firestops or fire-rated opening protection devices. In older buildings, the fire inspector has to figure out which walls are fire resistant. This can be accomplished by examination of the door openings and knowledge of the building code tables on types of construction. Fire doors must be self-closing and rated to 75 percent of the wall fire resistance rating so that a 2-hour fire separation wall is protected by Class B labeled 1½-hour fire doors. The exception is the exit access corridor, which may be protected by 45-minute doors in mechanical areas and 20- to 30-minute doors in office areas. Typically, corridors have 1-hour fire-resistant walls unless the building is protected by automatic sprinkler systems, in which case, the walls may be nonrated.

The fire inspector must also examine the horizontal fire protection systems or floor/ceiling assemblies. In multistory buildings, the floor/ceiling may be part of a fire-resistant envelope. The fire inspector needs to make sure that the ceiling has no openings that allow fire spread. In most buildings with drop ceilings, the drop ceiling may be or may not be part of the fire-rated assembly. To determine whether the drop ceiling is rated, the fire inspector needs to examine the interstitial space above the ceiling. Fire-resistant ceiling tiles are identified by labels and are required to be clipped in place against the grids to prevent the tiles from being dislodged by thermal pressure. Sometimes, these clips are removed by contractors to service equipment above the ceiling and may not be replaced. By at least examining some of these interstitial spaces, the inspector will be able to determine whether the ceiling is part of the rated floor ceiling assembly. Another clue to a fire-rated drop ceiling is that all penetrations, such as light fixtures, heating and cooling supplies, and return diffusers, must be protected by proper fire-rated boxes or fire dampers (see Figure 10.9). Fire-rated

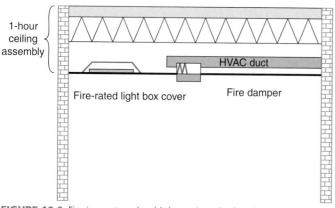

FIGURE 10.9 Fire inspectors should determine whether the drop ceiling is part of the rated fire-resistant assembly. Ceiling tiles should be clipped, the grids should be UL fire rated, and all openings in the ceiling for lights or heating should have proper fire-rated covers and fire dampers.

1-hour ceiling assembly

HVAC duct

Fire-rated light box cover

Fire damper

FIGURE 10.10 Fire-resistant walls must be properly taped and firestopped above the drop ceiling in the interstitial space. *Courtesy of J. M. Foley*

drop ceilings also have fire-rated metal grids that are of a heavier design than those in non-rated drop ceiling applications. If the drop ceiling is part of the rated assembly, all of these fire-restricting components must be present. If they are not, then the ceiling protects a non-rated concealed space. In these cases, any penetration of the floor assembly above the ceiling must be properly firestopped with fire-resistant materials matching the horizontal fire resistance rating of the floor assembly. Fire inspectors should examine building system areas carefully, such as electrical and telephone closets for penetrations. These areas are generally where conduit and pipes pass through rated floor/ceiling assemblies. Any open spaces, holes, or missing firestops should be noticed and corrected.

CEILING PLENUMS

In unfinished areas above ceilings, fire-resistant rated assemblies must be taped, spackled, and fire sealed to maintain their integrity. This includes the void or interstitial spaces (see Figure 10.10). The interstitial space above a drop ceiling may also be used as a return air plenum in some facilities. In these cases, all wiring must be in conduit or be fire rated for plenum use (see Figure 10.11). The void space should be free of combustible materials. The fire inspector should also note whether electrical fixtures are properly supported independent of the drop ceiling. In many cases, the fluorescent fixtures are dropped in the ceiling grid, which creates a potential entanglement hazard for firefighters should the ceiling collapse during a fire. In 1994, two Memphis firefighters were killed by such a situation during a high-rise fire in Tennessee.

In commercial kitchens, it is a good practice to examine the voids above the ceiling for two reasons. The first reason is that grease-laden vapors may escape from the range hood if it is not properly drafting, causing a grease buildup and fire hazard above the ceiling. Second, the inspector can see any changes in direction of the ventilation duct, which identifies the cleanout locations and any issues with clearance to combustible construction or access to cleaning the ducts. The fire inspector should identify where the duct turns upward, as this will be the bottom of a vertical shaft and is required to be a fire-rated enclosure.

FIGURE 10.11 Ceiling plenums used for return air supply must have properly protected wiring systems. Often, improper wiring occurs after certificates of occupancy are issued, such as that used for communication, television, and data line cables. Fire inspectors should look above ceilings whenever possible. *Courtesy of J. M. Foley*

VERTICAL SHAFTS

All shafts for the enclosure of exhaust systems, electrical chases, elevators, ventilation, waste removal, chimneys, and vents must be enclosed in fire-resistant shafts. Shafts generally require 1-hour fire ratings up to four stories and 2-hour fire resistance ratings above four stories. A fire-rated door or a damper assembly must protect any openings within the shaft wall. Fire inspectors should inspect shaft wall openings to ensure that protective closures are installed and maintained. This is especially critical in high-rise buildings with trash waste removal chutes. Trash chutes are usually protected by automatic sprinkler heads on every other floor but often, the fire doors become damaged by occupants due to overuse. Damaged trash chute fire doors should be replaced (see Figure 10.12). The fire inspector should also inspect the bottom of the shaft in the trash compactor room, as it must have a fusible link fire door at the bottom to close off the vertical shaft in case of a trash room fire. Vertical shafts for kitchen exhaust systems should have fire-rated access cleanout doors every 20 linear feet or usually every other floor. These doors allow the cleaning and inspection for buildup of grease in the vertical duct. The inspector should pay particular attention to access panels that must be frequently opened, as they often become damaged and may require replacement. Vertical shaft walls should be tight to the rated floor assemblies above and below and should be properly sealed for smoke and fire penetration.

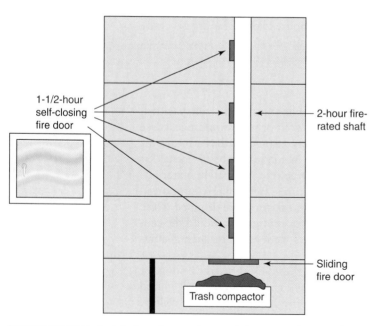

1-1/2-hour self-closing fire door

2-hour fire-rated shaft

Sliding fire door

Trash compactor

FIGURE 10.12 Vertical shafts such as trash removal chutes must be properly protected. Fire inspectors should examine fire-rated door closures and the sliding fire door at the bottom of the trash chute.

Summary

As with any fire protection system in a building, the passive systems are only as good as the building's maintenance. Access panels that are broken or left open, holes in the fire-resistant membrane, or the installation of plumbing, electrical, cable, computer, or HVAC systems after the building has been issued a certificate of occupancy often can compromise these passive fire protection features. The safety of the structure and the building occupants depends on these passive fire protection systems being properly maintained and in place at all times. Small compromises can have a devastating effect. As an example, in the 1980 MGM Grand Hotel fire, a fire-rated access door was left open in a high-rise stair tower that connected to the casino floor catwalk area. This 2 foot by 2 foot opening allowed smoke into the stairway, making it unusable for the hotel occupants above.[4] Most people would walk by that opening and never give it a second thought, but the impact it had during the fire was devastating to the occupants who needed that exit to escape. Fire inspectors must constantly improve their knowledge of the requirements for passive fire protection features in each type of construction so they can readily identify defects or potential fire spread issues that may have significant impact on occupant life safety.

Review Questions

1. Discuss the differences between early building code approaches to fire protection and current building codes as they relate to fire-resistant construction.
2. What is the relationship between fire load, fire severity, and fire endurance?
3. List several fire testing standards used in determining fire resistance requirements. What elements of the material do they examine?
4. List and identify the major differences between the five types of building construction. Which type do you believe would give the poorest performance under fire conditions?
5. Describe the differences between the following:
   a. Firewalls
   b. Fire separation walls
   c. Fire barriers
6. Describe the common fire resistance ratings for each of the following elements:
   a. Exit access corridor
   b. Stair enclosure of six stories
   c. Shaft enclosure of three stories
   d. A wall dividing two use groups
   e. An exterior wall
7. What test is used to determine the fire resistance rating of a fire door? List the failure points of the test procedure.
8. Identify the different types of wall openings and the appropriate opening protective devices that are required. Is glass permitted in every location?
9. When is the installation of a fire-stopping material required, and how is it tested?
10. Flame spread is an important facet of inspection of interior finishes. Do building codes restrict the application of interior finishes, and if so, how is that determined?
11. What does the term *noncombustible* mean? Are fire retardant materials noncombustible?
12. Why is the inspection of installation of decorative materials so important? What features should the fire inspector look for during the inspection?

Suggested Readings

Loyd, Jason B., and Richardson, James D. 2009. *Fundamentals of Fire and Emergency Services*. New Jersey: Brady /Pearson Publishing.

Smith, Michael. 2012. *Building Construction: Methods and Materials for the Fire Service*. 2nd edition. New Jersey: Brady/Pearson Publishing.

International Code Council. *2009 International Building Code*. Washington, DC: International Code Council.

International Code Council. *2009 International Fire Code*. Washington, DC: International Code Council.

Endnotes

1. Thomas M. Cunningham, *Historical Perspective: Plane Strikes the Empire State Building* (Annapolis, MD: U.S. Naval Academy Fire Department).
2. International Building Code – 2009, Table 601 (Washington DC: International Code Council).
3. International Building Code – 2009, Section 703.6 (Washington DC: International Code Council).
4. Richard Best and David Demers, *Investigation Report on the MGM Grand Fire, Las Vegas, Nevada, 1980* (USFA, NIST, NFPA 1982).

11

Fire Alarm Systems

Courtesy of J. M. Foley

OBJECTIVES

After reading this chapter, the reader should be able to:

■ Describe how fire detectors operate by identifying the signatures of a fire.

■ Know the general causes of false fire alarms.

■ Know the types of mode of operation of heat, smoke, and other special fire detectors.

■ Know the main components of an automatic fire alarm system.

■ Know the main circuits of a fire alarm system and their performance measures.

■ Know the appropriate listing and application of fire alarm components.

■ Understand fire inspection and maintenance requirements for automatic fire alarm systems.

Professional Levels of Job Performance for Fire Inspectors as Cited in NFPA 1031 and NFPA 1037

■ NFPA 1031 Fire Inspector I *Obj. 4.3.6 Determine the operational readiness of existing fire detection and alarm systems*

■ NFPA 1031 Fire Inspector II *Obj. 5.3.4 Fire protection systems and equipment*

- NFPA 1031 Fire Inspector II *Obj. 5.3.11 Verify compliance with construction documents.*
- NFPA 1031 Fire Inspector II *Obj. 5.4.4 Review the installation of fire protection systems.*
- NFPA 1031 Plan Reviewer I *Obj. 7.3.10 Evaluate plans for the installation of fire protection and life safety systems*

Introduction

A crucial element of a fire prevention inspection program is ensuring the adequacy and functionality of a fire alarm system. Fire alarms provide two important elements in building life safety. First, the fire alarm system must alert the occupants to a fire in the building and get them to evacuate; second, it must notify the fire department to respond to the fire at the building. The fire inspector must have a good understanding of how the components of a fire alarm system operate and how to identify potential fire code issues in the installation and maintenance of the fire alarm system.

The first identified fire alarm system was established in 1659 when Governor Peter Stuyvesant of New York instituted the "Rattle Watch." The "Rattle Watchers" were fire marshals that patrolled the city streets at night looking for the possible outbreak of fires. When fire marshals encountered a fire, they would use a wooden hand rattle to alert the neighbors and would yell to them, "Throw out your buckets" so that responding firefighters could gather them for use them in a bucket brigade. In America's colonial days, every homeowner had to have a leather fire bucket for use by the local volunteer firefighters.[1] Fire watches existed until the 1830s when the first telegraph was invented and two men, William F. Channing and Moses G. Farmer, successfully built the first operational fire alarm telegraph in the city of Boston in 1851. By 1859, additional advances in the fire alarm telegraph were made, and John N. Gamewell, an entrepreneur who purchased the patents from Channing and Farmer, began building fire alarm telegraph systems throughout the eastern United States.[2] Gamewell became the fire alarm industry leader. The next improvement was the development of the electronic fire sensor. The first electric fire sensor was invented in 1863 by William Watkins and was called the *Watkins thermostat*. By 1873, the Watkins thermostat was endorsed by the National Board of Fire Underwriters as a fire detector, and it became the forerunner of the modern heat detector.[3] These heat-sensing devices were extensively used through the late 1800s for property protection.

The basic concept of the fire alarm systems evolved through the nineteenth and twentieth centuries into a highly technological industry that is improving and innovating fire detection technology every year. Fire alarm systems are still based on the concept of early warning for occupants and notification of fire services; but now, they have also evolved into systems that can manage smoke movement in a building, close doors, give voice direction, allow firefighters to communicate, provide mass notifications, and institute many other actions to improve life safety. The inspection of the automatic fire alarm systems requires that the fire inspector understand fire detection principles and the basic provisions of the NFPA fire alarm for the installation, testing, and maintenance of the fire alarm system.

Fire Alarm Detection Methods

The first and most important concept that a fire inspector must understand is that fire detectors do not actually detect fire. Fire detectors react to environmental conditions that are characteristic of a fire, but may also be characteristic of nonfire phenomena. Fire detectors look for environmental signatures in the environment that may represent a fire such as flames, heat, smoke, or gas. These same signatures of a fire can also be caused by other environmental occurrences such as steam from a shower, gas effluent from a stove, or heat from a supply damper. The placement, selection, and type of fire alarm detector must be based on the most likely fire signatures that can occur in the building and must take into account any environmental conditions that may cause false alarms. Good fire alarm system design takes into account the type of fuel in the building and the most likely fire signature that fuel produces. As an example, in a flammable liquid storage room, we expect a fire to produce high heat and flame very quickly; therefore, heat or flame detection may be more appropriate than the installation of smoke detectors. Fire inspectors should examine each area of the building and determine if the fire detector is correct for the signature that is produced by the fuel. Often, in a building, the use of a space will change, but the fire detection equipment remains the same. These conditions may cause false alarms or delayed alarms depending on the type of detectors in the space.

The False Alarm Problem

Fire alarm systems are necessary for the early warning and evacuation of occupants during a fire. False fire alarms reduce the occupant's confidence in the fire alarm system and eventually train the occupant to disregard the alarm. When there are numerous false alarms in a building, the occupants normally disregard the signal as just another false alarm. This eventually leads to evacuation delays and the possibility of occupants becoming trapped in a real fire. Fire prevention inspectors must be attentive to these false alarm problems in their jurisdiction. Fire alarm systems can also be a source of aggravation to the local fire department if numerous false alarms are occurring. In fire department deployment studies conducted by the author, National Fire Incident Reporting System (NFIRS) fire report data demonstrated a 50 percent reduction in volunteer firefighter response to automatic fire alarm systems activations when there was a high rate of false alarms at the building. The volunteer firefighters' assumption is that likely it's just another false alarm. Eventually, however, the fire alarm system may detect a real fire, and the responding firefighters will be understaffed on the initial response because of numerous false alarms.

The NFPA 2009 report on false alarms claimed that 45 percent of false alarms were unintentional and 32 percent were caused by equipment malfunctions. Malicious or deliberate false alarms accounted for only 8 percent of all false alarms that year. The NFPA report also pointed out the impact on fire department responses that showed a total of 45 false alarm responses for every 10 structure calls that fire departments responded to.[4] That 4.5:1 ratio can take its toll on volunteer firefighters as well as putting firefighters at risk, responding to nonemergencies. Many career departments have reduced initial responses to alarm system activations as a result of excessive false alarms.

2

The NFPA reported that the major causes of most false alarm occurrences were smoking, dust, humidity, defective equipment, and high air velocities in the detection area. They also identified steam, construction, cooking fumes, and water intrusion into the alarm system as causes of false alarms. The majority of these conditions can be rectified with good fire inspection practices. The fire prevention inspector must be involved in reducing the false alarm problem by identifying incorrect applications of detection equipment, ensuring regular testing and certification of the fire alarm system. Fire inspectors must write notices of violations and orders to correct fire alarm problems before the false alarm frequency negates the effectiveness of the fire alarm in evacuating the occupants of the building.

Listing, Labeling, and Testing Laboratories

Students of fire protection should understand the importance of listing and labeling of fire alarm and fire protection equipment and how it relates to fire codes and standards. In its simplest terms, the code dictates when a fire protection system or device is required to be installed in a building. The applicable installation standard will then explain how the system of installation is to be performed as it relates to spacing or location of equipment. The listing and labeling requirement is to ensure that the manufacturer of the equipment has built it to the standard specification for performance. This process reduces the possibility of installing substandard equipment in a fire protection system. Building and fire codes generally require that all fire protection equipment comply with a nationally recognized testing laboratory's labeling and listing requirement. The two predominant laboratories in the United States are Underwriters Laboratories and Factory Mutual Research. The U.S. Department of Labor through the Occupational Safety and Health Administration (OSHA) maintains a list of other Nationally Recognized Testing Laboratories, or NRTL, for enforcement of OSHA safety regulations. Currently there are 16 recognized laboratories, including the ones mentioned on the list. Fire inspectors can go to the NRTL website at OSHA and see which of the ANSI, UL, or FM test requirements each laboratory is certified to conduct. Fire inspectors should recognize that, while UL and FM labeling are the most common, there are other nationally recognized laboratories that comply to those testing standards. For the purpose of simplicity the text will refer to the UL specific standards.

Fire Alarm System Components

All fire detectors and fire alarm system components must be listed and labeled for the specific applications that they are to be used for. Typically the labels will be from Underwriters Laboratories or Factory Mutual Research (*FM Global*). Certification and testing labels on fire detection devices ensure proper performance and testing under the fire alarm standard NFPA 72, *National Fire Alarm and Signaling Code*. Some important fire alarm detection equipment listings are as follows:

- UL 268 – System smoke detectors (no internal audible alarm)
- UL-217 – Single and multistation smoke detectors
- UL268A – Duct smoke detectors
- UL 521 – Heat detectors
- UL 864 – Fire alarm control panels
- UL 464 – Audible appliances
- UL 1480 – Speakers and amplifiers
- UL 1638 – Visual signals
- UL 1730 – Annunciator panels
- UL 1971 – Signaling systems for hearing impaired

The Underwriters Laboratories equipment listing and certification also tells the fire inspector whether the fire alarm equipment is intended for commercial or household applications. Underwriters Laboratories provides equipment listings on their website. Fire inspectors may search the UL's database for fire alarm equipment listings and fire alarm equipment services. UL provides a quick guide on how to search their database, which is a really useful tool to fire inspectors seeking equipment information.

alarm-initiating device ■ Fire detectors, manual pulls, stations, or water flow devices that detect heat, smoke, gases, light, or other signatures of the presence of a fire.

A fire alarm system consists of several interconnected components: **alarm-initiating devices** or fire detectors, alarm-indicating appliances or speakers and visual indicators, a fire alarm control panel, and a communication device to send a signal off site and call the fire department. Small fire alarm systems may be **non–power limited**, or powered by 120 volts, while larger alarm systems generally are **power limited**, or powered by 12–24 volts. The National Electrical Code®, NFPA 70, and the jurisdictional building code regulate all wiring methods employed in fire alarm system installations. There are two general categories of fire alarm systems: residential systems and commercial systems. Fire alarms for residential use detect and alert the occupants and may notify the fire department through an alarm monitoring service. Residential fire alarms in one- and two-family dwellings are installed in accordance with NFPA 72, *National Fire Alarm and Signaling Code*, Chapter 11, formerly NFPA 74, *Standard for Household Fire Warning Equipment*. These residential systems are self-contained, single-station detectors that are interconnected so that the initiation of any device sounds all the alarms on the circuit. They may be power limited (12–24 volts) or non–power limited (120 VAC) and can be combination burglar/fire alarm security systems. Household fire alarm systems are usually listed as residential under the UL equipment listings and may be used only in residential applications. Small residential occupancies like rooming houses or small apartment buildings may have multistation interconnected smoke detection systems of the non–power limited (120 VAC) variety installed throughout the building's common areas. These alarm-initiating devices must be installed according to NFPA 72 for spacing, wiring, and location. Generally, multistation smoke detection systems are nonmonitored and have no fire alarm control panel. The system consists of interconnected UL 217–listed single-station smoke detectors with an interconnected signaling wire that sounds the alarms in unison. Larger residential buildings such as hotels and motels normally have power limited (12–24 volts) fire alarm systems using a fire alarm control panel and monitoring service technology. These fire alarm systems must be listed for commercial application by UL and are installed in accordance with NFPA 72.

Fire alarm systems installed in industrial and commercial buildings must also detect and alert the occupants, but may also activate fire suppression systems to begin suppression of the fire or may interface with other life safety equipment in the building. These interconnections of systems may be a requirement of other NFPA standards or the building or fire code, but still must meet NFPA 72 requirements for supervision and monitoring by the fire alarm system. Ancillary fire alarm detection systems may be used in total flooding gas fire suppression systems under NFPA 12A, *Standard on Halon 1301 Fire Extinguishing Systems*; automatic fire sprinkler systems; preaction sprinkler systems under NFPA 13, *Standard for the Installation of Sprinkler Systems*; foam fire protection systems under NFPA 16, *Standard for the Installation of Foam-Water Sprinkler and Foam-Water Spray Systems*; or explosion suppression systems under NFPA 69, *Standard on Explosion Prevention Systems*, and they are installed under those NFPA standards and NFPA 72.

non–power limited ■ Current line power of 110 VAC to 600 VAC on wiring systems for automatic fire alarms.

power limited ■ The reduction of current line power on a fire alarm circuit to 12 or 24 volts AC.

Alarm-Initiating Devices

The alarm-initiating device under NFPA 72, *National Fire Alarm and Signaling Code*, is considered any device attached to the fire alarm control panel that either automatically or manually activates an alarm signal. Alarm-initiating devices include smoke detectors, heat and flame detectors, water flow switches, and manual fire alarm pull stations. Each type of device will initiate based on different signatures of a fire. The installation of each type of device is predicated on the fire load being protected and the predominant fire signature that will initiate an alarm the quickest.

HEAT DETECTORS

Heat detectors are by far the oldest and most reliable fire detection technologies going back to the original Watkins thermostat. These devices usually are uncomplicated in design and reliable in use because their operation is relatively simple. There are three common types of heat detectors: fixed temperature, rate of rise, and rate compensation. Some manufacturers

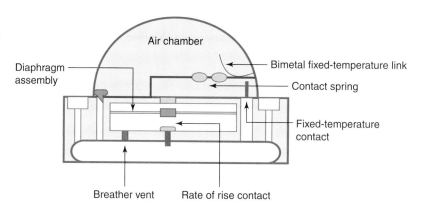

FIGURE 11.1 Fixed-temperature heat detectors operate by a fusible metal melting and releasing a contact spring that completes the circuit to a contact terminal and initiates the fire alarm.

of heat detectors also may use multiple methods of detection and may incorporate heat detection in other technologies like smoke detectors. These initiating detection devices are referred to as either **spot detectors** because they are located as single units and protect an area of the building, or **line detectors** because the detector is linear and protects a specific hazard such as an electrical cable tray. Let's look at each type of heat detector:

spot detector ■ Single fire detectors that cover a specific square foot floor area, usually 900 to 2,500 square feet.

line detector ■ A type of heat detector that runs linearly along a vessel or cable tray to detect heat.

- *Fixed Temperature:* The fixed temperature detector operates using a fusible link that melts at a specific temperature (see Figure 11.1). The fusible link's melting temperature is usually 135°F. The link releases a spring contact inside the detector that closes an electrical circuit and sends the alarm signal to the fire alarm control panel. Fixed temperature heat detectors are nonrestorable. NFPA 72 requires that fixed temperature heat detectors be inspected annually and be replaced at 15 years of service life. At the 15-year benchmark, at least two heat detectors out of every 100 detectors installed must be removed and tested by an independent laboratory. If the detectors fail to operate within the manufacturer's specification, then additional heat detectors must be removed and tested or all heat detectors should be replaced. If an owner decides to continue to perform the required laboratory testing instead of replacement, then the testing must continue to occur at an interval of every 5 years beyond the fifteenth service year.

- *Rate of Rise:* The rate of rise heat detector senses a rapid change of temperature over a 60-second time period. The temperature change is usually 12–15°F over the course of 1 minute. Rate of rise detectors usually are also combined with a fixed temperature element and can be tested annually with a hair dryer or heat gun or a manufacturer's testing apparatus. Care must be taken, however, not to concentrate any heat on the fixed temperature element, as that part of the heat detector is nonrestorable. The principle of operation of rate of rise detection again is relatively simple. In the unit, there are two air chambers separated by a rubber diaphragm (see Figure 11.2). One air chamber is sealed, and the other has an air vent to atmosphere. As the surface

FIGURE 11.2 Rate of rise heat detectors operate by the expanding heated air in the chamber not being able to escape fast enough through the breathing vent. The diaphragm then expands upward, pushing the contact spring against the contact screw and initiating the alarm.

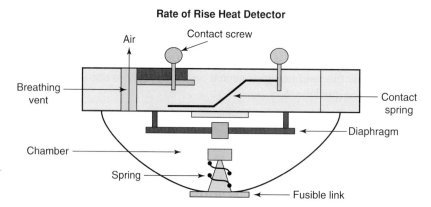

Rate of Rise Heat Detector

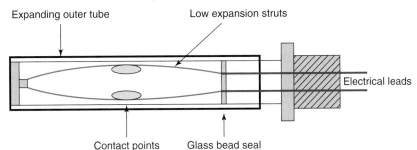

Rate Compensation Detector

Expanding outer tube

Low expansion struts

Electrical leads

Contact points

Glass bead seal

FIGURE 11.3 The rate compensation detector works at a fixed temperature based on the expansion of two dissimilar metals heated at a constant temperature. The detector will operate at a given temperature regardless of the temperature rate of rise.

of the detector is heated, the air in the sealed space expands, forcing the air on the other side of the diaphragm out of the vent. This air expansion closes an electrical contact and initiates the alarm. Typical fixed temperature and rate of rise heat detectors are UL listed at 900-square-foot spacing for the purpose of coverage. Some heat detectors, however, may also be rated for 50-foot spacing or 2,500 square feet.

- *Rate Compensation:* Rate compensation heat detectors are slightly different technologies than rate of rise detectors. Rate compensation detectors use the difference in the expansion and contraction of dissimilar metals to activate a signal. The rate compensation detector looks like an aluminum tube and is often used in kitchen fire suppression systems instead of the traditional fusible links in the hood plenum (see Figure 11.3). The advantage of the rate compensation detector is that it operates at a fixed temperature point without any thermal lag. Thermal lag is the time necessary for the heated air to get a fusible link to its melting point. The general operating principle of the rate compensation detector is that the two metal components of the detector have different expansion ratios. The outer metal case of the detector is one element, and inside the case are two metal expansion struts that expand slower than the outer case. If there is a rapid buildup of heat on the outer case, it expands faster and the inner struts close the electrical contacts mounted on the struts, sending the fire alarm signal. Rate compensation heat detectors operate in the temperature range of 135–194°F. Rate compensation detectors generally are listed for 2,500-square-foot spacing.

- *Line Heat Detectors:* Line heat detectors are long, twisted wires that are often used in the protection of electrical cable trays or flammable liquid storage tanks as well as other specialized pieces of industrial equipment. There are two predominant types of line heat detectors:
  - *Heat-sensitive cable:* Heat-sensitive cable is a very simple technology that consists of two wires with a heat-sensitive outer wrapping twisted together. The heat detector is either laid in a cable tray or suspended in close proximity to the piece of equipment to be protected. If a fire occurs, the heat-sensitive wrapping melts and allows the two twisted internal copper wires to short together, initiating an alarm signal (see Figure 11.4). Heat-sensitive cables can run up to 250 linear feet on a single detector unit. These linear heat

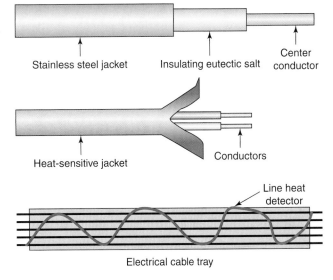

Heat-Sensitive Cable – Line Detectors

Stainless steel jacket

Insulating eutectic salt

Center conductor

Heat-sensitive jacket

Conductors

Line heat detector

Electrical cable tray

FIGURE 11.4 Line heat detectors may be thermistor type, such as mineral insulated cable or heat-sensitive cable, which has a meltable plastic jacket. Line-type heat detectors are used to monitor electrical cable trays or other applications such as aboveground storage tanks.

detectors can be used in many different applications including protection of aircraft hangers, tunnels, conveyor systems, and electrical cable trays or floating roof flammable liquid storage tanks.

- *Mineral Insulated Cable:* Mineral insulated or MI cable is a slightly different technology that uses an inorganic material, magnesium oxide, to transfer electrical current from an outer copper jacket to an inner copper or nickel wire. When the mineral insulated cable is heated by the fire, the magnesium oxide allows the flow of free electrons between the two conductors, thus initiating the alarm. The advantage of MI cable is that it is much more weather resistant and durable than heat-sensitive cable. MI cable is also nonflammable and resists damage by fire; therefore, it ensures the integrity of the circuit under fire conditions.

SMOKE DETECTORS

Smoke detection technology was actively developed through the late 1960s and early 1970s. According to Dr. John Hall of the NFPA, the first single-station smoke alarm was invented in 1965 using photoelectric technology. The first ionization smoke detectors were introduced in 1970, and by the mid-1980s more than three-quarters of U.S. residents had at least one smoke detector in their home.[5] Most smoke detectors operate on one of four main principles: the obscuration of light, the scattering of light, the ionization of air, or the sampling of air for particulate matter. Smoke detectors can be single or multiple stations and are listed under either UL 217 for single-station smoke detectors or UL 268 for system smoke detectors. The key difference between the listings is that UL 217 detectors can both detect and alarm, whereas UL 268 detectors can only detect and must be connected to a fire alarm control panel to initiate an alarm. Smoke detectors generally have a UL-listed spacing of 900 square feet per unit and a useful life expectancy of about 10 years. The life expectancy is based on NFPA analysis that demonstrated a decrease in smoke detector performance for units over 10 years of age.[6] Smoke detectors, unlike heat detectors, are under constant power conditions, and the internal electronics tend to fail over time from heat and dirt buildup. The NFPA and United States Fire Administration recommend that smoke detectors be replaced every 10 years. Many smoke detector manufacturers now install 10-year sealed lithium batteries in their smoke detectors so that the purchaser doesn't have to replace batteries during the detector's useful life span. At the end of the battery life, the detector is simply replaced with a new smoke detector unit.

photoelectric smoke detector ■ A fire detector with a photocell receptor that detects a fire by light scattering increasing energy to a photocell or reduction in light by smoke obscuration.

Smoke detectors may be mounted upon the ceiling or the walls in accordance with NFPA 72 and should not be closer than 4 inches to a wall–ceiling intersection nor more than 12 inches down from the ceiling if mounted on a wall (see Figure 11.5). Let's examine each type of smoke detection and the method of operation and inspection points for the fire inspector to apply.

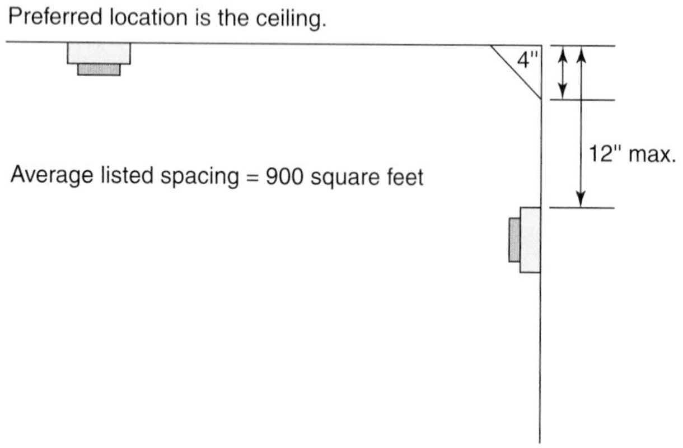

Photoelectric Smoke Detectors

The **photoelectric smoke detector** operates on the principle of light scattering. Inside the detector is a light-emitting diode (LED) and a photoreceptor cell. The LED and the photocell are located 90 degrees from each other so that the light beam cannot hit the photoreceptor cell (see Figure 11.6). When smoke enters the smoke chamber, the light is reflected off the smoke and onto the photoreceptor, which at a certain intensity initiates the fire alarm signal. Photoelectric smoke detectors are faster at detecting smoldering fires than flaming fires, but in the big picture, the time delay is relatively negligible. Photoelectric

Preferred location is the ceiling.

Average listed spacing = 900 square feet

4"

12" max.

FIGURE 11.5 All smoke detectors should be mounted either on the highest part of the ceiling or on the wall between 4 and 12 inches to the leading edge of the smoke detector for maximum effectiveness. Most smoke detectors are UL listed to protect 900 square feet of space.

smoke detectors generally are more reliable and less prone to false alarms than are ionization smoke detectors. Often, an ionization smoke detector that is generating false alarms due to environmental conditions can be replaced with photoelectric smoke detectors to resolve the false alarm problem. The fire inspector should inspect the following key points for photoelectric smoke detectors:

FIGURE 11.6 Photoelectric smoke detector uses reflected light to initiate an alarm. The LED is placed 90 degrees from the photocell, and as smoke enters the chamber the light is reflected to energize the photocell and initiate an alarm.

- The smoke detector body should be clean and dust free.
- The smoke detectors should be located to favor the air return diffuser in the room.
- The smoke detector should be at least 3 feet from any air supply diffuser in the ceiling that would push smoke away from the detector.
- The smoke detector should be located at the highest point in the space, such as at the top of stairway, or within 3 feet of the ceiling's peak.
- The smoke detector should have an undamaged insect guard in the outer case (see Figure 11.7). Spiders and gnats often are attracted to the LED light source and cause false alarms if the insect guard is damaged.
- The smoke detectors should be serviceable from the floor area without high-reach equipment. Frequently, spot smoke detectors are placed in locations where they are impossible to maintain or properly service.

Ionization Smoke Detectors

The **ionization smoke detector** is the most common smoke detector and operates on the principle that ionized air can carry electrical current. Located inside the ionization smoke detector is a dual ionization chamber that air moves through. A small radioactive source containing americium 241 ionizes the air molecules as they move through the chamber, allowing electric current to flow through the air between two electrode

ionization smoke detector ■ A fire detector that ionizes air and measures the disruption of current flow when ionized gases enter the ionization chamber.

FIGURE 11.7 Photoelectric smoke detector insect screen keeps insects from creating false alarms by being attracted to the LED light and heat. *Courtesy of J. M. Foley*

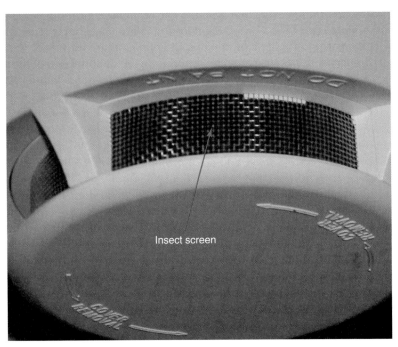

Insect screen

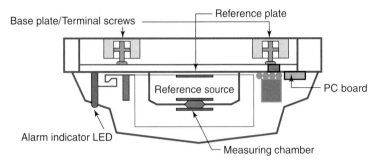

Ionization Smoke Detector

Base plate/Terminal screws

Reference plate

Reference source

PC board

Alarm indicator LED

Measuring chamber

FIGURE 11.8 The ionization smoke detector uses a radioactive source to ionize air in the measuring chamber. Smoke particulate entering the chamber reduces current flow and causes the alarm to initiate.

plates (see Figure 11.8). When smoke or other molecular gases enter the ionization chamber, the air's ability to ionize is reduced, and at a given point, the reduction in current initiates a fire alarm condition. Often, ionization smoke detectors are referred to as "dual chamber." This means that there are two ionization chambers in the unit; one chamber is open to atmosphere and the other is not. The purpose of this construction is to prevent atmospheric pressure and humidity from causing false alarms. If humid air enters one chamber, it is compared to the other; if they are the same, the alarm is canceled out. The dual chamber helps reduce possible environmental false alarms. Ionization smoke detectors are by far the most prevalent in the marketplace today, mostly because of their low cost. These smoke detectors are affordable and in most cases are given away by many fire departments through community smoke detector programs. Ionization smoke detectors are more prone to environmental false alarms than photoelectric models are. The key element in eliminating false alarms with ionization smoke detectors is "location, location, location." These smoke detectors must be kept away from kitchen areas or high-humidity areas like outside of bathrooms. They should not be located near dressing tables where aerosols may be discharged. The fire inspector should look at the smoke detector and use the same inspection criteria as discussed for photoelectric smoke detectors.

Buildup of dust holds heat in the smoke detector and can shorten the electronic life of the unit. If ionization smoke detectors become false alarm problems, they should be replaced with new units, photoelectric detectors, or other detection technologies.

SPECIALIZED DETECTORS

- *Beam Photo Electric Detectors:* Beam detectors are photoelectric detectors that operate on the principle of light obscuration. The application of beam technology is to protect large, open spaces such as atriums. The beam detector can cover an area of approximately 15 feet wide by 60 feet in length with a single-beam detection unit. Beam detectors consist of either a beam transmitter and a beam receiver or a single-beam transmitter/receiver that uses a reflector on the opposite side of the area being protected. Beam smoke detectors are more expensive than conventional smoke detectors; however, they cover far more area and require less expensive maintenance. The method of operation for the beam detector is a light beam projected across a space to a beam receiver or reflector. If smoke penetrates the beam's light path, the light becomes obscured. At a predetermined obscuration level (usually 25 to 50 percent obscuration), the reduced light causes a fire alarm condition to occur. Beam smoke detectors normally have a remote annunciate panel located in an accessible area that identifies both trouble and alarm conditions for fire inspection and maintenance purposes. Beam smoke detectors are field adjustable for the obscuration alarm settings and can be adjusted to prevent false alarming. The advantage of using beam detectors is that less equipment is required to protect a large area, and the system maintenance is more easily accomplished.
- *Duct smoke detector:* Duct smoke detectors are used in heating, ventilating, and air conditioning systems (HVAC) to detect fan motor or filter fires within the duct. Duct smoke detectors are listed under UL 268A, *Smoke Detectors for Duct Application.* Typically, duct detectors are standard smoke detector technology fitted with air sampling tubes that insert into the ductwork. The detector is mounted on the outside of the duct. Duct detectors that are located in enclosed ceilings should have a remote

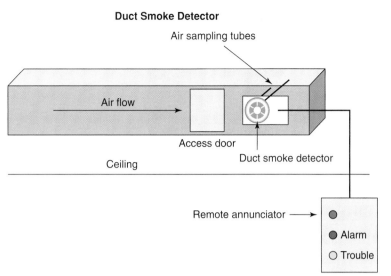

Duct Smoke Detector

Air sampling tubes

Air flow

Access door

Ceiling

Duct smoke detector

Remote annunciator →

○

● Alarm

○ Trouble

FIGURE 11.9 Duct smoke detectors use air sampling tubes inside the duct to detect smoke. Detectors in remote locations above ceilings have annunciators to tell when they are in alarm or trouble.

indicator provided below the ceiling so it can be determined whether the detector is in trouble or alarm condition (see Figure 11.9). Duct smoke detectors are typically installed in accordance with NFPA 90A, *Standard for the Installation of Air-Conditioning and Ventilation Systems* and the *International Mechanical Code* for air handling systems. The codes require that return air ducts moving over 2,000 cubic feet per minute have a duct smoke detector installed ahead of the return air fan. Air supply ducts moving over 15,000 cubic feet per minute must have duct smoke detectors installed ahead of the supply fan motor and air filters. The duct smoke detector is installed for the purpose of shutting down the ventilation system if smoke enters the duct. Duct smoke detectors shall be cleaned and maintained annually with air sampling tubes being examined for obstructions and the sensitivity of the smoke detector being checked. Inspection schedules should comply with NFPA 72, Chapter 10.

■ *Ultraviolet Detectors:* **Ultraviolet** or UV detectors are specialized detectors that use the ultraviolet spectrum to detect the spectral signature of a flame. Ultraviolet radiation is below 4,000 angstroms on the spectrum of visible light. Humans can see only visible light in the spectrum from 4,000 to 7,000 angstroms. Light wavelengths that are above 7,000 angstroms are called infrared light wavelengths. Ultraviolet detectors operate on the lower spectrum wavelengths of 150 to 300 nanoseconds and can detect flame radiation very quickly. These specialized detectors are most often used in industrial applications such as conveyors or pneumatic transfer systems that have a potential for a dust explosion or fires. These detectors may also be used in applications involving flammable liquid fires. UV detectors usually are connected to a fire suppression system that can quickly react and suppress a fire or explosion once flames are detected. UV detectors must be in close proximity to the expected fire problem in order to be effective. The UV sensor provides a cone of protection out to about 45 degrees from the centerline of the fire detector element (see Figure 11.10). At

ultraviolet ■ Ultraviolet is on the lower end of the spectrum below visible light in wavelengths 4,000 angstroms and less. Ultraviolet light is a fire signature that can be detected by ultraviolet fire detectors.

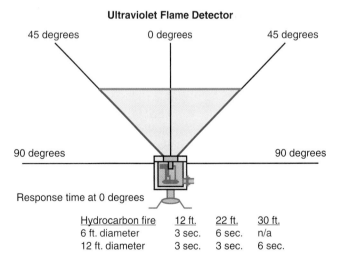

Ultraviolet Flame Detector

45 degrees   0 degrees   45 degrees

90 degrees   90 degrees

Response time at 0 degrees

	12 ft.	22 ft.	30 ft.
Hydrocarbon fire 6 ft. diameter	3 sec.	6 sec.	n/a
12 ft. diameter	3 sec.	3 sec.	6 sec.

FIGURE 11.10 Ultraviolet detectors must be close to the potential source of fire. The greater the distance from the sources, the larger the source must be for the detector to see it. UV detectors are most effective within 45 degrees of center.

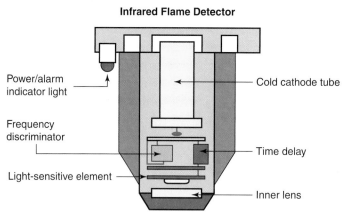

FIGURE 11.11 IR or infrared flame detectors are often used in the protection of aircraft hangars or areas of hazardous materials storage where open flame would be the predominant fire signature.

Infrared Flame Detector

Power/alarm indicator light

Frequency discriminator

Light-sensitive element

Cold cathode tube

Time delay

Inner lens

Detects infrared emissions in a single, narrow wavelength band using a photocell arrangement

45 degrees, the UV detector's sensitivity must be no lower than 50 percent of its design detection rating. There are newer UV technologies that may have sensitivity as high as 70 percent at 45 degrees. Another important design consideration with the use of UV detectors is their optical range or detection distance, which is determined by the *square law*. The square law states that as distance doubles, the target or flame must be four times larger to be effectively detected. This means that if the flame is farther from the UV detector, it must be larger to initiate an alarm response. The maximum effective distance for UV detectors is about 30 feet on the centerline of the unit to the anticipated heat source. In industrial applications, many UV detectors may be used to protect a single piece of equipment.

infrared ▪ Infrared is the high end of the spectrum above visible light at 7,000 angstroms. This energy wavelength is produced in fires and can be detected by infrared fire detectors.

▪ *Infrared Detectors:* **Infrared** flame detectors operate on the higher wavelengths of the spectrum above 7,000 angstroms (see Figure 11.11). IR detectors are often used in aircraft hangers and areas where high amounts of flammable liquids may be used or stored. Early versions of IR detectors would sometimes generate false alarms from sunlight. Newer models can discriminate and filter out sunlight as well as other environmental conditions like welding. The IR detector can identify the visible flame and compare it to a computer chip library of possible responses to reduce the potential of false alarms. It is critical for IR detectors to not produce false alarms because they are usually initiating the operation of high-expansion total flooding foam systems that are very expensive to replenish if discharged. IR detectors are also very effective in detecting non-carbon-based fires like hydrogen or alcohol fires that burn beyond the visible spectrum. These materials produce light waves that are generally not visible to human eyesight.

▪ *VESDA:* The *very early smoke detection aspirator,* or VESDA, is a special application smoke detection system used in high-end protection of sensitive electronic equipment like microchip production facilities (see Figure 11.12). VESDA is used in nuclear plants, financial record facilities, power generating plants, museums, and many other facilities where even a small quantity of smoke would critically damage valuable or historic contents or documents. The way the VESDA system works is through a series of air sampling tubes placed throughout the protected space that are connected to the VESDA detector. Air samples are drawn into the VESDA detector through the tubes where they are filtered and subject to a laser light source. Any particulate in the air will scatter the laser beam and initiate an alarm. Generally, multiple alarm levels can be established for the VESDA system to reduce any false alarm potential. The laser can detect very small levels of particulate that are not visible to the human eye. Maintenance on VESDA systems is required every 6 months to calibrate the detector and also annually to ensure that air-sampling tubes are clear and the filters are replaced.

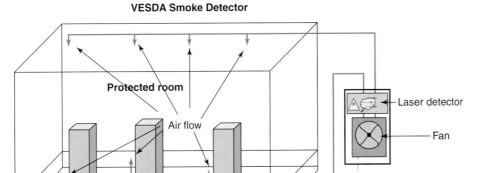

VESDA Smoke Detector

Protected room

Air flow

Laser detector

Fan

Exhaust

FIGURE 11.12 The Very Early Smoke Detection Aspirator (VESDA) detector operates by drawing air samples from the space and passing them through a laser to detect products of combustion.

Alarm-Indicating Appliances

Alarm-indicating appliances are the fire alarm system's output devices and may consist of fire bells, speakers, chimes, or strobe lights for the hearing impaired. Audible devices must provide a minimum sound pressure level of 75 decibels or 15 decibels above the average ambient sound levels in the area or 5 dBA over any sound lasting longer than 30 seconds within the space. The maximum sound pressure level of fire alarm speakers must not exceed 110 decibels at 5-foot minimum hearing distance in accordance with NFPA 72 (see Figure 11.13).

Fire alarm systems traditionally have had a multitude of signaling devices to alert occupants to evacuate a building. The building and fire codes have no standards for the signal output; therefore, different alarm signal equipment providers use different sounds to alert occupants. This creates ambiguity to occupants if they do not know what the alert signal indicates. In older buildings, fire inspectors will find traditional bells, coded bells, as evacuation signal outputs. The NFPA and several other industry groups worked on a standardized fire alarm evacuation signal for many years, and the result was the temporal signal. The temporal signal was adopted by NFPA technical committees in 1997 as a requirement of NFPA 72. Alarm-indicating appliances may utilize voice alarm, bells, chimes, or horns as long as they meet the temporal signal requirement. The temporal signal requirement is a slow temporal whoop that is on ½ second, off ½ second, and on again for ½ second. The sequence of signals must be repeated three times within 180 seconds. The NFPA 72 standard also requires that this signal be used only for total building evacuations; therefore, an alternate signal is required for staged evacuations. When authorities having jurisdiction allow the partial evacuation or relocation of occupants during a fire, the evacuation signal information should be given over the voice alarm system so as not to confuse occupants about when they should fully evacuate the building. Existing nonconforming signaling equipment is still allowed to be used until the equipment is no longer serviceable. All new fire alarm systems are required to use the temporal output signal for occupant evacuation. It is

alarm-indicating appliance ▪ Notification appliances that notify occupants by sound or visual light.

FIGURE 11.13 Audible and visual alarm-indicating appliances must be mounted 80 to 96 inches above the floor and must produce intelligible voice communications and alarm signals at 15 dba above ambient sound or 5 dba above loud sounds lasting more than 30 seconds. Maximum sound should not exceed 110 dba at a 5-foot hearing distance. *Courtesy of J. Foley*

interesting to note that the recognition of the temporal evacuation signal is still not totally clear to the public. In a study conducted for the National Research Council of Canada by Guylene Proulx in 2001, occupants were asked to identify different sounds including a car horn, a car backup alarm, a fire alarm bell, a slow temporal whoop fire alarm, and a buzzer. Only 6 percent of those sampled identified the temporal three signals as an emergency evacuation alarm, while 98 percent recognized a car horn and 50 percent recognized a bell as an emergency signal.[7] Over time, people will become more aware of the temporal signal as an evacuation signal through exposure. In the current version of NFPA 72, fire alarm voice communications also must comply with intelligibility requirements to ensure that communications over the fire alarm speakers are effectively understood by the building occupants. Some of these measures are human factors such as language, speed of talking, and annunciation or articulation of words, while other factors are building and equipment related such as sound attenuation, including the speakers' bandwidth distortion, and building area echoes that may distort the voice. The general idea is not to have the voice alarm sound like the drive-up window speaker at a hamburger stand, which is often unintelligible.

BELLS AND CHIMES

Bells of the single-stroke variety may still be used in fire alarm systems for evacuation as long as they produce the temporal pattern signal upon activation. Because of bell resonance, the temporal pattern is 1 second on, followed by 2 seconds off, then 1 second on again. Chimes may also be used in areas where notification of an emergency may require a more subtle sound level such as in a hospital operating room. After all, you wouldn't want a surgeon to be startled by the fire alarm while using a knife near your vital organs. Chimes must also utilize the NFPA 72–required temporal pattern signal output.

SPEAKERS AND HORNS

Speakers and horns are the most common alarm-indicating appliances in use and may be combined with visual flashing indicators as required for alerting the hearing impaired. Speakers must be mounted in such a fashion as to avoid any mechanical damage and are required to be at least 90 inches above the floor or 6 inches below the finished ceiling. The NFPA 72 standard additionally requires that indicating appliances located within sleeping rooms shall produce a low-frequency sound wave to awaken and alert sleeping occupants. This sound is referred to as a "square wave" of 520 hertz, and it will be required for fire alarm speakers by 2014.

STROBE LIGHTS

Visual strobe lights are required for alerting the hearing impaired in conformance with the Americans with Disabilities Act and NFPA 72, *National Fire Alarm and Signaling Code*. These devices may be part of an integrated fire alarm speaker strobe or may be a separate strobe light device. Strobe lights are typically mounted 80 to 96 inches above the floor level so that they are in the line of sight for the building occupants. The flash rate of the strobe is a critical factor and must be within NFPA 72 criteria as follows:

- Minimum flash rate: one per second
- Maximum flash rate: two per second
- Clear or white lens cover
- Maximum flash intensity of 1,000 candela[8]

Strobes must be placed in all public areas and also in sleeping rooms. The strobe lights may be placed a maximum distance of 100 feet apart. This flash rate must not exceed 2 hertz to prevent photosensitive epilepsy in some occupants. In sleeping rooms a strobe must be located within 16 feet of the bed and within 24 inches of the ceiling to effectively wake up a hearing impaired person.

Fire Alarm Control Panels

The fire alarm control panel, or FACP, is the brain of the fire alarm system. The fire alarm control panel must be listed for fire alarm service by UL 864, *Control Units and Accessories for Fire Alarm Systems*, and must meet the requirements of NFPA 72 for the appropriate type of signaling system being used in the building. Fire inspectors should check the inside of the fire alarm panel cover for the appropriate UL listing and make sure the panel is for commercial use and not residential use. Only one- and two-family dwellings may use UL-listed residential burglar/fire alarm control panels. These residential control panels sometimes find their way into commercial use such as in small apartment buildings or small commercial buildings.

The fire alarm control panel contains several different fire alarm circuit types with different system functions. Each circuit is connected to the central processor unit of the fire alarm control panel. The typical circuits are as follows:

1. *Primary power supply:* The primary power circuit is 120 VAC and enters the control panel and connects to a step-down transformer. The transformer reduces the voltages to 12 or 24 volts for the operation of the fire alarm system. In small fire alarm systems, the transformer may be external to the fire alarm control panel and must be permanently attached to a power source so it cannot be dislodged. The primary power circuit must be installed by a licensed electrician and must be on a dedicated circuit that is unswitched. The fire alarm circuit breaker must be identified and should have a breaker lock to prevent the accidental turning off of the system. Primary power is indicated by a green LED on the control panel.

2. *Secondary Power Supply:* A gel cell battery or an emergency power generator may provide the secondary power for the fire alarm. Fire alarm systems must transfer to secondary power within 30 seconds of any primary power failure. Secondary power must support the system for at least 24 hours and have enough reserve power to operate all devices for 4 to 5 minutes. Secondary power supplies must be supervised for low battery conditions and must have audible and visual trouble indicators. The trouble indicator must be capable of producing a signal every 7 minutes for a period of 7 days.

3. *Trouble alarm:* The trouble alarm relay monitors both primary and secondary power supplies, alarm-initiating circuits, and alarm-indicating circuits and institutes an audible and visual supervisory alarm upon detecting failures in any circuit. The relay is powered from the secondary power supply so that primary power failure can be detected.

4. *Alarm-Initiating Circuits:* These circuits contain the fire alarm detectors in each alarm zone. Each circuit is connected to the fire alarm control panel at a specific set of terminal points for initiating devices. Initiating circuits may contain smoke detectors, heat detectors, manual pull stations, and other alarm-initiating appliances like water flow switches for automatic fire sprinkler systems. Smoke and heat detectors, water flow alarms, and manual pull stations must each have their own indicators for trouble and alarm signals.

5. *Alarm-Indicating Appliance Circuits:* The alarm-indicating appliance circuits contain the alarm output devices including the strobes, speakers, bells, and chimes necessary for distributing the evacuation signal. They activate only after an alarm-initiating device has detected a fire. Alarm-indicating circuits are required to be supervised and may have a silence switch; however, when the switch is activated, it must give both an audible and visual signal that the indicating appliances are off at the fire alarm control panel.

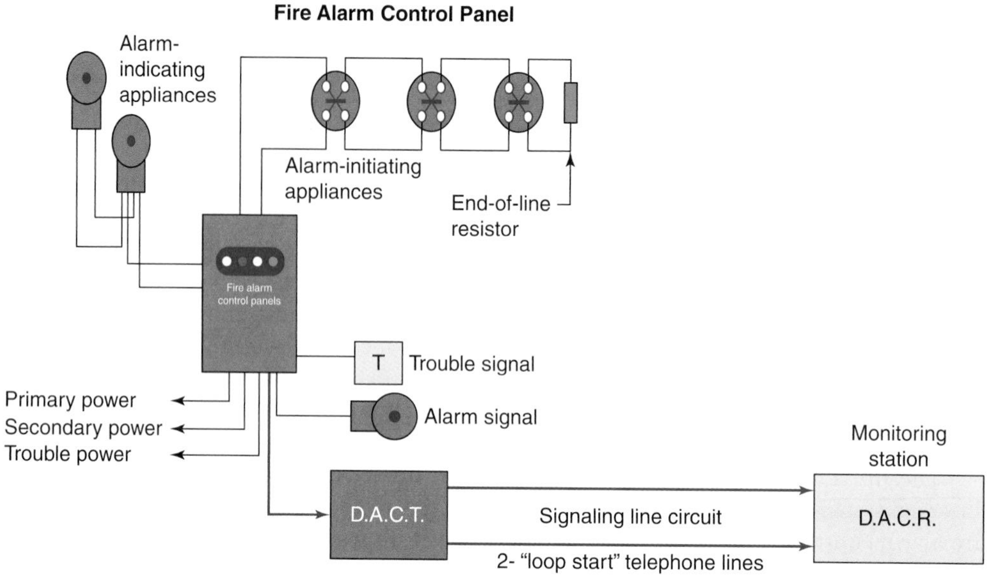

Fire Alarm Control Panel

FIGURE 11.14 The fire alarm control panel, or FACP, controls all functions of the fire alarm including primary and secondary power, trouble signals, initiating devices, and audible and visual appliances.

6. *Signaling Circuits:* The signaling circuits connect the control panel to outside monitoring services that notify the fire department. Typically, these circuits are connected to a digital access communicator transmitter (DACT), which connects to the local telephone utility, the Internet, or a radio frequency system and sends the alarm signal off premises to a monitoring station. The monitoring station has a digital access communication receiver (DACR) that processes the alarm and calls the fire department.

The fire inspector should also note that the fire alarm control panels must be protected within the building by an area smoke detector or by an automatic sprinkler system in accordance with NFPA 72 (see Figure 11.14).

Fire Command Centers

fire command center ■ A staffed or unstaffed location where the status of alarm systems, communication, and building control systems can be monitored or manually operated by the fire department.

Fire command centers are generally required in larger buildings and high-rise structures by building codes. The **fire command center**, or FCC, must be in a location as approved by the local fire department. The FCC will contain all of the necessary fire system controls and building operation functions for a fire department incident commander to manage a fire at the building. Fire command centers may also be staffed 24/7 by trained fire safety personnel or they may just be a room where the equipment is located. The IBC requires fire command centers to have at least the following fire protection features for the fire department's use:

1. Fire alarm annunciation panel specifying the device activated, smoke detector, duct detector, manual pull station, and water flow device by floor and by annunciator zone
2. Public address system and voice alarm controls with both general alarm and selective area capability
3. Fire department two-way communication systems
4. Sprinkler and water flow display panel
5. Elevator floor location panel
6. Fire pump status indicator
7. Emergency generator manual start and stop control
8. Emergency and standby power indicators

FIGURE 11.15 The building codes now require fire command centers to be 200 square feet in high-rise and large structures to provide adequate room for fire department incident commanders. *Courtesy of J. M. Foley*

9. Controls to automatically unlock egress doors
10. HVAC status and fan controls
11. Smoke evacuation control panel
12. Telephone for fire department use
13. Building plans and drawings
14. Work table
15. Room size – minimum 96 square feet by 8 feet wide
16. High rise – minimum 200 square feet by 10 feet wide[9]

After September 11, 2001, the IBC building code changed the requirement for the size of fire command centers to be more efficient for the fire department. The change was to increase the size of FCCs from a minimum of 96 square feet to 200 square feet (see Figure 11.15). Fire command centers also must have at least a 1-hour fire-resistant construction and must be accessible from the outside of the structure at a location approved by the local fire department. The current IBC code also requires the installation of a radio antenna in the building that meets the requirements of the local fire department for frequency and must meet 95 percent effective coverage for radio communication throughout the building. It is an important function for fire inspectors to know how all of the fire and building systems in the FCC operate so that the local firefighters can be effectively trained on the capabilities and operations of the different systems.

Voice Alarm Communication Systems

Voice alarm communication systems are by far the most effective means of evacuating occupants or supplying information and directions to the occupants in a building during an emergency. Studies conducted by John L. Bryan of the University of Maryland demonstrated that most people become aware of a fire in a building by human communication.[10] The modern temporal signal, while currently required by NFPA 72, may still be ambiguous to many people who have never been involved in a fire or had an evacuation experience. The voice alarm communication system (see Figure 11.16) allows a fire alarm operator or the fire department to communicate to all areas of the building in either a selective or general fashion throughout the building. The voice alarm system initially activates using the temporal signal for three cycles and then may initiate a prerecorded tape

FIGURE 11.16 The voice alarm communications panel may broadcast its signal to a fire area or throughout the building as a general alarm. New voice alarm systems must be intelligible systems for mass notification.
Courtesy of J. Foley

message to provide evacuation information to the occupants. The tape message is typically as follows: "Attention, attention; there is a fire emergency reported in the building. Please evacuate by the closest stairway and do not use the elevators." The message usually will repeat three times. In a fully staffed FCC, the fire alarm operator may read other scripts over the voice alarm system to provide additional direction to the occupants. For example, he or she may provide directions to the occupants to move to a certain floor or away from a certain area of the building while the nature of the alarm is determined. Typically, single initiating device events such as a single smoke detector or a manual pull station generally turn out to be false or accidental fire alarms, while multiple device events usually indicate that a fire is in progress in the building. The NFPA 72 standard permits alarm verification in fire alarm systems provided that the local authority having jurisdiction approves it. The advantage of a voice alarm communication system is that rapidly changing information may be broadcast to specific floors or rooms within the building very quickly from the fire command center. In some cases, selective evacuation and relocation of occupants will be more efficient than a total building evacuation. The local fire official determines the parameters for selective evacuation based on local fire department policy. For example, some cities have policies that buildings with full automatic sprinkler protection and voice alarm systems are permitted to use selective evacuation during fire alarm activations, while nonsprinklered buildings must fully evacuate on alarm.

Fire Areas and Annunciation Zones

fire area ■ The aggregate floor area of a building enclosed by firewalls, fire barriers, exterior walls, and horizontal fire rate assemblies.

The building codes require that the evacuation signal be automatic and sound throughout each affected **fire area**. A fire area can be a floor or multiple floors that are interconnected by atrium spaces. Fire-resistant construction, firewalls, or vertical and horizontal fire barriers forming a compartment determine the building's fire areas. A fire area may also have multiple fire alarm annunciation zones located within it. The purpose of the annunciation zone is to identify the location of the initiating device for the fire department. The IBC requires a minimum of at least one annunciation zone per floor. The annunciation zone may not exceed 22,500 square feet per zone, with the exception of water flow switches installed under NFPA 13, *Standard for the Installation of Sprinkler Systems*. Annunciation zones may not extend more than 300 linear feet in any direction. In high-rise buildings, the annunciator panel must indicate activation of smoke detectors, manual pull stations, water flow switches, and any other approved device separately. While the annunciator indicates the location of the initiating device, the indicating appliance throughout the fire area must activate to alert occupants. This is an important concept as some designers may send the evacuation signal only to the annunciation zone and not the fire area. Another important inspection point for the fire inspector is to make sure the annunciation zones make sense to the responding firefighters. If ZN-23C lights up on the fire alarm annunciator panel, there should be a corresponding document that tells the fire department where ZN-23C is located such as "ZN-23C – 23rd floor in the café." Remember, the annunciator has to make sense to the responding firefighter to be effective.

Alarm Verification

Alarm verification is a process permitted by NFPA 72 by which the building owner may provide an additional delay of time to verify the cause of the fire alarm activation before the alarm sounds within the fire area and calls the fire department. The NFPA 72 defines this delay process under **positive alarm sequencing** (PAS).

The requirements for positive alarm sequencing are as follows:

1. Upon activation of a fire alarm-initiating device, the operator has 15 seconds to acknowledge the alarm at the FACP. Failure to acknowledge the alarm results in activation of the audible indicating appliances.
2. Upon acknowledgment, the operator has 3 minutes or 180 seconds to verify and clear the initiating device. If it cannot be cleared in 180 seconds, the alarm-indicating appliances operate.
3. If during the 180-second verification period a second alarm-initiating device activates, the verification cycle is cancelled, and the alarm-indicating appliances immediately begin to operate.[11]

Newer fire alarm systems with addressable smoke detectors may also allow discrimination at the device level to avoid nuisance false alarms. Typically, the detector has a verification circuit that resets the smoke detector for a short period of time to determine whether the initiating cause has dispersed, such as cigarette smoke below a smoke detector. Some addressable smart smoke detector technologies actually have libraries of information on environmental alarm indicators that the detector can use as comparison for discrimination of false alarms. Compliant fire alarm verification helps to eliminate unnecessary false alarm responses by the fire department; however, fire inspectors must make sure they operate within the code requirements.

Manual Fire Alarms

Manual fire alarm pull stations have been around since the late 1800s. The manual pull station was an original component of street box fire alarm systems that were predominantly manufactured by the Gamewell Company. These street box systems were placed on street corners in cities throughout the country and some are still in service today. Street boxes were incorporated into building fire alarm systems using what was known as the **McCulloh loop** circuit. This circuit allowed building fire alarm systems to connect and report a fire over the municipal auxiliary street box fire alarm system. New York City had over 15,000 manual street boxes, and they are still in service today; today, however, voice communications are used from the box.

Manual pull fire alarm systems were installed in buildings in the early part of the nineteenth century as a method to alert occupants in high-occupancy buildings and schools. Most early systems were a series of manual pull stations either interconnected or connected to a fire alarm control panel. The pull station activated a series of alarm bells on each floor to notify

positive alarm sequence ■ A permissible method to delay the activation of fire alarm-indicating appliances for up to 180 seconds to allow owners an opportunity to reduce false alarms. The sequence requires a 15-second acknowledgment, and a second device activation will immediately cause indicating appliances to operate.

McCulloh loop ■ Patented by Chauncey McCulloh in 1882, the McCulloh loop allowed central stations to monitor and repair breaks in the circuits using a two-wire loop. This allowed greater dependability and monitoring of customers attached by McCulloh loops to the central alarm station. Its use revolutionized the alarm industry.

FIGURE 11.17 A double action manual pull station. Note the directions on the clear cover. Covers can reduce false alarms caused by pranksters.
Courtesy of J. Foley

the occupants. Early systems in schools and hospitals were called coded bell systems. The coded bell would sound a specific sequence of rings indicating the location of the activated manual pull box that was activated. Manual pull stations typically were installed in the common hallway of buildings along the path of egress so that the occupant could initiate a local alarm while exiting the buildings. Early manual pull systems, except for those in schools and hospitals, rarely were connected to the fire department. The activation of the manual pull fire alarm would also activate the local street box on the municipal alarm system to summon the fire department.

Building codes require the installation of manual pull stations in buildings of assembly, business, institutional, factory, mercantile, residential educational, high-hazard, and high-rise use groups. Each specific use group has a threshold or trigger at which point the manual pull stations must be installed. Typically, the thresholds are initiated by either the number of building occupants or the building's height. The IBC eliminates the need for manual pull stations in some buildings when smoke detection and automatic sprinkler systems are installed. In these cases, a single manual pull station must be installed at an approved location by the authority having jurisdiction. Typically, that location will be one that is constantly staffed like a front desk or fire command center. The manual pull station, once activated, is required to sound the temporal three signal throughout the affected fire area or building.

Manual pull stations are to be located within 5 feet of all exit doors and should be no more than 200 feet apart in corridors. Pull stations must be mounted between 42 inches and 48 inches from the floor to be compliant with both the building code and the Americans with Disabilities Act. Pull stations shall be red in color, and if they only sound a local alarm signal, they must have signage posted next to the pull station stating, "WHEN ALARM SOUNDS, CALL THE FIRE DEPARTMENT." Manual pull stations may be single, double, or triple action to activate (see Figure 11.17). Single action means you simply pull a handle down to activate the fire alarm. Double action boxes require two actions to operate, such as break glass and pull handle down. Typically, triple action boxes are double action pull stations protected by an alarmed cover. The cover must be clear to expose the pull station underneath and must have directions on how to properly operate it. Double and triple action manual pull boxes are used in locations prone to malicious false alarms. Manual pull fire alarms should not be installed in locations prone to false alarms such as elevator lobbies. Often, pranksters riding on elevators pull false alarms. If pull boxes must be in this location they should be double or triple action pull stations. Building operators should also provide reset keys or tools for the fire department's use to reset a manual pull box.

Automatic Fire Detection Systems

Automatic fire detection systems are required to perform five main functions under the NFPA 72 standard:

1. Provide notification of an emergency or nonemergency supervisory signal both on premises and off premises.
2. Alert all of the affected occupants in the building.

3. Call the fire department.
4. Control other fire safety functions in the building such as HVAC or elevator movement.
5. Monitor the fire alarm system for off-normal or trouble conditions.

The automatic fire detection system is comprised of a fire alarm control panel, fire alarm-initiating device circuits, fire alarm-indicating appliance circuits, and primary and secondary power supply circuits.

FIRE ALARM CONTROL PANEL

The fire alarm control panel, or FACP, controls all of the functions of the fire alarm system as previously mentioned. The FACP must be UL listed in accordance with UL 824, *Control Units and Accessories for Fire Alarm Systems*, and must be used within its listing for the appropriate type of signaling system in NFPA 72. Control panels may also be listed as *household* under NFPA 72, Chapter 11 for one- and two-family dwellings or *commercial* under NFPA 72, Chapters 6–8, and they are also listed for the monitoring service use as Auxiliary (A), Remote (RS), Proprietary (P), or Central Station (CS).

NFPA 72, *National Fire Alarm and Signaling Code*, identifies five fire alarm reporting categories or methods for all fire alarm system. These methods may be required by either the building or fire codes of the local jurisdiction. Let's examine each reporting method.

LOCAL ALARMS

A local fire alarm system is one that simply detects a fire and sounds an alarm on the protected premises. The alarm is not retransmitted to the local fire service and depends on the occupants calling 911 to notify the fire department. All fire alarm systems must be local alarm systems and are required to sound the fire alarm at the protected premises, and they may also retransmit the alarm off site to a monitoring company and the fire department. The *UL Fire Protection Equipment Directory* designates local alarm equipment with an "L" in the equipment type column of the directory.

AUXILIARY ALARM SYSTEMS

The auxiliary fire alarm system connects the protected property with a municipal street box fire alarm system. Upon detection of a fire, a shunt relay activates the municipal street box that is closest to the protected property to summon the fire department. The transmission of the alarm may be shunt-type relay, where the circuit from the municipal alarm box extends into the protected property, or local energy type, where the municipal alarm box is wired directly to the fire department. The signal may be transmitted over parallel telephone lines, coded wire, coded radio signal, or series telephone lines to the alarm center. These auxiliary fire alarm systems were popular in large cities but are disappearing as municipalities eliminate the use of fire alarm street boxes due to their high maintenance cost and general lack of use by the public. With today's enhanced 911 and cellular telephone communication, the municipal fire alarm street box is going the way of the telephone booth in most communities. In the *UL Fire Protection Equipment Directory*, the fire inspector will find auxiliary fire alarm equipment listed under the equipment type column. All auxiliary fire alarm equipment is identified by an "A" in the listing for the type of equipment.

REMOTE STATION ALARM SYSTEMS

The remote signaling system stations usually is a municipal police or fire department dispatch center that monitors fire alarms but may also be a private monitoring center providing alarm service that is constantly staffed by trained personnel. Remote station signaling systems are the most common type of fire alarm system monitoring. Remote stations may be UL listed under UUJS, but are not certificated the same as central station fire alarm systems. Remote signaling stations do not provide either runner services or system maintenance. The remote signaling station must comply with NFPA 72 for retransmission of fire alarm signals to the local fire department. The signal from the protected premises to the remote station may be transmitted by any of the following means: leased telephone lines, optic cables, private microwave radio systems, telephone company derived local channels, digital alarm radio systems, McCulloh systems, and two-way radio frequency. The key to communications is that the method must have redundancy and reliability. The NFPA 72-2010 standard also provides for communications using nontraditional systems such as cable access television through cable digital voice technology or Internet voice communications systems such as the Comcast or Verizon networks. The *UL Fire Protection Equipment Directory* lists equipment for remote stations under equipment type "RS."

PROPRIETARY PROTECTIVE SIGNALING SYSTEMS

The key difference between remote station signaling systems and proprietary signaling systems is the type of ownership. A proprietary system requires that the protected premises owner must also own the alarm receiving station. Many large companies use proprietary fire alarm systems such as Walmart, Vornado, and many large mercantile chains. This allows the owner to monitor events at all of their facilities by geographic regions or nationwide, if they desire. In a proprietary signaling system, the alarm receiving station may be either on site or remote; however, it must meet all of the requirements of NFPA 72 for staffing by trained personnel and the retransmission and recording of all fire alarms, supervisory, and trouble signal conditions at the protected premises. With the rapidly changing communications environment and the development of the Internet and other instantaneous communications links, proprietary fire alarms systems eventually will be replaced by these more cost-effective applications of advanced technology. Proprietary fire alarm equipment must be compatible throughout the system, and therefore it does not allow competition in the purchasing and maintenance of the alarm equipment. The *UL Fire Protection Directory* lists proprietary fire alarm equipment under type "P."

CENTRAL STATION SIGNALING SYSTEMS

Central station signaling systems are certificated by Underwriters Laboratories and appear in the *UL Fire Protection Equipment Directory* as "UUFX." A UL-listed central station must provide eight specific services to the customer in order to be certificated. The central station must provide system installation, maintenance, testing, and runner services for both fire alarm and supervisory signal conditions at the protected premises. The central monitoring station itself must have effective system management for monitoring of all signals from the protected premises. All signals must be retransmitted to the local fire department, and a written record of all alarm activities must be maintained.

Central station signaling system providers are permitted to just provide the monitoring services under certain conditions specified by UL. The central station is permitted to subcontract the system installation, maintenance, testing, and runner services to other UL-listed service providers certificated as "UUJS" in the UL directory. A non-UL-listed fire alarm service company may not represent itself as a UL central station provider even if it is subcontracted by a listed central station company. The only way a nonlisted fire alarm company may be used in a central station application is if the UL-compliant central station contracts directly with the nonlisted company and assumes all responsibility for the work.

UL-listed central stations are required to have a signed contract to provide maintenance, installation, and testing of all fire alarm system equipment with the property owner. The central station must provide "runner services" that require a certified fire alarm technician to respond within 1 hour of any fire alarm activation and within 4 hours of a supervisory or trouble signal or alarm. A UL-listed central station provides the highest level of assurance that the fire alarm system is properly maintained and serviced. Insurance providers of highly protected risks usually require the owner to subscribe to a UL-listed central station monitoring company. The International Building Code (IBC) requires that fire alarm systems be monitored but leaves the specific type of monitoring up to the authority having jurisdiction. Fire inspectors should request a copy of the certificate for a central station. Approved central station signaling system equipment is identified under equipment type as "CS" in the *UL Fire Protection Equipment Directory*.

Residential One- and Two-Family Dwelling Fire Alarm Systems

Generally, fire inspectors get only limited opportunities to inspect fire alarm systems in residential one- and two-family dwellings. These inspections usually are conducted during construction, property rental or resale, or perhaps as a courtesy inspection to a homeowner. Many fire departments may also institute smoke detector installation and giveaway programs in their communities. NFPA 74, *Standard for Household Fire Warning Equipment*, used to be the applicable standard for residential fire alarm systems. NFPA 74 is now incorporated in NFPA 72, *National Fire Alarm and Signaling Code*, Chapter 11. The chapter applies to one- and two-family dwellings, sleeping rooms in lodging or rooming houses, dwelling units in apartments, guest rooms in hotels and motels, dormitories, day care centers, and residential board and care centers. These requirements may be modified by local jurisdictional codes. The fire alarm may be a series of interconnected single-station smoke and heat detectors, or they may be connected to a FACP or residential burglar/fire alarm control panel. The activation of any detector shall sound the alarms throughout the dwelling unit. All single-station smoke detectors must also have a secondary backup battery power supply. The smoke detectors shall meet a minimum installation requirement of at least one smoke detector on every level of the home. Smoke detectors are also required outside each sleeping room as well as within the sleeping room, usually within 10 feet of the bed. The NFPA requires an effective sound level of 75 dba at the pillow to wake a person out of a sound sleep. Smoke detectors may be either ceiling or wall mounted and should be no closer than 4 inches to the wall ceiling joint nor more than 12 inches from the ceiling when wall mounted. Single-station smoke detectors should be tested monthly for operation and the batteries replaced annually unless they are warranted for greater periods of time (e.g., lithium ion 10-year batteries).

Carbon Monoxide Detection Systems

Carbon monoxide is known as the silent killer, and it kills or injures hundreds of homeowners every year. Carbon monoxide, or CO, is a chemical compound that when inhaled inhibits the transfer of oxygen to red blood cells, which eventually causes asphyxiation. The leading cause of carbon monoxide buildup in a home is cooking appliances. Carbon monoxide detection systems are to be installed under the requirements of NFPA 720, *Standard for the Installation of Carbon Monoxide (CO) Detection and Warning Equipment*. The installation of carbon monoxide detection systems is mandatory under the current building codes for new construction containing CO-producing appliances such as gas stoves, gas water heaters and furnaces, or fireplaces. The alarms are also required if the building has an attached garage next to the dwelling unit. Carbon monoxide detectors are listed by Underwriters Laboratories under UL 2034 for stand-alone detectors and UL 2075 for systems connected to a control panel. The NFPA requires that carbon monoxide detectors be installed outside bedrooms on every level of the home. CO detectors are mounted in accordance with the manufacturer's directions and generally at eye level. Unlike carbon monoxide in a fire, CO effluents from a leaking fuel-fired appliance attain room temperature quickly and mix with air. As required by UL, CO detectors have alarm levels that activate on CO concentration of 30 ppm within 30 days or 400 ppm within

15 minutes. Fire inspectors must look at the label on the device, as CO detectors do have a life span. Earlier models lasted about 2 years, while newer CO detectors may have a useful life of 5 to 7 years. Like smoke detectors, the CO detector "chirps" when the battery fails or the sensor has reached its end of life. The CO detector should not be placed within 10 feet of any fuel-burning appliances, or near air conditioner or heater supply vents. Fire marshals and fire officials should ensure that the fire department has a CO detector response procedure and that firefighters are properly protected with CO detection meters and self-contained breathing apparatus.

Fire Alarm System Circuitry

Fire alarm systems have a number of circuits that perform different functions for the system's operation. Five types of circuits are connected to the fire alarm control panel:

- Power circuits, both primary and secondary
- Alarm-initiating circuits
- Alarm-indicating appliance circuits
- Alarm signal notification circuits
- Trouble signal notification circuits

Each type of circuit has a specific classification under NFPA 72, which determines the performance measures that it must achieve. Let's take a look at each type of fire alarm circuit.

POWER CIRCUIT WIRING

Power supply wiring for a fire alarm or any electrical equipment must comply with the wiring methods specified in NFPA 70, *National Electrical Code®*. The NEC classifies power circuits as power limited or non–power limited in article 760-15.

Power-Limited Circuits

These circuits operate on 12 to 24 volts alternating current and must use specific low-voltage fire alarm wire in the circuits. Power-limited wire may be in conduit if it is within 7 feet of the floor. Wires above 7 feet may not require protection unless they could be subject to mechanical damage. Fire alarm system wiring is UL labeled as follows for power-limited systems (see Figure 11.18):

- FPLP – Fire power-limited plenum cable (used in return or supply air plenums)
- FPLR – Fire power-limited riser cable
- FPL – Fire power-limited circuit cable

Fire alarm wiring must be separated from all electric cables by at least 2 inches and may not be run in the same electrical conduit systems. Fire alarm riser cable must be protected by conduit in vertical shafts and elevator hoist ways. Fire inspectors can identify the proper fire alarm wire by the UL imprint on the wire jacket for the proper use location.

Non-Power-Limited Circuits

These circuits are 120 VAC or higher and require much larger conductors in the wiring. Non-power-limited wire must have rated insulation up to 600 VAC. Non-power-limited

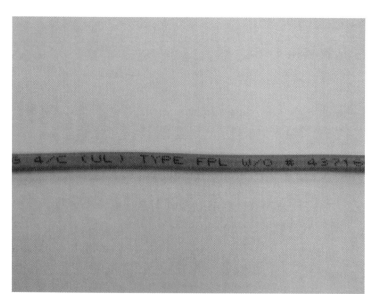

FIGURE 11.18 Both power-limited and non-power-limited cable are marked on the wire jacket as to their appropriate use and location. *Courtesy of J. M. Foley*

wire is also specifically designated for fire alarm system application and is labeled as follows by UL:

- **NPLFP:** Non-power-limited fire plenum cable (used in return air plenums)
- **NPLFR:** Non-power-limited fire riser cable
- **NPLF** Non-power-limited fire circuit wire

A fire alarm installer must use either fire alarm protective signaling wire or follow the NEC-required article 300 electrical wiring methods for system installation. Other types of non-power-limited electrical wiring may be substituted for listed fire alarm wire provided that it is permitted by article 760-61(d) of the NEC.

PRIMARY POWER SUPPLY

The primary power for the automatic fire alarm system may come from one or two sources: a commercial electric supplier via the local public utility, or an engine-driven electric generator with a trained operator on duty at all times, or an engine-driven generator with a storage battery having a 4-hour minimum rated capacity. All primary power sources must be supervised by the fire alarm control panel and there must be a dedicated branch circuit identified as "Fire Alarm Control Circuit" in the electrical panel box.

SECONDARY (STANDBY) POWER SUPPLY

Every fire alarm system must also have a secondary reliable power source capable of transferring power within 30 seconds of the primary power supply's failure. The new NFPA 72, 2010 edition, also requires standby power transfer within 10 seconds, which is consistent with the IBC for emergency electrical load transfer to secondary power supplies. The secondary power supplies must provide power as indicated in Table 11.1 for each type of fire alarm signaling system.

TROUBLE ALARM POWER SUPPLY

The NFPA 72 standard requires trouble alarm power to be separated from primary power supplies so that the trouble supervisory signal can be sent to the monitoring station if the primary power supply should fail. Typically, the secondary power supply is a gel cell battery that will provide trouble signal power. Trouble signals indicate a disruption in the circuit by loss of power and are different from supervisory alarms that indicate an off-normal condition in the alarm system. Off-normal conditions are the devices where a change has occurred such as a closed valve, a fire pump that is running, a dry pipe that

TABLE 11.1	Secondary Power Supply Requirements	
TYPE SYSTEM	**MAXIMUM LOAD (HOURS)**	**MAXIMUM ALARM LOAD (MINUTES)**
Local System	24	5
Auxiliary System	60	5
Remote Station Signaling Systems	60	5
Proprietary Signaling Systems	24	Normal traffic
Central Station Signaling Systems	24	Not specified
Emergency Voice Alarm Systems	24	5

Reprinted with permission from NFPA 72® 2013, 2010, 1999 *National Fire Alarm and Signaling Code*, Copyright © 2012, 2009, 1999, National Fire Protection, Quincy, MA. This reprinted material is not the complete and official position of the NFPA on the referenced subject, which is represented only by the standard in its entirety.

has low air pressure, or silent alarm switches that have been activated. Both trouble and supervisory alarms produce audible and visual alarms at the fire alarm control panel.

INITIATING DEVICE CIRCUITS

Fire alarm systems' initiating device circuits serve two very important functions. First, they connect the heat, smoke, or other special detectors to the fire alarm control panel; and second, they provide a path for supervising the fire alarm system for component failure or circuit integrity. The principle of alarm signal, supervisory, and trouble signals is pretty basic: The fire alarm control panel can detect three distinct electrical conditions from the devices attached to the initiating circuits. The three conditions are normal system power, reduced system power, and no system power. Each electrical condition provides a different signal function:

- Reduced power condition on one leg of the circuit is considered normal.
- A full power condition on both circuit legs is considered an alarm condition.
- No power on one leg of the circuit is considered trouble or ground fault.

end-of-line device ■ A resister or diode placed at the end of a circuit that reduces current flow back to the fire alarm control panel to supervise the continuity of the circuit.

The fire alarm system determines these changes in power by the application of an **end-of-line device**, which allows a slight power reduction on the return leg of the circuit. The end-of-line devices are placed at the farthest distance of the circuit loop from the control panel. This is usually at the last smoke detector in the loop.

NFPA 72 identifies how the initiating device circuits operate by circuit styles. Originally, NFPA 72 identified circuit styles as wiring classes A or B. Class A consisted of four-wire initiating circuit loops, and Class B consisted of two-wire initiating circuit loops. Fire inspectors may still hear fire alarm circuits being referred to as Class A or Class B. In today's NFPA 72, circuit styles for initiating device circuits are Style A or B for two-wire circuits (old Class B) and Style D for four-wire circuits (old Class A). The difference between the two styles is basic circuit reliability. In the Style A and B circuits each device is attached to a pair of wires, and at the last smoke detector in the loop, an end-of-line device is attached. The end-of-line device provides the circuit supervision. If a device fails or a wire becomes disconnected in this style circuit, all devices beyond the wire break no longer function. In Style D circuits with four wiring pairs, one loop of two wires is used for power supply and another loop of two wires is used for power return. These circuits use relays or diodes instead of end-of-line resistors to measure current flow and to provide operating power to the alarm and trouble signals in the control panel. The relays operate on power loss to one side of the relay coil, causing the other side's dry contacts to close, completing a circuit on the second pair of wires in the circuit loop. The diodes are like one-way doors that permit electric current flow only in a single direction in the wire (see Figures 11.19, 11.20, and 11.21). The control panel contains an alarm signal relay and a trouble signal relay, and under normal conditions the electrical contacts of both relays are open. In alarm mode, the current flows to the alarm relay closing it, causing the indicating appliance circuits and signaling line circuits to operate. If a fault occurs, the power is directed to the trouble relay, which closes and sends the supervisory signal to the signaling line circuits as well as local trouble audible and visual alarms on the control panel. The key difference between the four-wire circuit and the two-wire circuit is system reliability. This is important for the fire inspector to understand because a trouble signal in Style A or B circuits means that part of the fire alarm system is no longer functioning to protect the property, while in Style D circuits, there is a system problem, but the complete fire alarm system can still operate.

INDICATING APPLIANCE CIRCUITS

The indicating appliance circuits contain the audible and visual alarm devices as well as additional circuits for other types of control circuits such as automatic door closers, stair

Style D and E Circuits (Old Class A)

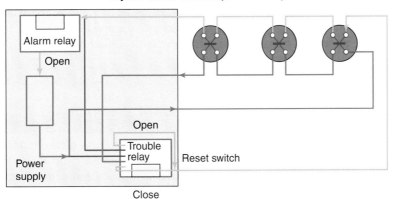

FIGURE 11.19 Four-wire style D circuit in normal mode with no power to alarm or trouble relays.

<u>Normal mode</u> - Current flow through two open relays and resistor to indicate normal power condition

Style D and E Circuits (Old Class A)

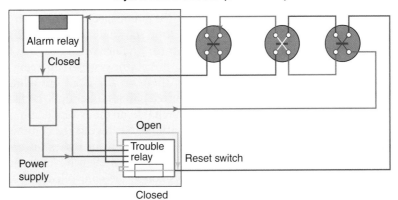

FIGURE 11.20 An alarm causes a direct short across the smoke detector, allowing full power to travel to the alarm relay and close the circuit.

<u>Alarm mode</u> - Current flow direct wire-to-wire short in initiating device, completing the circuit outside the resistors and closing the alarm relay

Style D and E Circuits (Old Class A)

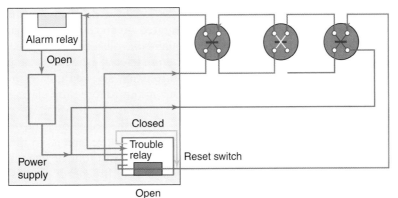

FIGURE 11.21 The advantage of style D four-wire circuits is that all devices are still powered even through a wire break or disconnection. The reverse polarity on the wire causes the trouble relay to operate and send a trouble signal.

<u>Trouble mode</u> - Open fault gives trouble signal. Alarm receipt condition is maintained during abnormal condition.

pressurization fans, and smoke damper controls. Indicating appliances use Style W and X circuits in accordance with NFPA 72 and are attached to the control panel. These indicating appliance circuits also contain diodes and the end-of-line device to supervise the integrity of the wiring circuit. These circuits may use either two wires or four wires, depending on the system circuit style.

SIGNALING LINE CIRCUITS

Signaling line circuits go from the protected property to the central station, remote station, proprietary station, or municipal auxiliary fire alarm system. Signaling may be accomplished through a multitude of communication methods including the plain old telephone service (POTS) lines, optical fiber circuits, radio frequency channels, microwave links, two-way and one-way licensed radio systems, managed facility-based voice systems, and the Internet. In all cases, the signaling path must be reliable and must have multiple communication paths to comply with NFPA 72. Most existing fire alarm systems employ redundant telephone lines with digital access communicator transmitters (DACT) installed in the control panel. The signal transmission system requires two **loop start telephone lines**, which are provided from the telephone company. This ensures a live phone line to the DACT. The DACT uses the plain old telephone lines to connect to the alarm monitoring point where a digital access communication receiver (DACR) is installed to receive the signal.

An important inspection point for the fire inspector is to ensure that the DACT has two redundant telephone lines that are loop start and not connected to another telephone system such as a private branch exchange (PBX) or switch board in the building.

In the newest edition of NFPA 72, fire alarm systems are being expanded to include mass communication systems. In fact, the term *fire* is all but eliminated from the new standard. New technologies like cellular telephones, fiber optic communication cable, and wireless Internet have introduced new reliable methods of signaling offsite locations for emergencies. Signaling line methods continue to evolve as the communications technology continues to innovate. The new NFPA 72 standard increases the scope of the fire alarm infrastructure to become an all-hazard warning system for mass notification. This also provides the capability for building systems to be controlled from remote locations via cell phone and Internet applications. Eventually, buildings' environmental and safety systems will have the ability to be managed from remote computers, iPad® tablets, or cell phones. These systems eventually will replace the older communication methods currently found in the fire codes.

Technological Changes in Fire Alarm Systems

In the 1950s and 1960s, fire alarm system technology was relatively rudimentary and mostly non-power-limited systems. In the 1970s, with advancements in smoke detection technology and development of the integrated circuit chip, the fire alarm industry dramatically changed, and power-limited computer-controlled fire alarms became the norm. These systems were called *multiplexing fire alarm systems*, as multiple signals could be sent out on a single communication path. Multiplexing allowed greater numbers of detection devices to be placed on the systems. This meant that the fire alarm system could protect much larger structures. The way multiplexing was accomplished was by connecting several secondary control panels called *transponders* to the fire alarm control panel, connecting all the initiating device circuits, and indicating appliance circuits to the transponder panels. This allowed the control panel to communicate with four or five transponders, and each transponder would communicate with about 40 fire alarm circuits attached to it. This approach gave the system the capability to monitor thousands of devices efficiently in a short time frame. Multiplexing fire alarm systems used two different methods of supervision called *time march multiplexing* and *frequency march multiplexing*.

loop start telephone line ■ A telephone line powered directly by the utility company and independent of the building power supply used to connect communication transmitters to a proprietary, remote, or central station fire alarm monitoring center.

- *Time march:* In time march multiplexing systems, each detection zone or group of devices is assigned an address on the transponder panel. The control panel calls up the transponder and pulses each address, starting with device 1 and working sequentially through the circuits. As each circuit is pulsed, it is compared to the previous scan, and if nothing changes, the system is normal. If a change of status does occur after the pulse, the system will not see the change until it pulses the transponder address again. It should be noted that a fire alarm signal takes precedence over all other types of supervisory signal from the transponder.
- *Frequency march:* The frequency march system is different in that each type of device or group of devices is assigned a specific frequency, which is activated upon a change in status for the device. This allows multiple transmissions of changes without sequentially waiting to be pulsed by the control panel. Frequency march allows for a quicker response to the initiating device status change.

In either type of system, the control panel must receive the change in fire alarm status within 90 seconds of the change occurrence and must be capable of receiving 10 percent of all alarm devices or a minimum of 50 alarm signal changes within 90 seconds, whichever is greater.

The next step in fire alarm evolution was the development of intelligent smoke detection or addressable fire alarm systems. This technology provides a logic computer chip inside the smoke detector that is assigned a specific analog or (now) digital address. This technology allows the control panel to communicate directly with the individual fire alarm devices and also allow better location identification of the specific detector that is in alarm. The other distinct advantage is that the logic chip can perform other maintenance functions in the smoke detector as well as help to discriminate between environmental and actual fire signatures to reduce possible false alarms. Addressable smoke detectors can adjust detector sensitivity and notify the control panel when cleaning is necessary.

Other Fire Protection Systems

Automatic fire alarm systems also perform many other fire-related functions. An automatic fire alarm system may connect many other types of fire suppression and fire protection equipment to the fire alarm control panel. Typically, the most common system connected to the fire alarm is the automatic fire sprinkler system. The fire alarm monitors several functions of the sprinkler system including water flow switches, valve tamper alarms, low-air alarms in dry systems, fire pumps when provided, and preaction and deluge sprinkler systems. These functions are monitored using the following types of devices:

- *Water flow Switch:* The water flow switch is an electric paddle switch inserted into the sprinkler pipe that indicates when water is flowing. The device can be adjusted to compensate for water surges by delaying the activation of the paddle for 15 to 30 seconds. Sprinkler systems may also have pressure switches that can detect the drop in pressure when water begins to flow through the sprinkler pipes.
- *Tamper alarm switches:* Tamper switches are electrical contacts that are attached to either a rod and cam or a plug-in strap device attached to a sprinkler control valve. The rod-type device is spring loaded and attaches to the branch line sprinkler control valves or main system OS&Y valves. The rod is held open by a cam placed on the valve wheel stem. If the valve handle is closed or turned, the cam pushes the rod to the closed position and a signal is sent to the control panel that the valve is off normal. The plug-in type devices use a wire rope loop through the valve handle that attaches to a plug. If the valve handle is turned, the plug dislodges and cannot be plugged back in without a special tool.
- *Pressure alarms:* Pressure alarms or pressure switches are often used on dry pipe sprinkler systems and fire and jockey pump applications. The pressure switch either

senses a pressure increase by water pushing against a diaphragm to close an electrical contact or senses a drop in pressure indicating that water is flowing in the system. These switches are normally located at the sprinkler dry pipe or alarm check valve on top of the alarm retard chamber.

■ *Kitchen suppression systems:* Another type of fire protection system that may be monitored by the fire alarm system is kitchen hood systems. Many kitchens use pre-manufactured hood wash-down systems that are connected to the automatic sprinkler system. These kitchen systems are attached to both an independent surface fire protection system (usually wet chemical) and the fire alarm. Activation of the fire wash-down system sends a signal to the control panel that there is a fire in the hood.

Building Management Systems

The fire alarm may control or monitor other building system functions in large structures and high-rise buildings. These system controls are usually related to emergency egress door releases, recall of the elevator, and smoke control in the building.

STAIR PRESSURIZATION

Stair pressurization is required in buildings over 75 feet in height if the building does not have smokeproof stair enclosures. Stair pressurization provides a positive pressure in the stairway of 0.15 inch of water column to keep smoke from entering the stairs when doors are opened and closed during an evacuation. To perform this function, a signal must be sent from the FACP to an electric relay to start the fan motor. The signaling path will be supervised to the connecting alarm relay switch. The fan motor and electrical supply for the fan are not supervised, as they are non-power-limited 120 VAC or higher and cannot be attached to a power-limited fire alarm system.

SMOKE CONTROL SYSTEMS

Controlling the migration of smoke in a building is a difficult and complicated process, to say the least. Smoke control systems are installed under NFPA 92A, *Standard for Smoke-Control Systems Utilizing Barriers and Pressure Differences*, and as required by building and mechanical code regulations. Smoke control systems may be designed as dedicated smoke removal systems or they may be integrated systems with the existing HVAC system for the building. The general objective of smoke control is to maintain the integrity of the egress system for a reasonable period of time so that the building occupants can safely escape via the exits and exit stairways. Smoke control systems may exert a negative air pressure on the fire floor by going to 100 percent outside exhaust and air supply. The systems also create positive air pressure on the floors immediately above and below the fire floor to prevent smoke migration. This usually is accomplished by 100 percent intake supply air on the surrounding floors with no exhaust on those floors causing air pressure to build up. Smoke control systems generally have to supply four to six air changes per hour, or an average of one air change every 10 minutes. To perform these complicated functions, the control panel must send signals to multiple fan motor relays to start or stop and must send signals to electrical damper relays to open or close based on the location of the activated fire detection device. These fan and damper actions are specified in the fire alarm's logic matrix provided by the system designer. The fire alarm matrix describes each function of the fire alarm and the appropriate equipment responses that should occur. The fire alarm system matrix can be very helpful in understanding the operation of the system. In addition, the fire command center must also have the capability for smoke control fans to be stopped or started manually from that location by the fire department (see Figure 11.22). The smoke control system panel shows the location and status of each fan and allows the fire department to secure the fans, if necessary, to control fire travel and further remove smoke from the area. The operation of smoke removal systems should be preplanned by the fire department.

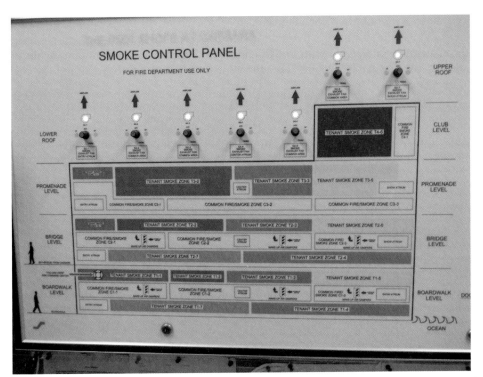

FIGURE 11.22 The smoke control panel in a fire command center must have the capability to manually start and stop fans. *Courtesy of J. Foley*

FIREFIGHTER COMMUNICATIONS SYSTEMS

In buildings over 75 feet above the lowest level of fire department access or high-rise building firefighters, communications systems are required to be installed. These communications systems are a component of the automatic fire alarm and are installed under NFPA 72. Firefighter communications must be a protected system for fire survivability; all wiring must be installed in fire-rated enclosures. Under the current NFPA 72, the two-way communications systems are considered level 2 or 3 for fire-resistant protection of the communication wiring. The level 2 protection consists of a 2-hour fire-rated barrier protecting the wiring, and level 3 protection requires both automatic sprinklers and a 2-hour fire-rated barrier for wiring protection.

Firefighter warden stations may be either fixed telephone locations or portable handsets that the fire department personnel take with them from the fire command center. The phones are plugged into communication jacks located in specific areas of the building. The communication jack locations are the stairwells on each floor, the elevator lobbies on each floor, the elevator car, the machine rooms, and the fire pump room. Existing firefighter communications systems have a common talk feature, meaning that once connected to the fire command center, up to five phones can operate together. This feature has been eliminated in the current NFPA 72 standard as it may cause miscommunication or possibly confusion on orders. Firefighter communication systems should be tested on an annual basis to ensure proper operation and proper identification of all jack termination points.

Elevator Recall The recall of elevator cars is required once the elevator travels beyond 25 feet. Elevator recall requirements are established in ANSI 17.1, *Safety Code for Elevators and Escalators*, and apply to all elevators regardless of whether a fire alarm system is present. Phase I recall requires that upon activation of the fire alarm system, at least one elevator car must return to the lobby for the fire department's use. If the fire is in the lobby, the elevator car returns to a designated alternate floor. The designated floor is determined by the fire department. Other elevator cars will continue to run unless the elevator lobby smoke detector is activated on any floor; in that case, the car recalls to the ground floor and stops.

When a smoke detector connected to a preaction sprinkler system operates in the elevator machine room or the elevator shaft, then phase II recall occurs, shutting down the elevators completely before the discharge of the sprinkler in the shaft. Again, this system receives its signals from a relay between the elevator car control and the fire alarm system circuits.

Door Control   Door access control systems may also be connected to the automatic fire alarm system. The fire alarm sends signals to special locking devices to turn off power and cause the locks to open, making them accessible to evacuating occupants. In the case of smoke containment, the alarm signal is sent to electromagnetic door holders, turning off the power and releasing the doors so that they will automatically close. Door-releasing devices must be approved and listed for fire alarm use under UL 228, *Door Closers-Holders with or without Integral Smoke Detectors*, standards.

Other Functional Controls   In some jurisdictions and in special amusement buildings, the fire alarm may be interconnected with other emergency systems to turn lights on, stop amusement rides, shunt music, or operate emergency lighting equipment. Some jurisdictions have adopted local fire code amendments requiring the shunting of electricity to live or recorded music performances in places of public assembly upon the activation of the fire alarm system. Fire inspectors must remember that the fire alarm signal must be 15 dba above ambient sound or 5 dba above any sound louder than ambient lasting over 60 seconds.

Testing and Inspection Methods

Automatic fire alarm systems are required to be inspected and tested in accordance with NFPA 72, Chapter 10. This standard requires monthly, biannual, and annual inspections of the fire alarm system components. The monthly test is a simple visual examination of the system noting any obvious deficiencies, such as broken or damaged equipment, supervisory or trouble alarm indicators at the control panel, or equipment in need of cleaning. Monthly testing requires certain components to be operated, including the audiotape message and at least one initiating device on every circuit. This test verifies that the speakers and indicating appliances are fully operational. The two-way digital access communicator, or DACT, must also be tested to ensure a working connection to the alarm-monitoring agency. During the semiannual test, at least 25 percent of the speaker circuits shall be tested and the primary power supply shall be disconnected and the secondary power supply tested to ensure the required performance. At least 10 percent of all the heat detectors shall be tested and all of the smoke detectors are also to be tested. Remote annunciators and all fire extinguishing system interconnection relays shall be tested. During the annual fire alarm inspection, all components shall be tested including all functions of the fire alarm control panel, including the lamp test, primary and secondary power test, battery charger, bells, horns, strobe lights, and speakers. All of the fire alarm system fuses shall be removed and checked for proper fuse ratings. Smoke detectors must be calibrated in accordance with the manufacturer's recommended practices or calibration methods. The fire prevention code stipulates that the calibration testing of smoke detectors must occur 1 year after initial installation and every 2 years after. If smoke detector calibrations during the second-year inspection are within 4 percent obscuration for gray smoke, then calibration testing may be performed every 5 years. All system heat detectors shall be physically tested upon the fifteenth year in service. At least two heat detectors out of every 100 installed must be destructively tested. If the heat detector fails in the laboratory test, then all heat detectors are required to be replaced. Typically, fire alarm system components will be visually inspected weekly or monthly and will be physically tested either semiannually or annually in accordance with NFPA 72, *National Fire Alarm and Signaling Code* (see Table 11.2).

TABLE 11.2	Fire Alarm Testing Requirements				

FIRE ALARM COMPONENTS	WEEKLY	MONTHLY	QUARTERLY	SEMI ANNUAL	ANNUALLY
All fire alarm equipment					X
Monitored fire alarms (CS),(RS),(P),(A)					X
Trouble signals				X	
Unmonitored fire alarms local	X				
Supervisory alarm transmitters					X
Voice alarm communications systems				X	
Batteries depending upon type (see NFPA 72)		X		X	
Remote annunciator				X	
Initiating devices: Smoke, duct, air sampling, manual pull, beam detectors				X	
IR/UV detectors			X		
Water flow and tamper switches			X		
Combination systems and CO monitoring				X	
Fire alarm control panel interface				X	
Guard tour equipment				X	
Notification appliances both visual and audible				X	
Two way communication systems					X
Mass notification systems					X
Mass notification system control equipment				X	
Public emergency reporting transmission equipment				X	X
Supervisory alarm station receivers	Daily				X

Reprinted with permission from NFPA 72® 2013, 2010, 1999 *National Fire Alarm and Signaling Code*, Copyright © 2012, 2009, 1999, National Fire Protection, Quincy, MA. This reprinted material is not the complete and official position of the NFPA on the referenced subject, which is represented only by the standard in its entirety.

Typically, fire alarm inspection and testing is conducted by certified fire alarm technicians and witnessed by the fire inspector. These private fire alarm system testing companies should meet the local or state requirements for industry certification. Typically, fire alarm technicians are certified by the National Institute for Certification of Engineering Technologies, or NICET. The certification from NICET demonstrates competency in the field of fire protection, and many states now require these certifications for technicians who may work on these life safety systems. The fire inspector should review any fire alarm reports from the testing agency and verify that they comply with NFPA 72 testing requirements.

Fire Alarm System Impairments

The fire prevention codes require local fire marshals and fire officials to be notified anytime a fire protection or alarm system is tested, repaired, altered, or is inoperative. The responsibility to perform this notification function is the duty of the impairment coordinator for the facility. System impairments fall into two general categories: those that are *planned* and those that are *unplanned*. The impairment coordinator must authorize any planned disruptions, and eight steps must be taken before a planned impairment of the fire protection system, including contacting the authority having jurisdiction and the local fire department officials:

1. The extent and duration of the system disruption must be known.
2. The affected area must have a risk analysis performed to identify and reduce any fire hazards
3. Safety recommendations must be made to the building owner, and this may include instituting a fire watch or the evacuation of the area.
4. The fire department and fire marshal or fire official must be notified.
5. The insurance company and fire alarm monitoring company must be notified.
6. Supervisors working in the area effected must be notified.
7. Impairment tags are in place on all affected equipment.
8. All the necessary tools and repair parts are on site to accomplish the repair.[12]

Any unplanned fire system disruptions require the same eight steps to be completed immediately after the impairment has occurred. Before a fire alarm system can be returned to service, the new system components must be tested and verified to be operational by the service technician. The impairment coordinator then shall make all the necessary contacts to restore the system and shall remove any system impairment tags.

Finally, any fire protection alarm components that may be recalled either through voluntary or mandatory circumstances shall be replaced at the owner's expense, and the fire marshal or fire official shall be notified when all components have been repaired. It is an important aspect of fire prevention inspection to identify the building impairment coordinator and to make sure that he or she is aware of the fire code responsibilities necessary to perform that function of building management.

Summary

Automatic fire alarm systems are a vital component of the building's life safety system. Fire alarms must be reliable, effectively heard, capable of controlling certain building functions, and provide an emergency communications path for the fire department to use to direct fire operations and evacuate occupants. The maintenance and inspection of the fire alarm system is a key aspect of an effective fire prevention inspection program. Fire inspectors must carefully review all certification and testing documents and ensure that all identified deficiencies are corrected. The fire inspector should witness some or all of the fire alarm testing, if possible, to learn more about the fire alarm system components and test method requirements of the system. Fire alarm systems should not be disrupted without the knowledge of the building impairment coordinator and the proper notification of the fire marshal or fire official. Fire alarm systems are vital to the life safety of the occupants of a building and may be considered an imminent hazard when they are not functioning in accordance with the NFPA 72 standard.

Review Questions

1. Describe how an automatic fire alarm system detects the signatures of a fire. List the environmental conditions that may lead to false alarms.
2. False alarms are a large problem, and they place firefighters and the public at risk. The NFPA reported that 45 percent of false alarms are human error and 32 percent are equipment malfunctions. Develop and describe an inspection program to deal with the false alarm issue. What would the key inspection points be?
3. Describe the operation differences between ionization and photoelectric smoke detectors.
4. Describe how rate compensation heat detectors operate. In what application would they be used?
5. Contrast the differences between infrared and ultraviolet fire detectors. What are the fire hazard applications for these devices?
6. How would you test alarm-indicating appliances? What are the minimum outputs that should be measured?
7. List and describe the major components of an automatic fire alarm system.
8. List and describe the major circuits of an automatic fire alarm system. Describe their function and operation.
9. What fire protection features are required in a fire command center?
10. Describe the differences between the following methods of fire alarm system monitoring:
    a. Local
    b. Remote signaling station
    c. Proprietary signaling station
    d. Central signaling station

Suggested Readings

International Code Council. 2006. *2006 IFC Fire Protection Systems*. Washington, DC: International Code Council.

National Fire Protection Association. 2008. *Fire Protection Handbook – 2008*. Quincy, MA: NFPA.

National Fire Protection Association. 2010. *NEC – 2010*. Quincy, MA: NFPA.

National Fire Protection Association. 2010. NFPA 72, *National Fire Alarm and Signaling Code 2010*. Quincy, MA: NFPA.

National Fire Protection Association. NFPA 90A, *Standard for the Installation of Air-Conditioning and Ventilating Systems*. Quincy, MA: NFPA.

National Fire Protection Association. NFPA 92, *Standard for Smoke Control Systems*. Quincy, MA: NFPA.

Endnotes

1. William Greer, *A History of Alarm Security,* 2nd edition (Bethesda, MD: National Burglar & Fire Alarm Association), 7.
2. Ibid., 25.
3. Ibid., 68–69.
4. National Fire Protection Association, *Unwanted Fire Alarms, 2011* (Quincy, MA: NFPA), ii.
5. *White Paper: Home Smoke Alarms & Other Fire Detection Equipment* (Public/Private Fire Safety Council, April 2006), 9.
6. http://www.usfa.fema.gov/downloads/pyfff/smkalarm.html
7. Guylène Proulx, Chantal LaRoche, Fern Jaspers-Fayer, and Rosanne Lavallée, *Fire Alarm Signal Recognition* (National Research Council of Canada, IRC-IR-828, 2001), 15–16.
8. *The National Fire Alarm Code* (Quincy, MA: NFPA), 123.
9. *International Building Code – 2009* (Washington, DC: International Code Council), 203.
10. John L. Bryan, *Smoke as a Determinant of Human Behavior in Fire Situations* (College Park, MD: University of Maryland, Department of Fire Protection Engineering, 1977), 85.
11. *The National Fire Alarm Code* (Quincy, MA: NFPA), 98, section 6.8.1.3.
12. *International Fire Code – 2006 – New Jersey edition* (Washington, DC: International Code Council), 54.

Courtesy of J. Foley

KEY TERMS

air accelerator, *p. 292*

air exhauster, *p. 292*

acceptance pump test curve, *p. 311*

alarm check valve, *p. 290*

commodity classification, *p. 287*

deluge system, *p. 294*

dry pipe system, *p. 291*

dry pipe valve, *p. 291*

ESFR sprinkler head, *p. 300*

fire department connection, *p. 291*

fire pump, *p. 310*

flow switch, *p. 305*

fusible link, *p. 295*

hazard classification, *p. 285*

hose allowance, *p. 297*

hydraulic calculations, *p. 285*

inspector's test valve, *p. 290*

K-factor, *p. 301*

large droplet sprinkler head, *p. 299*

master pressure-reducing station, *p. 314*

microbiological influenced corrosion, *p. 307*

pipe schedule, *p. 285*

preaction system, *p. 295*

pressure-reducing valve, *p. 312*

pressure-restricting device, *p. 312*

residential sprinkler head, *p. 299*

sprinkler head, *p. 299*

standpipe, *p. 311*

tamper switch, *p. 305*

water columned, *p. 292*

wet pipe system, *p. 289*

OBJECTIVES

After reading this chapter, the reader should be able to:

- Know the different National Fire Protection Association Standards that apply to water-based fire protection systems.
- Know the main types of automatic sprinkler systems and their associated components.
- Know the critical inspection points needing examination during a routine fire inspection.
- Understand the classification of hazards and commodities as they relate to automatic sprinkler system design and protection.
- Understand the basic classification of standpipes and the critical inspection points on these systems.

Professional Levels of Job Performance for Fire Inspectors as Cited in NFPA 1031 and NFPA 1037

- NFPA 1031 Fire Inspector I *Obj. 4.3.5. Determine the operational readiness of fixed fire suppression systems*
- NFPA 1031 Fire Inspector I *Obj. 4.3.15 Determine code compliance*
- NFPA 1031 Fire Inspector II *Obj. 5.3.4. Evaluate fire protection systems and equipment*
- NFPA 1031 Fire Inspector II *Obj. 5.3.11 Verify compliance with construction documents*
- NFPA 1031 Fire Inspector II *Obj. 5.4.3 Review the installation of a proposed fire protection system*
- NFPA 1031 Plan Reviewer I *Obj. 7.3.10 Evaluate plans for the installation of fire protection and life safety systems*

Introduction

Automatic sprinklers, fire pumps, and standpipes are critical active fire protection systems that require constant vigilance on the part of fire inspectors. Generally, the building codes allow reduced passive fire protection when automatic sprinkler systems are installed. Automatic sprinklers also allow the owner to increase the height and area of the building while reducing the type of construction by one class. These reductions may also occur in the hourly fire resistance ratings of firewalls, fire separation walls, and fire barriers. Automatic sprinklers allow increases in occupant loads without necessarily increasing egress capacity. In other words, the building code places a lot of "eggs in the basket" when automatic sprinkler systems are installed. The key safety concern for fire inspectors is proper maintenance of the system. These systems must work the first time and every time.

Automatic Sprinklers: A Brief History

The concept of automatic sprinkler systems goes back to 1812 in England, when the first rudimentary sprinkler system was installed in the Theater Royal in Drury Lane. The first sprinkler systems in America were installed in the New England Textile Mills in the mid-1800s. These systems were constructed of a series of perforated pipes that sometimes were operated using gunpowder and fuses to initiate the water flow through the pipes. Henry S. Parmelee, a piano maker from New Haven, Connecticut, is credited with the first patented automatic sprinkler head in 1874. He invented the sprinkler to protect his piano company. His invention revolutionized fire protection technology. Fredrick Grinnell, an industrialist and inventor from Massachusetts, licensed and improved on the Parmelee design and developed the technology of the glass bulb fusible link in 1890. John Freeman, an engineer from Factory Mutual, later presented a report on sprinkler protection to the committee on automatic sprinklers at the NFPA meeting in 1896, and this report eventually evolved into NFPA 13, *Standard for the Installation of Sprinkler Systems*, creating the first standard for the installation of automatic fire sprinkler systems.[1] Sprinkler technology at that time was used only for property protection and had little consideration for potential life safety value.

The next breakthrough came in 1953 with the invention of the standard sprinkler head. Until this invention, most of the water (40 to 60 percent) was directed up toward the ceiling and not down at the floor. The standard pendant and upright sprinkler heads began to improve on the effectiveness of applying water to the fire area.

Sprinkler technology improved through the 1960s, but it was in the 1970s and 1980s that vast improvements were made in providing greater life safety value to the sprinkler system. The invention of the early suppression fast response, or ESFR, sprinkler head, which was approved for listing by Factory Mutual Research in 1989, began to change the value of sprinkler systems to provide a greater level of life safety protection. The United States Fire Administration, the Marriott Corporation, and the International Association of Fire Chiefs worked together from 1976 to 1982 on the development of the quick response residential sprinkler head through a joint project called *Operation San Francisco*. This project eventually led to the formation of Operation Life Safety and in 1999 the creation of the Residential Fire Safety Institute.[2] The goal of this group is to encourage the installation of fire sprinkler systems in residential structures. Residential sprinkler head technology has since been approved by building codes for installation in light-hazard commercial occupancies such as hotels. Innovation is constantly changing sprinkler technology, and residential systems now may be installed using PEX-A flexible plastic tubing listed under UL 1821, *Standard for Safety of Thermoplastic Sprinkler Pipe*. This product reduces labor and material costs and makes the installation of fire sprinklers in new construction much more affordable to the consumer.

Water-Based Fire Protection Standards

Fire inspectors must be competent in applying automatic sprinkler system requirements that are found in both the building and fire codes. Every state or locally adopted code identifies the applicable standard in the code appendix. The edition of that standard is identified, such as NFPA-13-96, meaning the fire inspector should use NFPA Standard 13 and the 1996 edition. The fire inspector must make sure that they use only the adopted applicable standard identified in the code, as that is the only edition legally adopted by reference.

The following are the main NFPA standards applicable to automatic sprinkler systems that the fire inspector should be familiar with:

- NFPA 13, *Standard for the Installation of Sprinkler Systems*
- NFPA 13R, *Standard for the Installation of Sprinkler Systems in Low-Rise Residential Occupancies*
- NFPA 13D, *Standard for the Installation of Sprinkler Systems in One- and Two-Family Dwellings and Manufactured Homes*
- NFPA 14, *Standard for the Installation of Standpipe and Hose Systems*

These standards describe the basic installation, design, and features of automatic sprinkler systems and standpipes. There are also many additional fire protection standards related to automatic sprinkler protection that may apply to specific industries such as warehousing or other elements of water-based fire protection systems. These fire protection standards fall into four general categories: building fire protection, special facilities protection, fire protection equipment standards, and water-based system maintenance and testing standards. (See the complete list of the 21 water-based fire protection standards.)

Please visit Resource Central Additional Resources for a list of NFPA water-based standards.

Effectiveness of Automatic Sprinkler Systems

Automatic sprinkler systems are the first line of defense in building fires and are important in the protection of the community from large conflagrations. To be useful, automatic sprinklers must be cost effective and reliable in suppressing fires. The cost of installing commercial automatic sprinklers may run from 2 to 5 dollars per square foot.

The installation of specialized sprinkler systems can be as high as 10 dollars per square foot depending on the application. The United States Fire Administration estimates the cost of residential fire sprinkler systems to be under 2 dollars per square foot in most regions of the country, and this is far less than the cost of most floor coverings today.[3] An important life safety value of automatic sprinklers systems is that there have never been more than two fire fatalities in a property protected by working automatic fire sprinklers. According to the NFPA, automatic sprinklers are effective 95 percent of the time with the operation of a single sprinkler head. A 2006 NFPA study by John R. Hall examined the effectiveness of fire sprinklers. The results on the effectiveness and number of sprinkler heads that operated based on actual fires are listed below:

- 95% effective   One head operating
- 94% effective   Two heads operating
- 91% effective   Three heads operating
- 89% effective   4–10 heads operating
- 81% effective   More than 10 heads operating

Obviously, the fuel load present in the structure may alter the number of sprinkler heads that may operate; however, the effectiveness is still significant even when ten sprinkler heads are operating. The report also noted that multiple sprinkler heads operated in 35 percent of wet pipe system activations and 59 percent of dry pipe system activations. According to an analysis of NFIRS 5.0 fire incident reports from across the United States, from 1999 to 2002, the total identified failure rate of automatic sprinkler systems was only 7 percent of all properties protected. These sprinkler system failures were highest in public assemblies (19 percent), then storage facilities (15 percent), and educational (11 percent) buildings, and were the lowest in manufacturing (7 percent) and residential (3 percent) occupancies. While a 7 percent overall failure rate may seem high, the majority of noted failures of sprinkler systems were caused by human error or premature intervention shutting the sprinkler systems down. The NFPA analysis identified that in 65 percent of the system failures, sprinkler control valves were closed before the fire, and another 16 percent of failures were caused by premature closing of control valves. The study attributes 11 percent of sprinkler system failures to poor system maintenance, 5 percent to a mismatch between the commodities protected and the sprinkler system's design, and only 3 percent to sprinkler system component failure.[4] As we can see, four out of the five major causes of system failure are related to human intervention. Fire inspectors must be aware of these causes of failure and give careful attention to how sprinkler systems are operated and maintained. Fire inspectors must be able to identify the commodities being protected and how it matches with the sprinkler system's hydraulic design requirements.

Basic Sprinkler System Design

As a general rule, automatic fire sprinkler systems are designed and installed based on the degree of fire hazard that a building possesses. The building codes evaluate these hazards based on four common elements:

1. Building height: usually 3 stories above or below grade.
2. Building area sprinklers are generally required beyond 12,000 square feet.
3. Occupant load sprinklers are generally when occupant loads exceed 300.
4. Special hazard: the presence of specific quantities of hazardous materials or processes.

In general, the fire inspector should anticipate automatic sprinkler and standpipe systems for buildings exceeding three stories, 12,000 square feet, or that have 300 to 1,000 occupants. Sprinkler systems are also required by special hazards such as nonambulatory occupants; restrained occupants with physical or mental disabilities; penal institutions; high-rise structures; high-hazard buildings with chemical or flammable liquids storage;

and residential buildings in which occupants sleep, such as hotels and apartment buildings or dormitories. The fire inspector should be familiar with the local building and fire code requirements for the installation of automatic sprinklers. The fire inspector should also be aware of any local code or retro fitting provisions, as many states have additional requirements for installation of automatic fire sprinklers in existing buildings.

Automatic sprinkler systems' design was traditionally **pipe scheduled** up to the 1970s. A pipe schedule is a kind of "cookbook" approach to installing a sprinkler system in which the pipe sizes are determined by the number of sprinkler heads installed on each section of pipe. When material and labor costs were relatively inexpensive, pipe scheduling was the predominant method of sprinkler system installation. Fire inspectors usually can recognize pipe schedule sprinkler systems by the increasing pipe size on the branch lines. These systems resemble tree branches going from large-size pipes to small-size pipes at the end of each branch line. The components of a typical system include a sprinkler riser that could be located at the end, the center, or multiple locations in a girded and looped system design. The riser connects to a cross main, which is of equal size or slightly smaller than the riser, and the branch lines connect perpendicular to the cross mains. The sprinkler heads then are attached to each branch line, which decreases in size as you move away from the cross main. The sprinkler installer would use a pipe schedule table from NFPA 13 to determine the appropriate pipe configuration. For example, in the light hazard pipe schedule you can install only two sprinklers of 1-inch pipe; at the third sprinkler, the pipe must be 1-¼ inches; and at the fifth sprinkler, the pipe must be 1-½ inches.

The increasing pipe size toward the cross main compensated for the potential friction loss in the pipe but also used far more material than generally was necessary and added labor costs because of the weight of the pipe. In the late 1960s and early 1970s, automatic sprinkler system designers began to shift to the NFPA 13 alternative method of designing systems using **hydraulic calculations**. Almost all of today's sprinkler systems are hydraulically calculated because of the higher costs of labor and material. Hydraulically calculated fire sprinkler system design requires three things: first, that the water supply is sufficient to operate all of the sprinklers that are anticipated to operate within the specified design area for a specific duration of time; second, that the water supply has adequate flow and pressure; and third, that the **hazard classification** under NFPA 13 has been properly identified for the building based on the known fuel loads or the types of commodities to be placed inside the building. Three general hazard categories are specified in NFPA 13. Light-hazard buildings contain ordinary quantities of combustible materials and are considered to have a low heat release. Ordinary-hazard occupancies have moderate amounts of combustible materials and are considered to have a moderate to high heat release in a fire. Extra-hazard occupancies have appreciable combustible and highly flammable commodities and present a high to severe fire hazard and high heat release. Each hazard classification also is divided into groups based on the height of storage or the flammability and quantity of the materials in the building.

The NFPA 13 standard classifies buildings as follows:

- **Light Hazard:** low heat release, such as residential, assembly, educational, hospitals, museums, libraries, etc.
- **Ordinary Group I:** moderate heat release; storage up to 8 feet in height in facilities such as light manufacturing, warehousing facilities, parking structures, bakeries, electronic plants, and dairies
- **Ordinary Group II:** moderate heat release; storage from 8 to 12 feet in height in facilities such as mercantile stores, woodworking shops, metal fabricators, and large storage facilities
- **Ordinary Group III:** high heat release; these facilities have flammable liquids in limited quantities and combustible dusts; include feed mills, paper plants, repair garages, piers and wharves, and tire manufacturers

pipe schedule ■ A method of sprinkler design where pipe size and installation as well as the number of sprinklers permitted on a specific pipe size are determined by a table based upon occupancy hazard.

hydraulic calculations ■ A design method used in designing automatic sprinkler systems based on flow, pressure, and friction loss to the most remote location of the system.

hazard classification ■ Occupancy classifications under NFPA 13 used to identify potential fire hazard risk. Hazard class determines the sprinkler spacing requirements and the water supply demand.

- **Extra Hazard Group I:** high heat release; these are facilities that have a larger amount of combustible materials, which would include upholstery shops, die-casting, saw mills, and textile plants
- **Extra Hazard Group II:** Severe heat release; these facilities have appreciable quantities of hazardous materials and would include flammable liquid spraying, asphalt plants, flow coating operations, mobile or modular building assemblies, oil quenching, and solvent cleaning facilities

When inspecting warehousing facilities, the fire inspector must be aware that other standards besides NFPA 13 are required to be reviewed, such as NFPA 231, *Standard for General Storage,* to establish the proper hazard classifications. This standard provides the required knowledge for the fire inspector to correctly identify the types of commodities in the warehouse and determine the appropriate hazard classification that the automatic sprinkler system should provide. This is important because commodities, especially in transportation warehousing, change all the time, and the sprinkler system must be capable of meeting the fire challenges that the fuels present.

Once the appropriate hazard classification is determined, NFPA 13 established limits on the system size, the sprinkler head spacing on branch lines, and the coverage of floor area by each sprinkler head. Light- and ordinary-hazard sprinkler systems may not exceed 52,000 square feet in protected areas, with each sprinkler head covering from 225 square feet in light hazard to 130 square feet in ordinary hazard. This equates to approximately 400 sprinkler heads on a single-riser system. Extra-hazard fire sprinkler systems are limited to 40,000 square feet, with sprinkler heads spaced at 100 square feet per head. Extra hazard pipe schedule systems are limited to 25,000 square feet, or approximately 250 sprinkler heads on the system, and hydraulically calculated systems may go up to 40,000 square feet. The fire inspector may locate system design information on the hydraulic design plate located on the system riser above the main control valve (see Figure 12.1).

COMMODITY CLASSES

An important aspect of the fire inspector's role in automatic sprinkler system maintenance is ensuring that the sprinkler system adequately protects the building as it was originally designed. The most likely challenge to the system design is a change in the

FIGURE 12.1 The hydraulic design plate provides useful information to fire inspectors.
Courtesy of J. Foley

commodities being stored in the building. This is especially true in transportation warehousing, where commodities change on a weekly or daily basis. NFPA 231, *Standard for General Storage*, deals with storage and warehousing facilities and establishes seven different types of storage classifications that can challenge automatic sprinkler protection. Warehousing storage is usually arranged in one of four ways or some combination of the four: bulk storage, palletized storage, rack storage, or solid pile storage. The fire inspector needs to identify the type of storage and the **commodity classification** to determine whether the sprinkler density is adequate to protect the commodity. Fire inspectors must also realize that NFPA 231 also contains many requirements on sprinkler head clearances, storage pile height, partial storage areas, pallets, and cutoff rooms. Let's examine each commodity type:

commodity classification ■ Classification of storage under NFPA 231 for the purpose of identifying potential fire severity.

■ Class I – Class I commodities consist of noncombustible products on combustible pallets in ordinary corrugated cartons with or without a single-thickness cardboard divider or materials wrapped in ordinary paper wrapping with or without pallets. Class I commodities are moderately combustible and require sprinkler densities of 0.15 gallon per minute over 2,000 square feet in wet pipe systems and 2,600 square feet in dry pipe systems.
Examples: Food, beverages, frozen foods, dairy products, glass bottles, metal products with limited plastics, appliances, gypsum wallboard, cement bags, etc.

■ Class II – Class II commodities are products in slatted wooden crates, solid wooden boxes, multiple-thickness paper cartons, or equivalent combustible packaging with or without pallets. Class II commodities are also moderately combustible and require sprinkler densities of 0.15 gallon per minute over 2,000 square feet in wet pipe systems and 2,600 square feet in dry pipe systems.
Examples: Light bulbs, small Class I products placed in ordinary paper cartons, books, beer, and wine under 20% alcohol in wood boxes.

■ Class III – Class III commodities are wood, paper, natural fibers, and clothing materials or Group C plastics, with or without pallets, these products may also contain limited amounts of group A and B plastics. Class III commodities, because of the presence of plastics, are a moderate to high hazard and require sprinkler densities of 0.18 gallon per minute over 2,000 square feet in wet pipe systems and 2,600 square feet in dry pipe systems. The minimum sprinkler system design is ordinary group II.
Examples: Bicycles, shoes, clothing, books, magazines, plastic-coated food containers, wood products, door, windows, tobacco products, textiles, soaps, detergents, bleaches, pharmaceuticals.

■ Class IV – Class IV commodities are Class I, II, or III commodities containing an appreciable amount of Group A plastics in ordinary corrugated cartons or cartons with Group A packaging materials with or without pallets, packing materials like Styrofoam peanuts, or cocoon packaging. Class IV commodities are a high fire hazard risk and require sprinkler densities of 0.20 gallon per minute over 2,000 square feet in wet pipe systems and 2,600 square feet in dry pipe systems. The minimum sprinkler system design is Ordinary Group II.
Examples: Small appliances, upholstered furniture, mattresses, electrical wire on reels, insulating panels of Styrofoam or polyethylene.

Plastics

Plastics are divided into three groups in NFPA 231, *Standard for General Storage*. Commodities made of plastic generally have two to three times the heat release of ordinary combustible materials. NFPA 231 divides the different types of plastics into hazard groups based on their ability to burn rapidly and their heat release rate. Plastics in Group A burn the fastest, Group B burn slower, and Group C burn the slowest. Plastic groups A and B contain mostly thermoplastics that will soften, melt, drip, or liquefy when subjected to heat. Group C plastics are mostly thermoset plastics that burn slowly or char, such as phenolics and Bakelite plastics. Fire inspectors must be aware of the commodity classes as they can change very easily, especially in transportation warehousing. Class I commodities can also be shrink-wrapped and become Class IV commodities because of the plastic wrapping material. Automatic sprinkler protection must be up to the challenge of the potential burning fuel if it is to contain or control a fire.

Group A Plastics:

- Acrylic
- Acetyl
- Butyl rubber
- Ethylene propylene rubber
- Fiberglass reinforced polyester
- Natural rubber – expanded
- Nitrile rubber
- Polybutadiene
- Polycarbonate
- Polyester
- Polyethylene
- Polypropylene
- Polyutherane
- PVC
- Styrene acrylonitrile
- Styrene butadiene rubber

Group B Plastics:

- Celluloses (acetate, ethyl cellulose)
- Chloroprene rubber
- Fluoroplastics
- Natural rubbers – not expanded
- Nylon
- Thermoplastic polyester
- Silicone rubber

Group C Plastics:

- Fluoroplastics (polychlorinatedtrifluoroethylene)
- Polytetrafluoroethylene
- Melamine
- Phenolics
- Rigid PVC
- Polyvinyldiene chloride
- Polyvinyl fluoride
- Polyvinylidiene fluoride
- Urea

Pallet Storage: One last consideration the fire inspector must examine is the storage of pallets within the facility. It is recommended that pallets be stored outside and away from the building; but if they are stored inside, they are limited to 6 feet in height, and no more than four piles are permitted, which must be 8 feet apart. If those conditions are not complied with, then changes in the sprinkler design densities are required. Pallets stored over 8 feet would require sprinkler densities of 0.60 gallon per minute over 6,000 square feet.

FIGURE 12.2 Backflow preventors are installed between domestic and fire protection water supplies. *Courtesy of J. Foley*

Types of Automatic Sprinkler Systems

There are five main categories of automatic fire sprinkler systems that fire inspectors encounter in the field. They include wet pipe sprinkler systems, dry pipe sprinklers systems, deluge sprinkler systems, preaction sprinkler systems, and specialized sprinkler systems, which include fixed water spray, foam sprinkler systems, and water mist systems. Let's examine each type of system.

WET PIPE SPRINKLER SYSTEMS

Wet pipe systems account for the vast majority of sprinkler system installations. Wet pipe systems can only be installed in heated occupancies and are the most efficient systems because water is always available at the sprinkler head. The wet pipe fire sprinkler system must be attached to a reliable water supply, and a backflow preventer or dual check valve must be installed between the fire protection system and the domestic water supply (see Figure 12.2). The wet pipe sprinkler system is the simplest of all automatic sprinkler systems. The main components of the wet pipe sprinkler system start at the main control valve, which must be an indicating type valve (see Figure 12.3). Typically, this valve is an open stem and yoke, or OS&Y, valve or a flag-indicating valve. The stem of the OS&Y is a visual indicator of the valve gate position; when the valve stem is fully out, the valve is open, and when there is no stem showing, the valve is closed. The length of the stem is determined by the diameter of the pipe. A 6-inch

wet pipe system ■ A sprinkler system containing water at all times ready for immediate discharge on a fire.

FIGURE 12.3 OS&Y valve in the open position; the stem should equal the diameter of the pipe. *Courtesy of J. Foley*

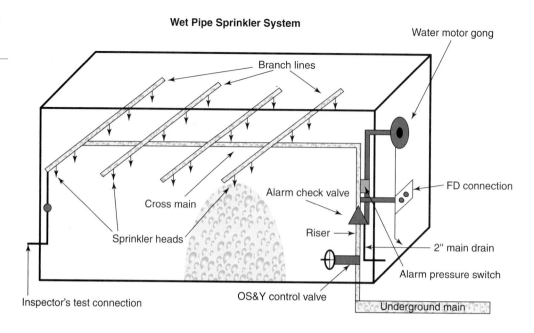

FIGURE 12.4 Wet pipe sprinkler system components.

Wet Pipe Sprinkler System

Water motor gong

Branch lines

FD connection

Alarm check valve

Cross main

Riser

2" main drain

Sprinkler heads

Alarm pressure switch

OS&Y control valve

Inspector's test connection

Underground main

alarm check valve ■ The part of a sprinkler system that transmits an alarm when the sprinkler system is first activated.

stem would indicate a 6-inch pipe. The main sprinkler control valve must be supervised electronically by the fire alarm system or by locking it in the open position. The valve must also have a sign indicating its control function. Moving up the riser, the next component is the **alarm check valve** (see Figure 12.4). The alarm check valve is a simple clapper valve that, when opened by a sprinkler activation, allows the water to flow to the alarm-indicating appliances attached to the valve. Between the alarm check valve and the indicating appliances, the fire inspector will find a large cylinder called the *alarm retard chamber*. The alarm retard chamber allows for slight fluctuations in water pressure without activating the fire alarm system; thus, it "retards" the alarm. On top of the retard chamber is usually an electric pressure switch that initiates the electronic portion of the fire alarm signal to the fire alarm control panel, reporting the activation to the alarm monitoring company and activating the building fire alarm system. The water continues to flow from the retard chamber through pipes to the water motor gong, which is a small paddle wheel bell mounted on the exterior of the building. This mechanical bell alerts people outside the building or neighbors to a fire in the building. The code requires the placing of a sign next to the water motor gong that reads "When Bell Rings, Call Fire Department" on the building's exterior. The water continues from the water motor gong, down a pipe to drain on the building's exterior. On the main body of the alarm check valve is the 2-½-inch main drain valve. This valve allows the entire system to be drained once the OS&Y valve is closed for restoration of the system and installing new sprinkler heads. The main drain valve also requires a sign indicating that it is the main drain. Beyond the alarm check valve, the water moves up a riser to a cross main, which then feeds a series of branch lines. The sprinkler heads are located at the appropriate spacing along the branch lines.

inspector's test valve ■ A valve located at the most remote location on a sprinkler system used by a inspector to test the activation of the sprinkler alarms. The flow simulates that of a single sprinkler head.

At the end of the most remote branch line from the system riser is usually the inspector's test valve. The **inspector's test valve** must also be marked with a sign and is used to flow water through a restricted orifice ½ inch in diameter. The inspector's test connection must go to the outside of the building or an approved drain location to prevent water damage in the building. In high-rise or multilevel buildings, the inspector's test valve is located at the cross main's connection to the sprinkler system riser or standpipe riser (see Figure 12.5). The inspector operating the valve will see water flowing through a site glass, which also has a ½-inch-diameter orifice fitting to simulate a single sprinkler head activation.

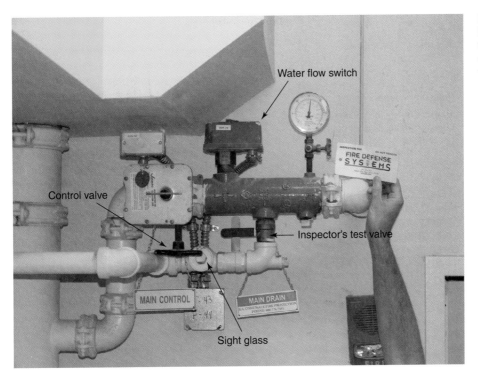

FIGURE 12.5 High-rise inspector's test valve location and site glass. *Courtesy of J. Foley*

The purpose of the inspector test valve is to ensure that water flow activates the fire alarm within 60 seconds to comply with NFPA 13. The **fire department connection** on wet pipe sprinkler systems is connected to the riser after the alarm check valve. The fire department connection is kept dry by a clapper valve in the line. The fire inspector should examine the exterior fire department connection for proper caps, signage, and ball drip valve. It is also a good practice to check the fitting threads for compatibility with the fire department's hose.

fire department connection ■ Exterior connection on an automatic sprinkler system for the fire department to use to supplement water quantity and pressure.

DRY PIPE SPRINKLER SYSTEMS

The **dry pipe system** is used in unheated buildings and areas where water may be subject to freezing. The dry pipe system has a different operating principle than the wet pipe system. Like the wet pipe, the dry pipe sprinkler system must have a reliable water supply and an indicating control valve located below the **dry pipe valve**. The dry pipe valve has a clapper inside that is held closed by air pressure on the system side of the valve. The principle of physics that operates the valve is Bernoulli's principle that *force = pressure × area*. The dry pipe clapper valve is unique in that the top of the valve or priming area is typically six times larger than the bottom side of the clapper facing the control valve. This is called a 6-to-1 dry pipe valve. Dry pipe valves come in 6:1 and 1.1:1 ratios between the top and bottom surfaces of the clapper. The 1:1.1 clapper is called a low differential valve, and it allows a faster response of the dry pipe valve when a sprinkler head operates. The 6:1 valve permits water pressure of 60 psi to be held back by 10 psi of air pressure, as illustrated in Figure 12.6. Dry pipe sprinkler systems are more for property protection than for life safety purposes because the air pressure must escape before water can reach the sprinkler head. In hydraulically designed dry pipe sprinkler systems, the designer must provide an additional 30 percent area increase in the remote operating area to compensate for the delay in water application.

NFPA 13 additionally requires that water must reach the most remote sprinkler head within 60 seconds of the dry pipe valve's operation. In large dry pipe systems (over 750-gallon capacity), quick opening devices must be installed to meet this 60-second time requirement.

dry pipe system ■ A sprinkler system filled with air pressure to be used in unheated spaces. Dry systems require 30 percent greater sprinkler operational areas.

dry pipe valve ■ A dry pipe valve is a 6:1 or 1.1:1 differential valve that uses air pressure to hold water pressure back in dry pipe sprinkler systems.

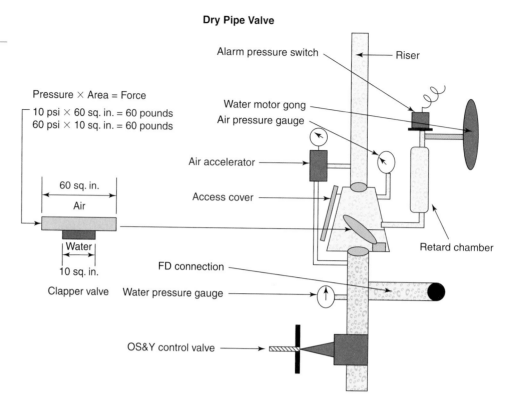

FIGURE 12.6 Dry pipe valve components.

Dry Pipe Valve

Alarm pressure switch — Riser

Water motor gong
Air pressure gauge

Pressure × Area = Force
10 psi × 60 sq. in. = 60 pounds
60 psi × 10 sq. in. = 60 pounds

Air accelerator —

Access cover —

60 sq. in.

Air

Water

10 sq. in.

Retard chamber

Clapper valve   Water pressure gauge —

FD connection —

OS&Y control valve —

air accelerator ■ A quick opening device installed on dry pipe sprinkler systems. Accelerators direct air below the dry pipe valve, forcing it to open quickly.

air exhauster ■ A quick opening device that releases air quickly from a dry pipe sprinkler system. Exhausters are located at remote ends of long runs of sprinkler mains.

water columned ■ A condition that occurs in dry pipe systems where water collects above the dry pipe valve over time. The condition causes additional weight on the clapper, delaying its operation. It can be corrected by draining the excess water off the dry valve or installing an anticolumn device.

There are two main types of quick opening devices, the **air accelerator** and the **air exhauster**. The air accelerator is the most common type and is found directly on top of the dry pipe valve. The accelerator is a three-chamber valve that operates when internal pressure changes between the internal chambers. The accelerator top and middle chambers are in equilibrium when the valve is closed. The operation of a sprinkler head causes air pressure to decrease in the middle chamber and creates an unbalanced pressure within the valve. The trapped air in the upper chamber pushes a diaphragm down and opens the accelerator to allow a path for air to move through a pipe and under the dry pipe clapper valve, forcing it open quickly (see Figure 12.7). The remaining system air then dumps to atmosphere until water pressure reaches equilibrium in the lower and middle chambers and the valve closes. Large dry pipe systems like those installed on piers and wharves may also have air exhausters installed at the remote end of the main. The air exhauster is also a three-chamber valve that uses pressure differentials to operate. The difference between the air exhauster and the air accelerator is the size of the air discharge. On an air exhauster, the discharge is 2 inches or larger to evacuate the air in a long run of piping very quickly. When water finally reaches the exhauster, it forces the air release valve closed so there is no loss of water pressure.

Important inspection points for the fire inspector include the examination of the air pressure and water pressure gauges located on the dry pipe valve. Typically, an additional 15 psi of air pressure is added above the valve tipping point as a safety factor on the dry valve to prevent false alarms. Here is an example: If the water main static pressure was 60 psi water pressure in a 6:1 valve, 10 psi air pressure holds the valve in equilibrium (60 psi × 1 = 60 psi and 6 × 10 psi = 60 psi). A safety factor of 15 psi is added, so the air pressure gauge should read 25 psi. To determine the tripping point on the clapper valve, simply divide the static water pressure by 6 (for example, 60 psi/6 = 10 psi tripping point). If the pressure on the air valve is higher than 25 psi, the system may be **water columned**. This condition occurs when residual water accumulates in the dry valve over time due to either improper draining or maintenance of low point drain valves. Each foot of water column adds 0.434 psi on top of the valve, delaying the proper operation of the valve. When

Air Accelerator

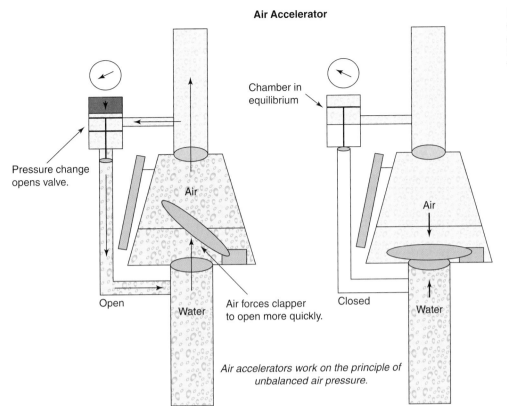

Chamber in equilibrium

Pressure change opens valve.

Air

Open

Water

Air forces clapper to open more quickly.

Closed

Air

Water

Air accelerators work on the principle of unbalanced air pressure.

inspecting dry pipe systems, low point valves, and drum drips (see Figure 12.8) should be examined for proper maintenance and that they are being properly operated. This is especially true if the system has had air leaks. All dry systems have air compressors to maintain the piping air supply. If the area is moist around the compressor and the compressor's water traps are not being maintained, the air in the system forms condensation that settles in low points on the systems. NFPA 13 requires low point drains in any pipe that can accumulate water. Pipes accumulating quantities below 5-gallon capacity require only a drain valve and a plug. Pipes accumulating over 5-gallon capacity require the installation of a drum drip at the low point. A drum drip is constructed with a 12 inch by 2 inch diameter pipe with nipples and valves located at both ends. The top valve should be open during normal use to allow water condensation to enter the drum. To drain the drum, the top valve is closed and the bottom valve is opened to let the accumulated water out. The process is then reversed to restore the drum drip. In cold climates, the draining

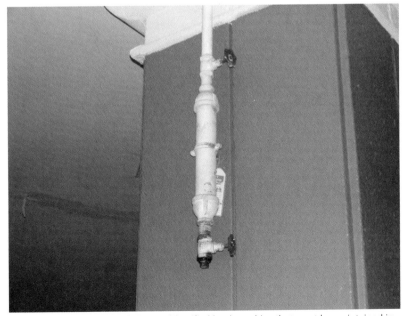

FIGURE 12.8 Low point drains are identified by drum drips that must be maintained in the system. *Courtesy of J. Foley*

of these low point drains and drum drips may need to occur on a weekly basis to prevent pipes from freezing. Low point drains are not supervised, so checking to ensure they are properly operating is essential. Fire inspectors must also make sure that the dry pipe valve room itself is properly insulated and permanently heated to at least 40°F in cold climates. Electrical heat tracing of the dry pipe valve is not permitted. The last important difference in the dry pipe sprinkler system is the placement of the fire department connection. In a dry pipe system, the connection is located above the indicating control valve but below the dry pipe valve. This is important for the fire department to know because if the dry pipe valve fails to open, water cannot be pumped into the system by the fire department connection. Dry pipe sprinkler systems are more complex than wet pipe systems, therefore, maintenance is more expensive. The dry pipe system is also subject to more corrosion because the pipes are empty, which allows oxidation and rust on the internal pipe surfaces. Dry pipe systems also have longer response times for water to be applied to the fire for at least 60 seconds, which means more sprinkler heads will operate or fuse.

DELUGE SPRINKLER SYSTEMS

deluge system ▪ A special type of sprinkler system used to protect high-risk areas, deluge systems have open-sprinkler heads, which all flow during activation.

Deluge systems are usually installed in areas with flammable liquid or chemical storage, where a large fire can develop very quickly that requires significant amounts of water to control. Deluge systems are usually limited in size because of the water capacity they must deliver by the system, which can be up to 3.0 gpm/sq.ft. The sprinkler heads in a deluge sprinkler system have no fusible links and are open so that water flows through all sprinkler heads in the system. Deluge systems require a fire alarm detection system to initiate the deluge valve's operation (see Figure 12.9). In typical deluge system installations, either infrared or ultraviolet flame detectors may be used as initiating devices. Deluge systems should not be used in areas where water damage is a top priority. Deluge systems are complex and require regular maintenance to ensure proper system operation. All deluge valves will also have manual activation capability by an integrated pull station.

FIGURE 12.9 Deluge systems have open-sprinkler heads and are connected to a fire alarm system.

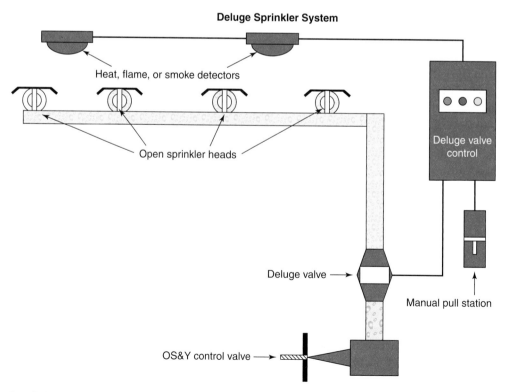

Deluge Sprinkler System

Heat, flame, or smoke detectors

Open sprinkler heads

Deluge valve control

Deluge valve

Manual pull station

OS&Y control valve

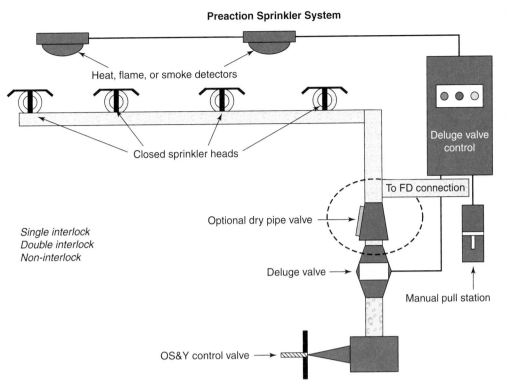

Preaction Sprinkler System

Heat, flame, or smoke detectors

Closed sprinkler heads

Deluge valve control

To FD connection

Optional dry pipe valve

Single interlock
Double interlock
Non-interlock

Deluge valve

Manual pull station

OS&Y control valve

FIGURE 12.10 Preaction systems provide the advantage of a dry system with the actions of a wet system. Preaction systems are connected to a fire alarm system that operates the deluge valve.

PREACTION SPRINKLER SYSTEMS

Preaction systems are, in some regards, the best of both worlds. They can be utilized in areas without heat but still provide water on the fire with the quickness of a wet pipe sprinkler system. Preaction sprinkler systems use a combination of a fire detection systems and a deluge valve with closed fusible link sprinkler heads. The concept of operation is based on testing of fire detectors, which must operate and initiate before the operation of a **fusible link** sprinkler head. The fire alarm initiating devices detect the fire and open the deluge valve to flood the system with water so that when the sprinkler head fuses, water is already at the sprinkler head with no 60-second delay. Preaction sprinkler systems are divided into three categories: single interlock, double interlock, and noninterlocking types (see Figure 12.10). In single interlock systems, the alarm-initiating device operates the deluge valve and floods the system directly. In a double interlock system, both the fire detector and the sprinkler fusible link must operate to open the deluge valve and flow water into the system. Double interlock systems may also employ a dry pipe valve but require the additional 30 percent increase in sprinkler remote operation area. Double interlocking systems are generally used in large freezers where accidental tripping of the system could cause pipes to freeze quickly. This would place extreme costs on the building owner to drain or remove the frozen pipes to restore the system. The redundancy of double interlock ensures that the system is more fail-safe. Noninterlocking systems are the most common preaction systems and are very reliable. In a noninterlocking system, either the sprinkler head fusing or the initiating device detecting a fire causes the system deluge valve to operate. In general, preaction systems are limited in size to about 1,000 sprinkler heads. Preaction sprinklers are also commonly used in elevator shaft protection under ANSI 17.1, the elevator safety standard for phase II recall of the elevator car. While these systems are efficient, they are also complex and require additional maintenance as opposed to wet pipe sprinkler systems.

preaction system ■ A dry pipe sprinkler system attached to an automatic fire alarm system that fills with water upon activation of the fire alarm, sending water to the sprinkler head before the head activation.

fusible link ■ Mechanical link soldered with a eutectic metal that melts at a specific temperature, releasing water from the sprinkler head.

FIGURE 12.11 Specialized foam firefighting systems often are installed in occupancies such as hospitals to protect helipads. These systems should be inspected for operational readiness by the fire inspector. *Courtesy of J. Foley*

SPECIAL SYSTEMS

There are a number of specialized water-based sprinkler systems that may be installed to address specific fire protection challenges in buildings. These systems are installed under the appropriate NFPA standards identified in the building code. Specialized systems are usually installed primarily for property or equipment protection as opposed to life safety.

Foam Sprinkler Systems

Foam-distributing sprinkler systems may be installed to protect flammable liquid loading racks, storage tanks, aircraft hangers, and other types of industrial applications. Foam sprinkler systems are installed in accordance with NFPA 16, *Standard for the Installation of Foam-Water Sprinkler and Foam-Water Spray Systems*, and use special foam application heads to distribute Class B foams on the fire hazard (see Figure 12.11). Low-, medium-, and high-expansion foam systems are installed under NFPA 11, *Standard for Low-, Medium-, and High-Expansion Foam*, and NFPA11A, *Standard for Medium- and High-Expansion Foam Systems*; and fixed water spray systems used on electrical transformers and substations are installed under NFPA 15, *Standard for Water Spray Fixed Systems for Fire Protection*. Each of these types of water-based fire protection systems is employed only in special applications and requires special knowledge for the fire inspector to review before conducting the fire inspection.

Water Mist Systems

A water mist system is a relatively new sprinkler technology that was developed in the 1990s and is considered an alternative replacement for halogenated fire suppression gases. Water mist systems are installed under NFPA 750, *Standard on Water Mist Fire Protection Systems*. Water mist requires special distribution heads, and the water is atomized by the introduction of high-pressure air or nitrogen gas into the piping. The water droplets must be no larger than 1,000 microns at the distribution nozzle. These systems produce very little water damage and are very effective on suppressing fires. This is because the finely divided water droplets convert quickly to steam. Water misting systems are usually preengineered and may be installed as wet, dry, deluge, and preaction systems. The systems have redundant banks of nitrogen cylinders, and the water is discharged at a rate between 175 and 500 psi at the sprinkler head. Water mist systems may be used in areas where valuable or irreplaceable Class A storage is maintained, such as museums, archives, and libraries, as well as computer rooms and areas traditionally protected by halogenated suppression systems.

Limited Area Systems

Limited area sprinkler systems are those that have 20 sprinkler heads or less and are attached to the domestic water supply or a standpipe, if available. Limited area sprinkler systems are usually installed to protect small hazardous areas within buildings such as laundry areas, storage rooms, or heater rooms. Limited area systems must be hydraulically calculated in accordance with NFPA 13, *Standard for the Installation of Sprinkler Systems;* NFPA 13R, *Standard for the Installation of Sprinkler Systems in Low-Rise Residential Occupancies;* or NFPA 13D, *Standard for the Installation of Sprinkler Systems in One- and Two-Family Dwellings and Manufactured Homes,* whichever is applicable to the building. The domestic water supply must be sufficient to provide both the fire protection demand and the domestic water requirements. Limited area sprinkler systems attached to domestic water supplies shall not have control valves unless they are supervised or secured in the open position.

Retrofit Sprinkler Systems

Many states have retrofit sprinkler provisions in their fire prevention codes requiring sprinklers to be installed in specific use buildings retroactively. These requirements are usually state specific and apply only to existing structures that may be below minimum fire safety standards. Often, these requirements may also be part of regulations for housing or health codes. Fire inspectors must be aware of all fire safety regulations that may exist within their state.

Residential Automatic Sprinkler Systems

Residential automatic sprinkler protection is installed under NFPA 13R for small residential structures up to four stories and NFPA 13D for sprinkler protection of one- and two-family dwellings.

NFPA 13R—RESIDENTIAL SPRINKLER SYSTEMS

The purpose of NFPA 13R, *Standard for the Installation of Sprinkler Systems in Low-Rise Residential Occupancies,* is to make sprinkler system installation more cost effective and attractive to small residential building operators. NFPA 13R systems do not fully protect the structure like the NFPA 13 requirements. The intent of NFPA 13R is to protect the areas of highest vulnerability and reduce the potential for flashover in a compartment fire. An NFPA 13R system does not receive the same fire resistance trade-offs under the building codes that are permitted for NFPA 13 systems. The NFPA 13R standard omits sprinkler protection in attics, concealed combustible spaces, bathrooms up to 55 square feet, and closets and pantries under 24 square feet. Balconies, porches, overhangs, and open exterior stairways are also exempt from sprinkler installation under 13R.

Another concern for the fire inspector is that of mixed use and occupancy. NFPA 13R applies only to the residential portions of the building and not to other use groups located within the building. The nonresidential use groups must be completely separated from the residential use by either firewalls or fire separation walls in order to utilize standard NFPA 13R. The other use groups would require compliance with NFPA 13 for complete sprinkler protection. Incidental uses such as a common kitchen or dining rooms that are used predominantly by the residents could be protected in accordance with a 13R sprinkler system. NFPA 13R also reduces the water flow requirement for the sprinkler system to 30-minute duration with no inside **hose allowance**. Fire inspectors should make sure that responding firefighters understand the difference between buildings protected by NFPA 13 and NFPA 13R for their safety. Structures under NFPA 13R should still be considered only partially sprinklered and can still have significant fires.

hose allowance ■ Additional water provided in the calculation of a sprinkler system demand for firefighting purposes. Sprinkler systems may have both inside and outside hose allowances that must be provided over a fixed period of operational time.

NFPA 13D—RESIDENTIAL ONE- AND TWO-FAMILY DWELLING SPRINKLER SYSTEMS

By far, the greatest threat to the public is fire in the home. One- and two-family dwelling fires accounted for 279,000 fires in 2010 according to the NFPA. These home fires led to 2,200 deaths and 9,400 injuries as well as 5.9 billion dollars in direct property loss.[5] In 2009, the IBC adopted residential sprinkler systems to be installed in all new one- and two-family dwellings in accordance with NFPA 13D. As each state has adopted the IBC codes, many have postponed or eliminated the requirement for 13D sprinkler system installations. Currently, California and Pennsylvania have adopted the requirements beginning in January 2011, New Hampshire will adopt it in 2012, Iowa in 2013, and South Carolina in 2014. The main distraction to the enactment of residential sprinkler regulations seems to come from the home building industry, which looks at sprinklers as an impediment to home construction instead of a selling feature. Homebuilders have gained allegiance in government trying to protect construction jobs, using the argument that sprinkler system costs will reduce home sales. An interesting report was produced by Bucks County, Pennsylvania, which had adopted residential sprinkler requirements in six municipalities between 1989 and 2000. During that time frame over 7,000 homes were purchased with NFPA 13D sprinkler systems. The builders spent over $815 million in capital investment. Bucks County produced a report, "Communities with Home Fire Sprinklers," in 2011 that provided several interesting facts based on their collective fire experience over the 11 years. In analyzing their NFIRS fire data, their fire experience showed significant benefits to residential sprinklers. Most significant was the fact that, of the 90 fire deaths that occurred in Bucks County, zero deaths were in sprinklered dwellings. Second, they identified that property loss in sprinklered dwellings averaged $14,000 while in nonsprinklered dwellings the average loss was $179,896 A third critical factor was that the water used to suppress the fires averaged 340 gallons in sprinklered dwellings as opposed to 5,914 gallons in nonsprinklered dwelling fires. Bucks County estimated the costs of installing sprinklers to be between $1.23 and $2.71 dollars per square foot.[6] The higher costs were in areas where there were no available water supplies. Homes in areas with water mains averaged construction costs of $1.10 per square foot. The National Institute of Standards and Technology (NIST) produced a report entitled "Home Fire Sprinkler Cost Analysis" in 2008 and determined that home fire sprinkler installation costs ranged from $0.38 to $3.66 per square foot with the average cost to builders being $1.48 per square foot. It is interesting to note that the $0.38 per square foot used newer flexible PEX-A sprinkler technology.[7] It should be obvious that the cost of sprinkler protection did not deter people in Bucks County, Pennsylvania, from buying sprinklered homes. This battle to sprinkler residences will continue and hopefully will become implemented on a national scale to significantly reduce home fire deaths and injuries in the United States. This is of critical importance to the firefighting community and is a key life safety initiative of the National "Everybody Goes Home—Line of Duty Death" program.

NFPA 13D systems are intended to be cost effective and simplistic in their design. They do not require significant piping or water supply. The typical single-story home under 2,000 square feet requires a water supply of 7 minutes duration, and all larger dwellings require a 10-minute duration. The water supply may be from municipal water, a well, or a storage tank. The sprinkler system is hydraulically calculated using two sprinkler heads with a design density of 0.55 to 0.5 gallon per minute per square foot. The minimum sprinkler operating pressure is 7 psi, and the system uses UL 1626–listed residential sprinkler heads.[8] NFPA 13D supply piping may be stand-alone or multipurpose providing both fire protection and domestic water. Piping may be copper, CPVC, PEX-A, or other approved materials.

Automatic Sprinkler System Components

To perform an effective fire inspection of the automatic sprinkler system, the fire inspector must be capable of identifying every component of the system. Each component's installation requirements are spelled out in the appropriate NFPA sprinkler standard. The fire inspector should review the standard and be familiar with the different component parts of all automatic sprinkler systems.

SPRINKLER HEADS

The most important component of the sprinkler system is the **sprinkler head**. Sprinkler heads come in several basic configurations and performance types, including those for special applications. The different types are as follows:

Upright Sprinklers

These have a large deflector with bent teeth that are located at the top of the sprinkler frame. The deflector is embossed "SSU," which stands for "Standard Sprinkler Upright," and the frame base identifies the sprinkler manufacturer. The frame color also identifies the operating temperature range for the sprinkler head's fusible link. Standard upright sprinklers have a ½-inch orifice and a K factor (flow coefficient) of 5.6. All ordinary-temperature sprinkler heads are uncolored on the frame, and the fusible link melts at 165°F.

Pendant Sprinkler Heads

The pendant sprinkler head is recognized by the small flat deflector attached to the frame. The deflector will be embossed "SSP" for "Standard Sprinkler Pendant." This head can only be installed in the pendant position with the deflector facing the floor. Pendant sprinklers are usually installed below drop ceilings and may be of the concealed variety with a special cover plate.

Sidewall Sprinkler Heads

The sidewall sprinkler head has a flat deflector extending out parallel to the sprinkler frame. On the top of the deflector, "sidewall" is embossed. Sidewall sprinklers come in two general varieties; either standard or extended throw. The throw describes how far water will extend from the sprinkler head. Standard throw is 14 to 16 feet, and extended throw is 20 to 30 feet. The standard sidewall sprinkler head is ½-inch orifice and the extended throw is ½-inch orifice with a larger deflector. The advantage to these sprinkler heads is the extended coverage area. They are typically used in hotel and apartment applications.

Concealed Sprinkler Heads

The concealed pendant sprinkler head is very common and used by architects to make the sprinkler head less visible for aesthetic purposes (see Figure 12.12). The concealed sprinkler has an ordinary pendent-style head that is covered by a flat plate attached by a fusible metal. The plates come in a variety of colors and may be custom colored, provided that the manufacturer performs the painting. The cover plate disengages from the sprinkler at 135°F and drops away. The sprinkler head fuses at 165°F. A key inspection element for concealed sprinkler heads is to make sure the caps are in place and not missing. If the cap cannot be properly installed, the sprinkler is not in alignment with the ceiling tile. The plane of the sprinkler deflector must be below the plane of the ceiling tile. Fire inspectors should also take note if the cap has been repainted, as this may affect the cap's ability to operate effectively and will require replacement of the cover plate.

Residential Sprinkler Heads

The **residential sprinkler head** is smaller than the standard sprinkler and comes in both pendent and sidewall varieties (see Figure 12.13). The residential sprinkler head has a round triple heat collector attached to the fusible link on the top of the sprinkler deflector. Residential sprinkler heads may be used in light-hazard occupancy applications under NFPA 13. The residential sprinkler head is considered a fast-response head due to the special heat collector arrangement.

Large Droplet Sprinkler Heads

The **large droplet sprinkler heads** are for special use in storage and warehouse applications and are identified by a large, smooth, round deflector plate. The deflector produces large water droplets at sufficient velocity to penetrate strong fire updrafts for high ceiling

sprinkler head ■ A water spraying head usually having a ½-inch discharge held closed by a fusible link.

residential sprinkler head ■ A specialized quick response sprinkler head used in residential and light-hazard occupancies.

large droplet sprinkler head ■ A specialized sprinkler head that produces large water droplets for specific high-challenge fire hazards. The large drop penetrates the fire plume more effectively.

FIGURE 12.12 Fire inspectors should examine the foam storage equipment for foam-water spray systems. *Courtesy of J. Foley*

FIGURE 12.13 Quick-response residential sprinkler heads must comply with UL 1626. *Courtesy of J. Foley*

sprinkler applications. These sprinkler heads are effective in areas where fire plumes can develop quickly and overcome standard sprinkler heads such as in a flammable liquid fire.

ESFR Sprinkler Heads

ESFR sprinkler head
■ A specialized sprinkler head used in the protection of flammable liquids and aerosols.

The **ESFR sprinkler head** (ESFR stands for "Early Suppression Fast Response") is used in warehousing and other areas where standard sprinkler protection is challenged, such as in the protection of aerosols (see Figure 12.14). ESFR sprinkler heads are designed to replace the need for in-rack sprinkler protection and can protect commodities such as Group A plastics, which burn rapidly and up to 40 feet in height. The ESFR sprinkler heads have

FIGURE 12.14 ESFR sprinkler heads are used for high fire-risk occupancies. *Courtesy of J. Foley*

large **K-factors** of about 14 and can flow up to 40–50 gallons per minute as compared to a standard sprinkler head that may flow 20 gpm. The ESFR head is used in warehousing, rubber tire storage, and high piled storage applications.

On-Off Sprinkler Heads

The on/off sprinkler head has the ability to turn itself off and reduce water damage once the fire has subsided or is controlled. The on-off sprinkler head uses a bimetallic snap disk that responds to temperature change (see Figure 12.15). As the disk heats, it snaps the valve open to flow water; when the disk cools, it then snaps back to its original position and closes the valve to stop the flow of water. These sprinkler heads have been a popular replacement for halogenated fire suppression systems in computer rooms and similar types of facilities with sensitive electronics.

K-factor ■ Is the relationship of a volume of water at a specific pressure that can pass through a sprinkler head's orifice. K-factors are standardized for sprinkler heads and allow designers to calculate flow and pressure using the formulas $Q = K (SQRT) P$, $K = Q/P$, or $P = (Q/K)^2$

Bimetallic snap disk

FIGURE 12.15 On-off sprinkler heads are often used to reduce the possibility of water damage. *Courtesy of J. Foley*

FIGURE 12.16 Fire inspectors should check spare sprinkler head cabinets for proper heads and wrenches based on NFPA 13. *Courtesy of J. Foley*

Spare Sprinkler Heads

The NFPA 13 standard requires that all automatic fire sprinkler systems maintain a stock of spare sprinkler heads and the properly sized wrenches need to change the sprinkler heads on site. Systems with 300 heads or less must have at least six spare sprinklers of representative types for the system. Systems with 300 to 1,000 heads must have 12 spares, and systems over 1,000 heads must have 24 spare sprinklers. Spare heads must be located in a cabinet, which should be placed by the system's main control valve (see Figure 12.16).

FIGURE 12.17 Temperature ranges are identifiable by the color of the sprinkler frame. *Courtesy of J. Foley*

SPRINKLER TEMPERATURE RANGES

An important aspect of the fire inspection is determining whether the sprinkler heads are of the correct temperature range for the location where they are installed. NFPA 13 makes identifying sprinkler temperature ranges easy by using different color sprinkler frames to identify the temperature by simple by observation (see Figure 12.17). Sprinklers with glass bulb fusible links also identify temperature ranges by the color of the liquid inside the bulb. The temperature ranges are established based on the maximum ambient ceiling temperature to which the sprinkler head may be exposed. Table 12.1 shows the classification and temperature ranges of typical sprinkler heads.

TABLE 12.1 NFPA Sprinkler Temperature Ranges

CEILING TEMPERATURE (°F)	SPRINKLER TEMP RANGE (°F)	CLASSIFICATION	SPRINKLER FRAME COLOR	GLASS BULB COLOR
100	135–170	Ordinary	Uncolored	Orange or red
150	175–225	Intermediate	White	Yellow Green
225	250–300	High	Blue	Blue
300	325–375	Extra High	Red	Purple
375	400–475	Very Extra High	Green	Black
475	500–575	Ultra High	Orange	Black
625	650	Ultra High	Orange	Black

Reprinted with permission from NFPA-2012, *Installation of Sprinkler Systems,* Copyright © 2012, National Fire Protection Association, Quincy, MA. This reprinted material is not the complete and official position of the NFPA on the referenced subject, which is represented only by the standard in its entirety.

SPRINKLER PIPE

Sprinkler distribution pipe must meet the Underwriters Laboratories (UL) and American Standards for Testing and Materials (ASTM) specifications for automatic sprinkler systems piping. Fire inspectors should be familiar with the different types of sprinkler pipes available and their applications. Sprinkler pipe may be steel with welded or grooved couplings in Schedule 40, Schedule 30, or Schedule 10, where the schedule describes the wall thickness of the pipe. Schedule 40 is the thickest wall, while schedule 10 pipe is the thinnest and is referred to as light wall sprinkler pipe. The pipe may be black iron or hot-dipped galvanized for corrosive protection. The schedule of the pipe translates into weight, and weight translates into installation labor cost factors. Steel pipe may be grooved and joined with mechanical fittings, commonly called *Vic* fittings. Sprinkler piping may be also welded, or screw flanged, especially if it may be subject to high operating pressures above the UL listing of standard mechanical fittings.

Sprinkler systems may also use copper pipe with a K, L, or M wall thickness. Copper pipe has better friction characteristics than steel but does cost more in installation. In light-hazard occupancies under NFPA 13, 13R, and 13D, CPVC plastic pipe is also permitted for use. The CPVC pipe can be used only in wet systems and must be protected by fire-rated construction in most applications; but in some areas, such as unfinished basements, it may be unprotected. CPVC pipe must meet the same hydrostatic testing requirements as steel pipe in any sprinkler system installation. Typically, that testing is 200 psi for 2 hours, or 50 psi for 2 hours over the highest expected operating pressure. During the test there may be no visible leakage of the pipe.

Another newer innovation permitted by NFPA 13D for one- and two-family residential sprinkler systems is the use of cross-linked polyethylene tubing, or PEX-A, which is made specifically for sprinkler system applications. The advantage of PEX-A is that the system can be multipurpose, serving both domestic water and fire protection requirements. The PEX tubing is flexible and requires fewer hangers, fittings, and labor to install, making it more cost effective. PEX tubing has been in application in home plumbing systems for more than 30 years with excellent results. National Institute of Standards and Technologies (NIST) conducted an analysis of PEX installations, in which it was determined that the sprinkler system installation costs may range from $829 for a typical one-story ranch-style home to $2,075 for installation in a two-story colonial home with a basement. This lower cost makes the installation of residential sprinklers even more attractive.[9]

SPRINKLER SYSTEM PIPE HANGERS

It is important for the fire inspector to examine the sprinkler pipe hangers during the inspection tour. NFPA 13 requires that at least one pipe hanger be placed on each section of pipe. Hangers must be capable of supporting the weight of the pipe filled with water and an additional 250 pounds at each hanger location. Hangers must be spaced no further than 15 feet apart on cross-mains and 12 feet apart on 1-¼-inch pipes. Branch lines may be unsupported for certain lengths based on the diameter of the pipe, 1-½ inches to 60 inches, 1-¼-inch pipe may be unsupported to 48 inches, and 1-inch pipe can be unsupported to 36 inches. The sprinkler system riser must also be supported using riser clamps at each floor level and hangers of the horizontal pipe within 24 inches of the riser cross-main connections. Fire inspectors should also be aware of the requirements for earthquake bracing, especially on the east coast and areas where earthquakes are not a predominant hazard factor. The IBC contains seismic tables that dictate the type of earthquake activity that may be present in a geographic area. Earthquake bracing must be both ridged and flexible in design. Flexible couplings may be required on risers at seismic joints and expansion joints in buildings. This can be accomplished with seismic separation assemblies specified in the standard. Earthquake sway bracing is installed in the longitudinal and lateral directions of the cross-mains. Lateral braces should be at about 30 to 35 degrees for maximum load handling (see Figure 12.18). The angles of the sway bracing and longitudinal braces must be calculated by the design professionals based on the anticipated movement of the system piping.

SPRINKLER SYSTEM VALVES

All system valves on automatic sprinkler systems must be indicating-type valves. They may be OS&Y, where the amount of stem projecting from the valve wheel indicates the position of the valve gate, or they may be flag-style valves that have an external flag that runs either parallel to the pipe to indicate it is open or perpendicular to the pipe, indicating the valve is closed (see Figures 12.19 and 12.20). All sprinkler control valves must be electronically monitored with tamper switches.

FIGURE 12.18 Earthquake lateral bracing must be installed based on seismic zones in jurisdiction. *Courtesy of J. Foley*

FLOW AND TAMPER SWITCHES

Automatic sprinkler systems valves and water flow conditions are required to be monitored and supervised by the fire alarm system. **Flow switches** are electrical contact switches attached to a vane or paddle that enters the pipe. Flowing water pressure pushes the paddle in the direction of the flow, and after a preset time delay the water flow switch activates and sends an alarm to the fire alarm control panel. Water flow switches can be adjusted from 15 to 60 seconds to accommodate surges in water in the system to reduce false alarm potential.

    Tamper switches are electrical contacts that send supervisory signals when a valve is being operated. Tampers are usually spring loaded rod-and-cam appliances that are attached perpendicular to the valve (see Figure 12.21). When the valve wheel is turned, the rod moves off the cam, and a signal is sent to the fire alarm control panel. All sprinkler system valves are required to be supervised by NFPA 13.

flow switch ■ An electrical switch that signals the fire alarm control panel when water is flowing in a sprinkler system.

tamper switch ■ An electronic switch installed on control valves to provide supervisory signals to a fire alarm control panel indicating the position of the valve as open or closed.

FIGURE 12.20 Indicating control valve with internal tamper switch. *Courtesy of J. Foley*

FIGURE 12.21 Tamper switch is attached to a control valve and the rod rides on a cam, indicating when the valve is moved to a closed position. *Courtesy of J. Foley*

Inspection of Automatic Sprinkler Systems

The fire inspector must be aware of the different types of sprinkler systems located at each property they may inspect. Each system will have different inspection points that should be observed. Keep in mind that the annual fire inspection is a visual inspection of the property supported by documentation and certifications that the fire protection system has been properly serviced and maintained to NFPA 25, *Standard for the Inspection, Testing, and Maintenance of Water-Based Fire Protection Systems*. The visual inspection points for the fire inspector should include the automatic sprinkler system control valve to ensure it is in the open position. The inspector should examine the alarm check valve on wet systems and the dry pipe valve on dry systems. The fire inspector should ensure that the fire alarm control panel is operational on either deluge or preaction systems. The fire inspector should be looking on the riser for the hydraulic design plate and should make sure all valve controls are properly identified. The inspector should note the water and air pressure gauges and, if the system is dry pipe, make sure it is not water columned. The sprinkler valve control room must be heated and free of combustible storage and may be required to be fire-resistant if it contains a fire pump. The fire inspector should examine the spare sprinkler stock for the proper number and types of sprinkler heads and make sure the wrenches are in place. If a fire pump is present, it should be examined for leakage or rust, and the pressure gauges should be inspected as well as the pump specification plate for capacity and head pressure ratings.

As the fire inspector moves through the building, in each area the condition of the sprinkler system should be noted. Are the right sprinkler heads in place for the ambient ceiling temperature? Are they unobstructed, or not in the right position? Is the storage or any other obstructions 18 inches below the sprinkler head? In flammable liquid store rooms, is there at least 36 inches clearance below the sprinkler head? The fire inspector should observe if sprinkler heads have been painted or if there are caps missing on concealed sprinkler heads. The fire inspector should take note of the fuel load within the space and determine if it is suitable for the sprinkler hazard class specified on the system riser information plate. If the ceiling is open, the inspector should note the system pipe hangers and their general condition. Are they rusted, damaged, or broken? Is the earthquake bracing in place on the cross-mains, and are there flexible connections at the building expansion and seismic joints? In the exit stairway, the inspector should examine the branch line connection to ensure the branch line control valve is open and properly supervised and labeled. The fire inspector should obtain a copy of the sprinkler system certification and pump test report before leaving the property and should review it later for any deficiencies that were noted or may need correcting. Often, contractors will note deficiencies on certification reports that are not always corrected unless the fire inspector questions them.

The building owner should keep accurate records of weekly, monthly, and semiannual testing in accordance with NFPA 25, *Standard for the Inspection, Testing, and Maintenance of Water-Based Fire Protection Systems*. Every quarter, the owner or their sprinkler system contractor is supposed to conduct a main drain flow test to ensure that the flow activates the fire alarm. The inspection, testing, and maintenance of the sprinkler system is usually conducted by a private fire protection company or contractor. Each state may have rules on licensing or certification of these third-party fire protection contractors to ensure proper and complete inspections to the NFPA 25 standard. The fire inspector should review the test reports from both the sprinkler contractor and the fire alarm contractor, as there is a crossover of components. The fire protection contractor will inspect the water flow and tamper switches, and the fire alarm contractor must ensure that they are properly reporting to the fire alarm control panel. Some fire departments and states require standardized contractor reports to ensure that all of the requirements of all applicable NFPA standards are addressed and have been inspected. The advantage of the standardization of report forms for the fire inspector is ease of finding information when reviewing the report for code compliance. The fire inspector should have knowledge of the local requirements for third-party contractors to service fire protection equipment and should ensure that those standards are complied with. It is also very beneficial for the fire inspector to witness third-party fire protection system testing whenever possible. This will help the fire inspector gain hands-on knowledge of how the tests are conducted and the actual workings of the fire protection systems.

Flow Testing

Every 3 years, a full system flow trip test is required on all wet and dry pipe valves, and they should be examined internally for corrosion every 5 years. The fire inspector should keep a record on these tests to ensure they are completed.

Internal Pipe Inspection

NFPA 25, *Standard for the Inspection, Testing, and Maintenance of Water-Based Fire Protection Systems*, requires that a 5-year flush test be performed on all sprinkler system pipes to determine if there are any internal obstructions in the pipe. If the flush test is clear, no further action is required until the next 5-year flush test period. During the flush test, signs of a problem include plugging of the test flow connection, discolored water, and discharge of foreign materials like sand, rocks, pipe scale, or microbiologic materials. If there are any of these indicators, a more complete pipe investigation needs to be conducted on the sprinkler system. Currently there are new methods available for the examination of sprinkler piping for these types of problems. Contractors can utilize ultrasound technology or internal cameras to determine the pipe's internal condition. Remote branch lines should be removed and evaluated for materials in the pipe such as scale, which is rusted metal flakes or microbiologic growth. The problem with **microbiological influenced corrosion**, or MIC, is that it may cause piping failure under critical fire conditions of flow and pressure. Microbiologic materials' presence in the pipe can be determined by

microbiological influenced corrosion (MIC) ■ The buildup of living microorganisms in sprinkler pipe, causing corrosion and deterioration of the pipe.

water analysis and can be chemically treated. Sand and silt can be flushed from pipes, but other biologic materials such as zebra mussels are far more difficult to deal with. The presence of zebra mussels may show up in a flow test when shells and live mussels are discharged. These live organisms can be killed with chlorination or the addition of potassium to the water; however, the debris remaining still needs to be removed from the system by extensive flushing. Fire inspectors should make an attempt to be present during flow tests and fire pump tests so that any MIC or other obstructions are identified and corrected before they become problematic and cause the system to fail.

Common Sprinkler System Violations

The overall goal in inspecting the automatic sprinkler and standpipe system is to ensure that it will operate properly and contain or suppress a fire in the protected property. Fire inspection helps to ensure the maintenance of the sprinkler system in compliance with the NFPA 25 and other NFPA installation standards. We have discussed many inspection points; however, here are the most common types of violations of the fire prevention code that can be easily identified even by new fire inspectors.

OBSTRUCTED SPRINKLER HEADS

Sprinkler heads must be unobstructed in order to function properly. The standard sprinkler head clearance is 18 inches below the sprinkler deflector. This allows the head to form the umbrella pattern necessary to cover the floor area. Sprinkler spacing on branch lines is determined by the square footage of the floor area that each sprinkler head must protect. Sprinkler heads in light-hazard systems can protect up to 225 square feet in noncombustible construction and 168 square feet in combustible construction. Sprinkler heads in an ordinary-hazard system protect 130 square feet per sprinkler head, and extra-hazard sprinkler heads protect 100 square feet per head. NFPA 13 also established a maximum distance between sprinklers on the same branch line or opposing branch lines at 15 feet for light and ordinary hazard and 12 feet for extra hazard and high piled storage. Sprinklers must have adequate clearance to vertical obstructions that are common such as lighting fixtures or signs. The clearance distances are shown in Table 12.2.

Sprinkler heads must have minimum clearances to horizontal obstructions that may include bar joists, pipes, fixtures, electrical conduits, fans, or banners. In older editions of NFPA 13, this was known as the "beam rule," but in today's code, the rule applies to any horizontal obstruction to the sprinkler head. While many of these concerns are addressed in the construction phase of a building, when furniture such as office partitions or privacy curtains and cubicles is added to the building, the vertical distance of the top of the partition must be a certain minimum horizontal distance away from and below the sprinkler

TABLE 12.2	Sprinkler Head Obstruction Clearances
MAXIMUM DIMENSION OF OBSTRUCTION	**MINIMUM SPRINKLER HORIZONTAL CLEARANCE**
½ to 1 inch	6 inches
1 to 4 inches	12 inches
Greater than 4 inches	24 inches

Reprinted with permission from NFPA-2012, *Installation of Sprinkler Systems,* Copyright © 2012, National Fire Protection Association, Quincy, MA. This reprinted material is not the complete and official position of the NFPA on the referenced subject, which is represented only by the standard in its entirety.

head. If all the partitions are 18 inches below the sprinklers, they are generally in compliance. Sprinkler heads must also be installed no closer than 4 inches to a wall or 3 inches to a pipe hanger, as this would impede their operation. Fire inspectors should look for temporary items installed at special events, such as drapes on pipe, banners hung from ceilings, or booths with solid roofs or tent tops, that may obstruct the sprinkler heads.

CLOSED VALVES

The closing of sprinkler system valves accounts for over 16 percent of sprinkler system failures. The closing of sprinkler system valves under the fire prevention code also creates an imminent hazard to the property. Remember that the building code's fire resistive construction features are reduced when automatic sprinklers are installed. When a closed valve disrupts the sprinkler system, the building falls below the minimum fire safety requirements of the building code. Fire inspectors should address closed valves immediately upon detection. If a valve is closed due to an emergency, then the impairment coordinator for the facility should have taken the necessary steps to notify the fire department, the insurance company, and the maintenance contractor to make the necessary repairs and rehabilitate the system. The impairment coordinator must institute all safety precautions, and a fire watch should be put in place in the affected areas. The building owner should not close any sprinkler system valves without proper notification of fire officials and all safety precautions being taken, including a fire watch.

The fire inspector should also work with the firefighters in the fire department so that they understand how to properly support a building with automatic sprinklers to avoid premature sprinkler system shutdowns during a fire. All fire departments should have operational guidelines that follow NFPA 13E, *Recommended Practice for Fire Department Operations in Properties Protected by Sprinkler and Standpipe Systems*.

This standard requires the fire department to support the sprinkler system with an engine through the fire department connection. In buildings with fire pumps, the fire department engine acts as a backup in case of primary pump failure. Engines assigned to the sprinkler support role should not support any other hose lines and should remain connected until the fire is declared extinguished. On arrival at a sprinkler-protected property, a firefighter with communications should be dispatched to the sprinkler control valve room and remain with that valve to secure it or reopen it as directed by the incident commander. Firefighters should allow some period of soak time if the sprinklers are operating; they should also keep in mind that the sprinkler operation causes smoke to bank down and reduces visibility in the fire area. Premature shutdown of the sprinkler system may allow the fire to grow and overcome the system if it is reactivated. Incident commanders should also have a good understanding of the water supply system in the area of protected properties and should avoid using water supplies for hand lines that decrease the water available to the automatic sprinkler system. Many sprinklered buildings have been lost as a result of premature system shutdown and inadequate water supplies due to fire department operations.

IMPROPER COMMODITY HAZARD CLASS

As we have discussed, automatic sprinklers must be properly matched to the type of commodities they are intended to protect. The fire inspector should determine the types of commodity classes present and whether they are segregated or intermixed. The inspector should look at the storage height and note if the commodities are in solid pile, on pallets, shrink wrapped in plastic, and what type of packaging is present. Extra storage pallets should also be noted if they are inside the building. Fire inspectors must pay close attention to the type and arrangement of commodities in warehouse facilities as well as other types of buildings and must compare types of commodity to the hazard classification of the sprinkler system identified on the sprinkler riser specification plate.

FIGURE 12.22 A typical centrifugal fire pump with an electric driver is installed to increase pressure for the automatic sprinkler and standpipe system. *Courtesy of J. Foley*

Fire Pumps

Fire pumps are installed to provide additional pressure in both sprinkler and standpipe systems. Fire pumps may be installed horizontally or vertically depending on the manufacturer of the pump (see Figure 12.22). Fire pumps must be operated weekly for 10 to 30 minutes, depending on outside weather, and may be powered by either electric motors or diesel engines. Electric fire pumps must be connected ahead of the main circuit breaker for the building to ensure power is reliable and uninterrupted (see Figure 12.23). Diesel fire

FIGURE 12.23 The main pump controllers must have an uninterrupted power supply independent of the building's electrical distribution system. Fire pumps must be installed in fire-resistant room enclosures. *Courtesy of J. Foley*

pumps must run 30 minutes per week, and oil pressure, water temperature, and engine speed must all be recorded by the operator to ensure the diesel engine will start and operate properly. All fire pumps require an annual flow test to 100 percent of their rated capacity. The results of the flow test should be compared to the fire pump's **acceptance pump test curve** and the previous year's flow test curve to ensure the pumps are still within the manufacturer's specification. Large buildings and high-rise structures may have multiple fire pumps at different levels of the structure to provide pressure. Often, insurance underwriters may require redundant fire pumps over and above the minimum building code requirements. These pumps must also be flow tested in accordance with NFPA 25, *Standard for the Inspection, Testing, and Maintenance of Water-Based Fire Protection Systems*. Accommodations must also be provided to conduct the flow test by the installation of a pump test header on the exterior of the building. During the annual fire pump inspection, all pressure maintenance pumps or jockey pumps should also be tested for correct pressure settings to ensure they are properly maintained.

acceptance pump test curve ▪ Initial pump performance curve demonstrating that the fire pump is in compliance with manufacturer's specifications. The curve demonstrates flow versus pressure and provides maximum pump pressure, maximum flow, and horsepower and efficiency of the pump.

Standpipes

Standpipes are required in buildings that exceed 30 feet in height above the lowest level of fire department vehicle access or 30 feet below the highest point of fire department vehicle access such as multilevel basements below grade. Standpipes are installed in accordance with NFPA 14, *Standard for the Installation of Standpipe and Hose Systems*, which has existed since the early 1900s. Traditional standpipes were designed based on the requirements to flow 250 gallons per minute at 65 residual pounds of pressure through 100 feet of 2-½-inch-diameter fire hose equipped with a 1-1/8-inch straight tip nozzle. In the ICC building code today, the nozzle pressure has changed to 100 psi at the topmost outlets to accommodate the modern variable-gallonage nozzles used by most fire departments. Local fire departments, however, must be aware of the older requirements, as these standpipes are predominantly used in older structures based on 65 psi nozzle pressure. Standpipes are divided into several operational categories and may be automatic wet or dry systems, semiautomatic dry pipe systems, or manual dry pipe systems.

standpipe ▪ A vertical wet or dry pipe used to supply fire hose valves in a building.

- **Automatic Wet Standpipes:** In this type of system, the standpipe is filled with water and has an automatic water source such as a fire pump to provide capacity and pressure.
- **Automatic Dry Standpipes:** The automatic dry standpipe is filled with pressurized air and uses a dry pipe valve and automatic water supply to provide capacity and pressure to all hose outlets.
- **Semi-automatic Dry Standpipes:** These standpipes are dry and require a secondary device such as a manual pull station to operate a deluge valve and flood the system. These systems are used in environments such as cold storage warehousing facilities, where draining a system may be problematic due to freezing temperatures.
- **Manual Dry Standpipes:** These are simple dry pipes with no attached water supply. The fire department must supply the water from an engine through hose connected to the fire department connection.
- **Manual Wet Standpipes:** These are systems that are wet, usually with a roof top storage tank, but have no automatic water supply and rely on the fire department to provide pressure and capacity through the fire department connection.

Standpipes are further classified, based on their intended users, as Class I, II, or III. Let us examine the three standpipe classes.

CLASS I

Class I standpipes are for fire department use only (see Figure 12.24). These connections consist of a 2-½-inch fire hose outlet with a 1-½-inch reducing cap for smaller-diameter hose. The Class I standpipe has a maximum allowable operating pressure at the connection

FIGURE 12.24 Class I fire department standpipe hose valve location. *Courtesy of J. Foley*

of 175 psi but may incorporate a **pressure-restricting device** at 150 psi. Pressure-restricting devices are usually either cam-and-wheel or cam-and-lever types that reduce the ability to fully open the valve unless they are removed by the fire department. These are not the same as the **pressure-reducing valves** that will be discussed later in this chapter. Class I standpipes are required to be spaced no greater than 200 feet in sprinklered properties and 150 feet in nonsprinklered properties. The IBC requires Class I standpipes in covered malls, heliports, helistops, marinas, high-rise buildings with automatic sprinkler protection, sprinklered public assembly Groups A-1 and A-2, nightclubs with over 1,000 occupants, and buildings with over 10,000 square feet of floor area per story. Class I standpipes must have hose thread compatible with the local fire department. Fire inspectors should check outlet threads with a used fire hose coupling of each appropriate size to verify thread compatibility.

Class I standpipes shall be located in every exit stairway on each floor level. They shall also be located at the building entrances and at the entrance to any horizontal exits. The maximum distance between standpipes is a 200-foot travel distance to all areas of the building.

CLASS II

Class II standpipes are for trained occupant use and provide a single 1-½-inch hose connection with a hose rack, 100 feet of single-lined cotton jacket hose, and a nozzle. The maximum operating pressure at the hose station is set to 100 psi. The hose is held in place on the rack by a series of pins that are designed to feed the hose out of the cabinet as the occupant takes the nozzle and moves toward the fire. Often, these hoses are pulled off the pins in the cabinet, creating a pile of hose. Class II standpipes are spaced at 100 feet of hose and a 30-foot hose stream to all portions of the building. Class II standpipes are required in uses A-1 and A-2 over 1,000 occupants when they are not sprinklered and they must be installed on stages over 1,000 square feet in area on both stage right and left and at the rear of the auditorium. Fire inspectors may also encounter Class II standpipes in hospitals, older elementary schools, warehouses, piers and wharves, and older high-rise buildings.

CLASS III

Class III standpipes provide both fire department Class I connections and the occupant Class II hose station. Class III standpipes shall be installed in all buildings greater than 30 feet above the lowest level of fire department access or 30 feet above the highest level of fire department access. In buildings with automatic sprinkler systems, the Class III standpipe may be downgraded to a Class I fire department–only standpipe. The pressure requirements on Class III standpipes are 100 psi maximum on the hose station and 175 psi maximum on the fire department Class I connection. Class III standpipes shall be installed in the same locations as Class I and II standpipes according to the IBC.

Pressure-Reducing Devices

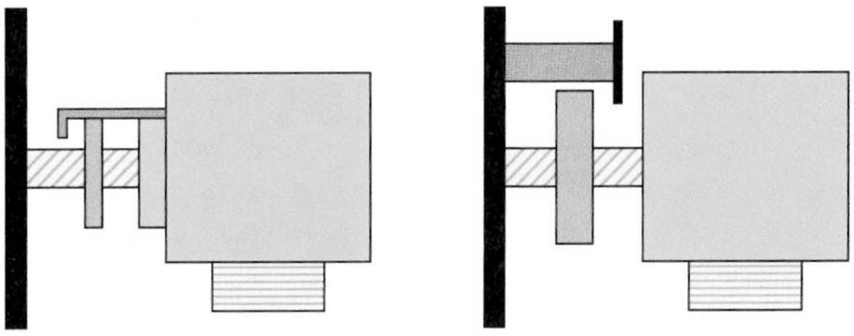

FIGURE 12.25 Pressure-reducing devices reduce pressure to 150 psi on Class I standpipes. These devices can easily be removed by firefighters if more pressure is required.

PRESSURE-REDUCTION DEVICES

Pressure-reducing devices, or PRDs, are required to be installed in Class I standpipe systems when operating pressure exceeds 150 psi and Class II standpipes at pressures exceeding 100 psi. The typical Class I fire hose valve is outfitted with a cam-and-lever or wheel restrictor attached to the valve handle that restricts the valve from being fully opened (see Figure 12.25). The PRD can easily be removed by firefighters from the hose valve to allow full pressure, if necessary. Full pressure may not, however, exceed 175 psi at the hose connection. In Class II standpipes, the PRD is usually a restricting orifice washer that is placed inside the 2-½ to 1-½ reducer cap. If the fire department is using this connection, they must remove the reducer completely and use their own fire department reducer fitting.

PRESSURE-REDUCING VALVES

Pressure-reducing valves, or PRVs, are required when standpipe hose connection pressures exceed 175 psi at the point of hose connection. These valves fall into two categories: those that are field adjustable, and those that are factory preset. Pressure-reducing valves are usually found in high-rise applications where operating pressures exceed 175 psi on the standpipe riser. The standpipe must provide 100 psi at the top outlet, and pressure-reducing valves are required once the building height exceeds 175 feet. This base pressure is determined by multiplying the building height by 0.434 and adding the minimum nozzle pressure at the topmost outlet:

$$175 \text{ feet} \times 0.434 = 75.95 + 100 \text{ psi nozzle pressure} = 176 \text{ psi}$$

The installation of pressure-reducing valves is generally not necessary until a building exceeds 19 to 20 stories. Buildings higher than 20 stories will require some form of pressure-reducing valves to be installed. Pressure-reducing valves came to the forefront of the fire service in 1991 after the One Meridian Plaza Fire in Philadelphia, where three Philadelphia firefighters died during a high-rise fire that began on the twenty-second floor. One of several major problems encountered by firefighters was that the non-field-adjustable pressure-reducing valves on the standpipes were incorrectly set. Firefighters could get no more than 40 psi from the outlets when they needed at least 65 to 100 psi to produce effective fire hose streams. Ironically, the same problem had been uncovered in California at the First Interstate Bank fire in Los Angeles several years before. The LAFD conducted a survey of all city high-rises and found that the problem existed in many of the buildings with improperly set pressure-reducing valves. (See the discussion on the Meridian One fire in Philadelphia in Chapter 1.)

PRV valves are dynamic valves that use spring pressure at the inlet to achieve the correct outlet pressure of the valve. The correct pressure setting is determined by the

FIGURE 12.26 A master pressure-reducing station may be installed to step pressures down to 175 psi, such as this station mounted before a high zone fire pump on the tenth floor of a hotel/casino. *Courtesy of J. Foley*

location of the valve within the building, and the valves on each floor will be set differently. PRV valves are located on the lower floors of the building where the pressure due to building height exceeds 175 psi. PRV valves may be field adjustable but often require a special tool to perform the adjustment. Newer PRV valves are shipped with the tools and charts to make proper adjustments to the valve, and it is good practice to keep tools and the chart by the spare sprinkler head cabinet in the sprinkler valve control room. Factory-set pressure-reducing valves are set at the factory, and metal tags are applied identifying the proper installation location in the building for the installer. Problems that can occur include tags that get accidentally removed or valves that are installed in different or wrong locations.

NFPA 14, *Standard for the Installation of Standpipe and Hose Systems*, provides other ways that excess pressure at the standpipe

master pressure-reducing station ■ A master pressure reducing station is usually located in the standpipe and sprinkler system to ensure pressure is reduced to below 175 psi.

hose connection can be addressed. One method is a **master pressure-reducing station** placed in the system at or near the high-pressure fire pump. NFPA 14 requires that master pressure-reducing stations be located no more than 7 feet 6 inches off the floor, and a bypass method for maintenance and service must be provided (see Figure 12.26). If the master PRV fails, the maximum pressure allowed must reduce to 175 psi. An alternate method of pressure reduction permitted in NFPA 14 is to install split-zone standpipes. The split zone eliminated the need for pressure-reducing valves because each zone is divided at a maximum pressure of 175 psi at the riser base (see Figure 12.27). Water is transported to the upper zones by express risers. Split-zone standpipes are identifiable by two sets of fire department connections that identify where each zone begins and ends.

Fire inspectors must be aware of the type of pressure-reducing appliances installed in buildings that they inspect. The testing of these appliances can be somewhat complicated, especially with PRV valves. NFPA 14 requires that all standpipe valves be flow tested every 5 years to ensure proper valve operation and settings. This requires a method to flow water to a suitable drain capable of handling the standpipe flow. On PRVs, both inlet and outlet pressure must be recorded in both flowing and static pressure conditions to ensure proper valve settings. Pressure-reducing valves that fail must be reset and retested or replaced.

Standpipes and Fire Inspections

During the course of the fire inspection, the fire inspector should be examining the standpipe visually in each stairway as he or she traverses

FIGURE 12.27 Another approved method of pressure reduction is the dividing of the building into zones. In this case the low zone feeds four stories of the building, and the high zone feeds a high-rise tower. *Courtesy of J. Foley*

the building (see Figure 12.28). The inspector should look to make sure that the standpipe or hose cabinets are properly labeled if they do not have a vision glass panel. The cabinets should be opened to check that the 2-½-inch valve cap is in place and that the valve handle is on the valve, as handles frequently are removed or missing. The valve should be properly positioned within the cabinet so that the fire department can properly connect a hose and open the valve easily. The cabinet door should have a handle of sufficient size to be opened with a gloved hand by a firefighter. The fire inspector should check the hose threads on the standpipe for correctness and compatibility, and any thread damage should be noted. All standpipes require a system control valve at the riser base, and the control valve on each riser should be visually inspected to make sure it is open and properly supervised. During the exterior inspection of the building, the fire department con-

FIGURE 12.28 Fire inspectors should examine all standpipe connections for proper threads, valve wheels and any signs of leakage. *Courtesy of J. Foley*

nections should be inspected for proper caps, and any obstructions, such as plants, should be removed. The fire department connection must be visible from the street and marked as to what systems it supplies. The fire department connection should also be checked for compatibility with fire department threads (see Figures 12.29 and 12.30). The fire inspector should carry a pocket spanner wrench to check all the caps and make sure they are not frozen or corroded in position or are too tight to be removed by firefighters. Lastly, before leaving, the fire inspector should review or obtain a copy

FIGURE 12.29 The fire department connection should be properly capped and should identify the system it supplies. This system has a bar code inspection tag attached. *Courtesy of J. Foley*

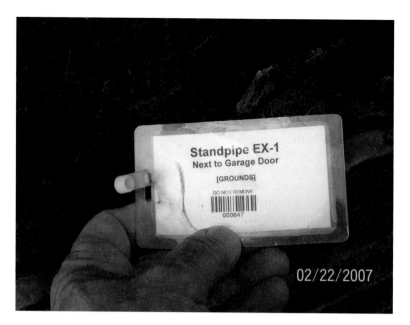

FIGURE 12.30 Bar code tag for inspector to verify condition of FDC during inspection. *Courtesy of J. Foley*

of the annual standpipe inspection report provided by the fire protection contractor to make sure the owner is conducting the appropriate system maintenance. The certification should include the 5-year annual flow test and a statement that the standpipe has been hydrostatically tested for 2 hours at 50 psi above the highest system operating pressure. If the system has a fire pump, the annual fire pump test certification should also be reviewed or requested. The fire inspector should note any deficiencies noticed by the certification agency or that he or she noticed during the inspection to ensure that the violations are corrected.

Summary

Automatic sprinkler and standpipe systems are effective fire containment and control systems provided that they are properly installed and maintained in buildings. Fire inspectors need extensive knowledge in the design and maintenance features of all different types of sprinkler system within their community. Inspectors need to evaluate sprinkler protection, as it compares to the products it is designed to protect, and they must determine whether there are deficiencies in sprinkler coverage or the ability of the system is compromised. Fire inspectors need to be well versed in NFPA standards to determine the proper installation and maintenance of automatic sprinkler systems and the proper operation of system components. Fire inspectors should also make efforts to witness sprinkler and standpipe system testing so they may develop a better knowledge of how these systems operate.

Review Questions

1. Describe and contrast the purpose and differences of the NFPA standards related to automatic sprinkler systems.
2. List and describe the four main types of automatic sprinkler systems and describe the key components and inspection points of each type of system.
3. NFPA 13, *Standard for the Installation of Sprinkler Systems*, establishes hazard classifications of buildings for automatic sprinkler protection. List the types of building in your jurisdiction that would fall into each hazard classification.
4. Describe the differences between automatic sprinkler systems that are installed as pipe scheduled as opposed to hydraulically calculated systems.
5. Describe how a dry valve operates and how a fire inspector can tell if the system is water columned.
6. List and describe the different types of quick opening devices that may be installed on dry pipe sprinkler systems.
7. When are low point drains and drum drips installed on dry sprinkler systems, and how are they maintained?
8. Describe the different types of automatic sprinkler heads and their applications.
9. Prepare an ordinance for the adoption of NFPA 13D sprinklers in one- and two-family dwellings. What would be the key elements to overcome community resistance to this ordinance?
10. Identify the common fire code violations that a fire inspector may encounter in the course of inspecting an automatic sprinkler system. What elements should the inspector examine?
11. List and describe the different classifications for standpipes and who is intended to operate them. What are the minimum and maximum operating pressures of the different systems?
12. Identify special types of automatic sprinkler systems and their applications in industrial fire protection.
13. All water-based fire protection systems must be maintained to NFPA 25, *Standard for the Inspection, Testing, and Maintenance of Water-Based Fire Protection Systems*. Create a matrix of necessary inspections to track compliance with NFPA 25 requirements. How could this be used in the course of a fire inspection?

Endnotes

1. "Fire Sprinkler History - NFSA, NFPA, & Tyco," *The Station House* (Lansdale, PA: Tyco Co., February 2005, Volume 4, Issue 1).
2. http://www.firesafehome.org/the-residential-fire-safety-institute-rfsi/
3. Newport Partners, *Home Fire Sprinkler Cost Assessment - Final Report* (Quincy MA: The Fire Protection Research Foundation, September 2008).

4. John R Hall, Jr., *An Analysis of Automatic Sprinkler System Reliability Using Current Data* (Quincy, MA: National Fire Protection Association).
5. Ben Evarts, *Trends and Patterns of U.S. Fire Losses in 2010* (Quincy MA: NFPA Fire Analysis and Research, September 2011).
6. *Communities with Home Fire Sprinklers: The Experience in Bucks County Pennsylvania* (Washington Crossing, PA: Home Fire Sprinkler Coalition, November 2011).
7. Newport Partners, *Home Fire Sprinkler Cost Assessment - Final Report* (Quincy, MA: The Fire Protection Research Foundation, September 2008).
8. *Water Purveyor's Guide to Fire Sprinklers in Single Family Dwellings* (National Fire Sprinkler Association, 2006)
9. David T. Butry, M. Hayden Brown, and Sieglinde K. Fuller, *Benefit-Cost Analysis of Residential Fire Sprinkler Systems* (Gaithersburg, MD: National Institute of Standards and Technology, NISTIR 7451, September 2007).

13

Plan Review of Water Supply for Fire Protection

Courtesy of J. Foley

OBJECTIVES

After reading this chapter, the reader should be able to:

- Identify and describe the water sources in your community.
- Describe the properties of water both at rest and in motion.
- Understand the importance of evaluating the water supply systems.
- Recognize specific requirements of ISO grading for water supply systems.
- Understand the responsibilities of a water supply officer.
- Understand how to conduct a flow test and review specific components of a water supply system.

Professional Levels of Job Performance for Fire Inspectors as Cited in NFPA 1031 and NFPA 1037

- NFPA 1031 Fire Inspector I *Obj. 4.3.16 Verify fire flows*
- NFPA 1031 Plan Reviewer I *Obj. 7.3.6 Evaluate code compliance for fire flows and hydrant location and spacing*

Introduction

Adequate water supply is an important factor to the fire defense of any community. The fire inspector has an obligation to take a lead role in improving the local water resources for fire protection purposes. Fire and building codes do not necessarily impose this obligation on the fire inspector; however, it is implied in the tasks that a fire inspector must perform every day. The proper performance of automatic sprinkler systems depends heavily on the reliability of the water supply system. Adequate and reliable water systems ensure the best protection from fire that a community may have. A city or town may be equipped with the best fire department in the world, but without water it will have little impact on suppressing fires.

Water has many characteristics that make it an excellent fire suppression tool. Water is plentiful, relatively inexpensive, and easy to store in reservoirs and storage tanks. Water is stable and does not react with many chemicals. Additionally, water is a **polar compound** and can absorb or dilute most polar liquids, making it a universal solvent. The most important property of water, however, is its ability to absorb great quantities of heat through vaporization when suppressing fires. Although many types of fire suppression agents exist, water will always be the predominant agent used in firefighting.

polar compound ▪ A compound (e.g., H_2O) in which the electric charge is not symmetrically distributed, so that there is a separation of charge or partial charge and formation of definite positive and negative poles.

The fire inspector must work closely with the local water utility manager, fire chief, planning or zoning officer, and construction and building officials when reviewing water supplies in the community. Fire inspectors need to ensure that proper suggestions are made on water system improvements that directly affect the fire department's ability to operate efficiently. Sometimes, the arrangement of fire hydrants or the size of water main may be insufficient for the protection of the community. The fire inspector must identify these deficiencies to ensure that the community's fire protection objectives are met.

There are many benefits to the fire department in having input into the water system's improvement. First, a better water supply system is established for both commercial and domestic use. Second, increased water capacity minimizes the effects of droughts and dry spells in the hot summer months. Water supply reliability improves the mitigation of fire disasters and reduces the operational time of the fire department in getting water to the fire. Water supply improvements also translate into a reduction of insurance costs within the community.

When fire inspectors consider water supplies within their jurisdiction, they also need to consider all potential water supply sources that may be available. Rural areas of a community may have to supplement firefighting water with natural or manmade water supplies. The fire inspector should inventory all potential water sources that may exist in these rural areas of the community. It is also common for the fire prevention inspector to provide support to the fire department by serving in the capacity of the **water supply officer** (WSO) during emergencies.

water supply officer (WSO) ▪ A designation to an officer to work with the water utility and develop water supplies for firefighting purposes.

Water Distribution Systems

Water distribution systems fall into two main categories of ownership. They may be owned by the government or they may be privately owned and regulated by the government. Government ownership is generally a municipal or county utility authority that maintains the infrastructure including all water towers (see Figure 13.1), water mains, pumps, and chlorination plants, and the fire hydrants. These utilities are publically owned and operated through fees and tax collection. Private companies also may own water utilities and are regulated by government for water quality and the services they provide to the community. Depending on how the water company is established, fire departments may be required to pay annual fees for fire hydrants in the community. Natural water supplies fall into the same two categories. In some cases, a lake or pond (see Figure 13.2) may be government owned, and in other cases, single owners or a homeowner association may privately own them. It is important for fire inspectors to identify the proper owner or authority having jurisdiction for each type of water source in their community. This is especially true when conducting flow tests of the system. The utility must be notified, and in some cases a permit may be required from the utility to conduct the test.

FIGURE 13.1 Water towers are utilized to provide reserve storage capacity and create water pressure on the distribution system. *Courtesy of J. Foley*

Flow Testing

All water-based fire protection systems are predicated upon the adequacy and reliability of the water supply. To properly design automatic sprinkler systems, the water flow and system pressure must be known at the point of connection to the municipal or private water supply. This data is often provided to the designer by the municipal water utility but should always be reconfirmed by actual flow testing of the system at the proposed building site. Often, flow data from the water utility may be dated or may not reflect recent improvements made to the system since the last flow test date. This water flow information is also important to the fire department

FIGURE 13.2 Not all communities have water distribution systems, and sometimes firefighters must depend on natural water supplies for fire protection. *Courtesy of J. Foley*

for the purpose of firefighting capability within a given area of a community. Conducting fire flow tests is especially necessary in older portions of cities or towns that may have underground supply piping that is centuries old (see Figure 13.3). A complete understanding of the water supply distribution system improves firefighting strategies and helps the fire department overcome any deficiencies that they identify in the water system. Fire inspectors should obtain water maps from the local utility that identify the size and connections of the underground water mains.

TOOLS AND EQUIPMENT

Like any task that is performed in the fire service, the fire inspector must have the proper tools and equipment to conduct a proper flow test. The necessary equipment should include

FIGURE 13.3 In colonial days, wooden water mains were used to supply domestic water. Many of these mains were in use until the 1960s.
Courtesy of J. Foley

a 6-inch ruler marked in 1/16-inch increments for measuring hydrant outlet openings and to draw the water supply curve on special 1.85n graph paper. The inspector will also need a 2-1/2-inch tapped hydrant cap with a water pressure gauge that reads up to 250 psi for determining the static and residual pressures on the water system (see Figure 13.4). The fire inspector should have a **pitot gauge** to measure velocity flow, a hydrant spanner wrench, a 90-degree elbow, and a play pipe with a smoothbore nozzle to complete the task. Many fire departments now use water diffusers with pressure gauges to conduct flow tests (see Figure 13.5). If you use a diffuser, make sure to follow the manufacturer's instructions and account for any friction loss from the additional hose. Additionally, a clipboard, pencils, and forms or graph paper also are required to record test information. Before conducting any flow tests, the water utility should be advised that a test is going to be conducted so they do not think there is a rupture in a water main or an active fire in the community.

pitot gauge ■ Hydraulic tools used to measure fluid velocity.

FIGURE 13.4 Flow testing equipment: pitot gauge and water flow charts. Inspectors also need a 1/16-inch ruler and appropriate forms to record their information.
Courtesy of J. Foley

FIGURE 13.5 Water diffusers reduce the pressure of flowing water from hydrants and standpipes and are connected to the hydrant by a short length of hose. The pitot gauge is inserted into a slot in the diffuser. *Courtesy of J. Foley*

The flow test procedure is outlined in National Fire Protection Association standard NFPA 291, *Recommended Practice for Fire Flow Testing and Marking of Hydrants* and should be followed by the fire inspector.[1] The test should be conducted under ordinary water demand conditions, and peak demand times should be avoided so that the public is not inconvenienced with brown water or low pressure from the test. Flow tests should also be conducted in warm weather to avoid creating ice hazards on roads or sidewalks whenever possible.

To properly conduct the fire flow test the fire inspector should review the utility maps and determine the surrounding water main sizes. In gridded systems water will always flow from larger water mains to smaller water mains. The fire inspector should determine the flow direction to establish the test fire hydrant (see Figure 13.6). Once the inspector has determined the flow direction, he or she should select the flow hydrant, which should be the

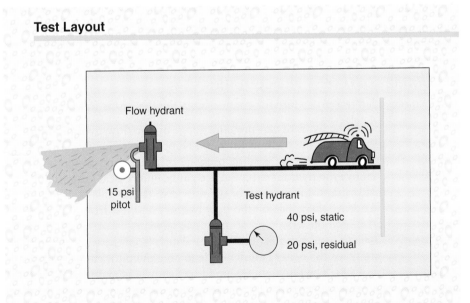

Test Layout

Flow hydrant

15 psi pitot

Test hydrant

40 psi, static

20 psi, residual

FIGURE 13.6 Example of a flow test using a flow hydrant and a test hydrant. The inspector should ensure that the hydrants are in the correct direction of water flow by reviewing the sizes of the water mains.

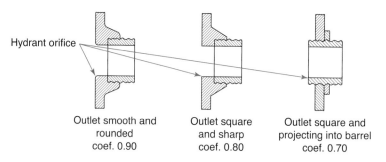

Hydrant orifice

Outlet smooth and
rounded
coef. 0.90

Outlet square
and sharp
coef. 0.80

Outlet square and
projecting into barrel
coef. 0.70

Cutaway of 2-1/2 inch hose outlet. By using your finger you can determine
the coefficient of roughness to be used in the flow calculation (NFPA).

FIGURE 13.7 The coefficient of roughness is determined by how the outlet is attached to the barrel of the fire hydrant. *Reprinted with permission from NFPA 291-2013*

one closest to the construction site; this is where the flow data will be recorded. The inspector should also identify the residual test hydrant downstream of the flow hydrant on the same water main. The residual test hydrant should be in the direction of the water flow. In some cases, only one fire hydrant may be available and will have to serve as both flow and residual test hydrant. Choosing the correct flow and test hydrant is very important in determining the fire flow.

Before beginning the flow test, the inspector should open the fire hydrant's 2-1/2-inch outlets and stick a finger inside the hydrant barrel to feel the outlet's edge inside the hydrant barrel. Fire hydrants usually have one of the three outlet configurations shown in Figure 13.7. Determine which outlet type is present on the test fire hydrant. The outlet type determines the proper coefficient of discharge for the fire hydrant flow calculation. The next step is to measure the inside diameter of the outlet, using the 1/16-inch ruler. It is important to make sure that you properly measure the outlet diameter, as 1/16 of an inch may cause a 5 percent error in the flow calculation. The fire inspector must be aware that not all fire hydrant 2-1/2-inch outlets are a true 2-1/2 inches. Once this data has been gathered, you are ready to begin the flow test. When conducting the test, it is best to avoid using the 4-1/2-inch steamer connection, as it often does not produce a solid flow like the smaller 2-1/2-inch outlets, and this will affect the accuracy of the flow measurements.

RECOMMENDED TEST PROCEDURE

The recommended test procedure for conducting a flow test is as follows:

Step 1. Install the hydrant cap-pressure gauge on the flow hydrant outlet.

Step 2. Slowly open the hydrant fully and record the static pressure on the pressure gauge. This is the static pressure measurement (water at rest).

Step 3. Remove the hydrant cap pressure gauge and relocate it to the residual test hydrant downstream of the flow test. If only one hydrant is available, then leave the gauge on the flow hydrant and test from the other 2-1/2-inch outlet.

Step 4. Open the hydrant fully and let any debris from the water main flush out. Once the water is flowing clear, take the pitot gauge reading (see Figure 13.8). The pitot gauge should be firmly held and inserted into the flowing water at the centerline of the outlet and at least one-half the diameter of the outlet away from the barrel. Once the gauge is properly inserted the correct distance into the flow, take the pressure reading on the pitot gauge and record it.

Step 5. While the water is still flowing, record the pressure decrease on the hydrant cap gauge downstream on the residual test hydrant. This records the remaining pressure on the system and should drop the static pressure by at least 25 percent or 15 psi to validate the flow data. If flow has not decreased the static pressure by 25 percent or 15 psi, open additional fire hydrant outlets until you get a sufficient drop in pressure. If additional fire hydrant outlets are used, record the pitot readings and the residual pressure at the test hydrant at the same time.

FIGURE 13.8 Flow testing hydrants is necessary to determine adequacy and reliability of the water system. *Courtesy of J. Foley*

Step 6. Take the data collected and, using 1.85n graph paper (see Figure 13.9), draw the water curve for the hydrant system. This will show the available water at the fire hydrant location.

Let's look at an example of a flow test in Figure 13.9. In this example, the static pressure is 40 psi. Once water flows, the pressure drops to 20 psi or decreases by 50 percent. The pitot gauge reading is taken, and it is 15 psi at full flow. We can determine the water flow in gallons per minute by using the Freeman formula.[2] In the formula, Q = the flow in gpm, c = the coefficient of roughness based on the outlet type in Figure 13.7, d^2 = the

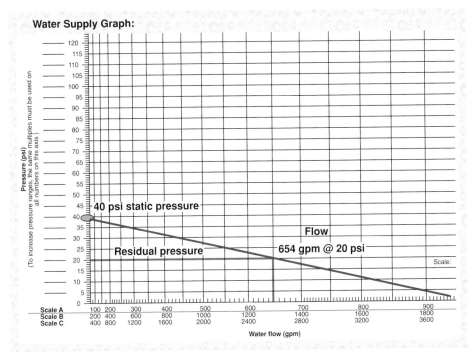

FIGURE 13.9 Water graph paper allows the water curve to be drawn as a straight line. Any point along the line will show pressure on the left and flow in gpm at the bottom on Scale A.

measured diameter of the 2-1/2-inch outlet in 1/16-inch increments, and p = is the pitot reading pressure from the flow test. The calculation is as follows:

$$Q = 29.83cd^2\sqrt{p}$$
$$Q = 29.83(0.9)(2.5)^2\sqrt{15}$$
$$Q = 654 \text{ gpm}$$

Once the flow in gallons per minute has been calculated, we have the three data points needed to construct the water supply curve in Figure 13.9. First, plot the static pressure at zero flow, which represents the system's available water. Next, select a scale at the graph paper and locate the flowing pressure at 15 psi with a total flow of 654 gallons per minute. The water curve can now be drawn on 1.85n graph paper using the ruler to connect the two data points. This special graph paper allows a curve to be drawn as a straight line. The area located below the line represents all of the available water from this fire hydrant. As we can see in Figure 13.9, the maximum flow for this hydrant is 654 gallons per minute at 20 psi residual pressure. The NFPA recommends that fire hydrant pressures not fall below 20 psi, as damage can occur to the water mains. The flow test demonstrates the available water at the test fire hydrant only, and the fire inspector must be aware of that. If more underground piping is to be added to extend the service onto the property, the friction loss of that pipe would have to be calculated to determine any additional pressure loss from the fire hydrant to the sprinkler system connection. If the sprinkler system hydraulic demand falls above the available water curve, then either a larger water supply main is required or a fire pump must be installed to elevate the pressure and increase the height of the water curve. On marginal systems, the fire inspector must also consider that the test data is not representative of the maximum daily demand for water in the area. Any sprinkler system demands that fall right on the line need to be carefully reviewed.

Urban Water Supply Systems

Urban areas are usually supplied water by a public utility through a distribution system of water mains for domestic consumption. The fire inspector should be aware of the many components of the water system and the key elements that control the system requiring maintenance either by the water utility or, sometimes, the fire department. Many fire departments play an important role in maintaining the fire hydrant system by conducting inspections of fire hydrants and performing routine annual maintenance. Types of maintenance may include painting the hydrants, greasing the caps and fittings, inspecting the hose threads, and identifying damaged fire hydrants.

Most urban areas have a well-developed fire hydrant system, and the fire department tends not to think in terms of other available natural water supplies for firefighting purposes. It is still important, however, to identify additional water sources, as these may be vital in emergency situations. Fire departments must always be prepared to convert their method of operation to other available sources of water in an emergency. Emergencies may include floods, broken water mains, failed pumps, frozen valves, earthquakes, or other types of service interruptions on the system. An emergency water contingency planning should be a part of every fire department's emergency management planning and is usually addressed through the fire prevention bureau.

Generally, water utilities supply and distribute water by the following two main methods or some combination of the two:

1. Gravity feed water systems
2. Direct pump pressure water systems

A gravity feed water system retains a water supply either naturally or by the construction of a dam. The dam may restrain a river or a lake and creates a manmade water

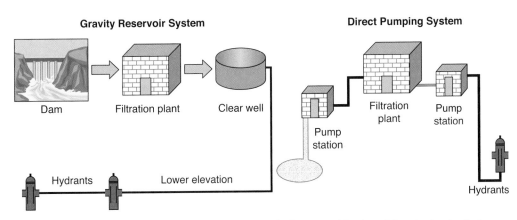

Gravity Reservoir System

Dam → Filtration plant → Clear well

Hydrants   Lower elevation

Pump station

Direct Pumping System

Filtration plant   Pump station

Pump station

Hydrants

FIGURE 13.10 Water utilities generally use these two methods or a combination of them to develop their water supply.

reservoir. Water is released from the dam, is filtered for impurities, and is pumped to a clear well to remove further sediment. The water then moves from the treatment plant to the distribution systems for domestic and fire protection use (see Figure 13.10). Many water utilities also use elevated storage tanks on the system to maintain an emergency water supply and to increase or balance the water system pressure on the distribution system. If water needs to flow in an up-hill direction, a tank is placed at the higher elevation to increase the static pressure on the system (see Figure 13.11).

In the second type of water distribution system, the water is pumped from deep wells from the aquifer below ground. The water is then chlorinated and filtered and moves into the distribution system for use. Direct pumping stations may chlorinate the water at the station or may move the water to a distribution

Pumping Station Gravity Storage

Pump station

Standpipe   Hydrants

FIGURE 13.11 Water tanks or standpipes are used to maintain a water reserve and equal static pressure on the system.

plant, where it is processed and then pumped by a secondary pumping station into the water distribution system. Plants may also incorporate additional high lift pumps to increase water pressure during high demand periods or fire emergencies.

As previously mentioned, most water supply systems use one or a combination of these two methods to distribute domestic water to the consumer. Additionally, water reserves must be maintained based on maximum daily domestic water consumption and the fire flow demands of the community that are established by the Insurance Service Organization for the purpose of fire insurance rating.

FIRE HYDRANTS

NFPA 291, *Recommended Practice for Fire Flow Testing and Marking of Hydrants* requires the color-coding of fire hydrants in accordance with the fire flows they can produce.[3] The bonnets on top of the fire hydrant should be painted to match the color in the flow charts provided in the NFPA 291 standard. This color-coding system enables firefighters to identify the fire flow range of a hydrant during emergency response and helps them select a fire hydrant of proper capacity based on the fire situation (see Figure 13.12).

The NFPA also recommends the use of reflective paint so that fire departments can easily locate and identify the flow range of the fire hydrant at night. Fire hydrants that have residual pressures normally lower than 20 psi should have the minimum effective

NFPA 291 Flow Color Chart	
BLUE	1,500 GPM OR MORE
GREEN	1,000–1,499 GPM
ORANGE	500–999 GPM
RED	LESS THAN 500 GPM

FIGURE 13.12 NFPA 291 identifies fire hydrant flow by reflective paint colors on the hydrant's top. *Reprinted with permission from NFPA 291-2013*

operating pressure stenciled onto the bonnet of the hydrant. This recommendation also applies to fire hydrants with very high capacities. This provides additional information to the fire department pump operator about the particular fire hydrant. Flush hydrants that are located at the ends of water mains should also be marked on the hydrant barrel for maintenance purposes. Flush hydrants need to be flowed annually to remove debris that may collect in dead-end water mains

Plan reviews performed by fire inspectors may also involve the installation of a new fire hydrant on a proposed construction site. NFPA 24, *Standard for the Installation of Private Fire Service Mains and Their Appurtenances* provides the recommended installation practices for fire hydrants on public or private water mains.[4] Fire hydrants should be installed so that the 4-1/2-inch steamer connection is at least 18 inches above the curb line to the center line of the fire engine pump connection. Proper installation of a fire hydrant requires a concrete pad and stones to be placed at the bottom bell fitting of the fire hydrant to allow the barrel to drain properly and not undermine the fire hydrant and cause it to sink or settle (see Figure 13.13).

NFPA 24 also requires the installation of a concrete thrust block at any change in direction of the underground piping connecting to the fire hydrant. Thrust blocks prevent the movement of pipe if water hammer occurs. A water hammer is caused by valves being shut too quickly, causing excess pressure to build up on the pipe wall, which can damage the pipe connections. The fire hydrant bell fitting must also be tied back to the piping by solid connector rods to ensure that the pipes do not separate from the bell during hydrant use. Many types of fire hydrants are used throughout the United States, depending on local climatic conditions. In warm climates, wet barrel fire hydrants may be used because freezing is not a issue. In the northeast and colder climates, the standard dry barrel fire hydrant is used, and sometimes the underground fire hydrant will be used in planned developments. The fire inspector or chief fire marshal should make recommendations on the location and spacing of all new fire hydrants as part of the plan review process. Spacing

NFPA 24 - Proper Fire Hydrant Installation

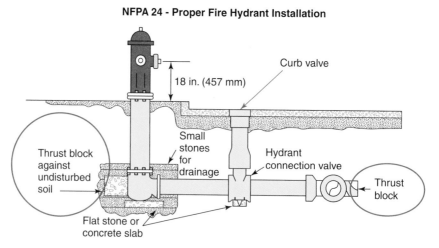

FIGURE 13.13 The NFPA 24 standard is used to install fire hydrants on private or public water mains. *Reprinted with permission from NFPA 24-2013*

should be in conformance with recommended practices of the American Water Works Association or other recognized standards or local fire code requirements. Engineering specifications for specific types of fire hydrants should be provided in the local planning and zoning regulations as well as the specification for system valve and shutoff requirements. Traditionally, fire hydrant spacing was based on population density and the expected numbers of fire hose streams that would be needed to control a major fire.

In urban areas, the fire hydrants are spaced closer than they are in rural areas. The American Water Works Association, or AWWA, recommended hydrant spacing at 500 feet in residential neighborhoods and 300 feet in industrial and commercial building areas. These spacings may be modified in commercial and industrial areas if the local fire department uses large-diameter hose to a distance of 500 feet.

The use of belowground fire hydrants also should be be carefully considered during plan review and discussed with the local fire service. Underground fire hydrants must have signage to identify their location, and if they are covered with snow or ice in the winter they are difficult to access or use. They must also be inspected regularly to ensure that debris is not causing drains to clog, which can cause icing problems and render the hydrant useless.

ELEVATED STORAGE TANKS

Elevated water storage tanks are used for pressure maintenance and emergency water supply reserves on gravity water distribution systems. Aboveground water storage tanks can range from 100,000 gallons to 4,000,000 gallons in capacity. All elevated storage tanks have some method of heating the tank during winter operations. Internal heating of the tank may be provided by electric heat coils, steam lines, or steam injection into the water tank. Bell-shaped tanks (see Figure 13.14) usually have an interior access ladder that runs from the base up through the center of the tank to the roof. Access to the storage tank for maintenance is by a hatch cover at the rooftop, which provides access to the tank's interior for maintenance. Aboveground storage tanks must be capable of withstanding both live loads of maintenance workers and dead loads of equipment, such as cell telephone antennas, as well as snow and ice. Local fire departments should preplan storage tanks so that they are aware of all of the hazards that may be present, including electrical hazards, steam hazards, and confined spaces. The internal tank heaters are usually gravity circulation systems of hot water or steam coils. In some cases, direct steam injection may be used if the climate is severe in the winter months. Storage tanks, like any other high structure, should be inspected by the fire inspector and preplanned by the fire department.

PRESSURIZED STORAGE TANKS AND ROOFTOP TANKS

Pressure storage tanks or rooftop storage tanks may also be used in protected properties for providing supplemental water capacity or pressure to the automatic sprinkler system. Typically, these tanks are installed inside the building, or they may be rooftop mounted. Pressure tanks may be installed instead of a fire booster pump if additional pressure requirements are marginal to operate the

FIGURE 13.14 Elevated water storage tanks must be provided with an internal heating system in cold climates. Storage may be from 100,000 gallons to 4 million gallons. *Courtesy of J. M. Foley*

FIGURE 13.15 The OS&Y valve in the center is fully open; these valves must also be electrically supervised on fire protection systems. *Courtesy of J. M. Foley*

automatic sprinkler system. The pressure tanks typically range from 2,000 to 5,000 gallons of water but may be larger in some special applications. Pressure tanks are similar to the wooden rooftop storage tanks used at the turn of the century. Pressure tanks should be inspected regularly as required by NFPA 25, *Standard for the Inspection, Testing, and Maintenance of Water-Based Fire Protection Systems*, Chapter 6, along with the other parts of the fire suppression system it supplies. Rooftop or internal supplemental water tanks may also be installed as secondary water supplies for fire protection systems. Typically, the tank must be elevated above the roof to provide 15 psi pressure at the topmost sprinkler head. These tanks must also have heat in cold climates to keep from freezing.

VALVES

All internal water supplies systems have control valves to isolate the system for maintenance. All control valves by NFPA 13, *Standard for the Installation of Sprinkler Systems* must be indicating types, such as the outside stem and yoke valve (OS&Y; see Figure 13.15). The amount of yoke extending beyond the valve wheel should equal the diameter of the pipe on which the valve is installed. The yoke indicates the position of the gate inside the valve. The amount of stem protruding from the valve wheel shows whether the valve is either fully or partially opened, or fully closed.

BACKFLOW PREVENTION

Typically the water supply main enters a building and will be separated into the domestic water supply and fire protection water supply. A backflow preventor or double check valve must be installed on the fire protection side of the system to prohibit fire protection water from entering the domestic water supply (see Figure 13.16). The reason for this is that the water in the fire protection system is stagnant and may build up microbiologic organisms over time. The backflow-preventing device must be maintained on a regular schedule as specified in NFPA 25, *Standard for the Inspection, Testing, and Maintenance of Water-Based Fire Protection Systems*, Chapter 9.

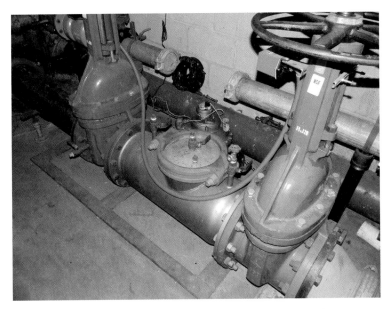

FIGURE 13.16 A backflow preventor must be installed between the fire protection water supply and the domestic water supply. *Courtesy of J. M. Foley*

FIRE BOOSTER PUMPS

Fire pumps or booster pumps are placed on automatic sprinkler systems to increase pressure due to building size or height. Additional booster pumps are located at the water distribution plant to increase distribution pressure, or they may be located at well points to extract water from aquifer and bring it up to the distribution piping. The two main types of pumps that are used are vertical pumps, which are used to lift water at well points, and the horizontal split case pumps used to increase pressure in fire protection systems (see Figure 13.17). All fire booster pumps must have both primary and secondary

FIGURE 13.17 Fire booster pumps are used in fire protection systems to increase pressure to the upper stories of a building. *Courtesy of J. M. Foley*

emergency power supplies. Fire booster pumps are installed and initially tested under NFPA 20, *Standard for the Installation of Stationary Pumps for Fire Protection*, and are to be maintained after acceptance testing in accordance with the inspection and testing schedules specified in NFPA 25, *Standard for the Inspection, Testing, and Maintenance of Water-Based Fire Protection Systems*, to ensure efficiency and reliability. The fire prevention codes also require fire pumps to be operated weekly and flow tested annually. The fire inspectors should obtain a copy of the booster pump's performance curve graphs on its initial acceptance test to ensure that the performance measures are within acceptable standards from the pump manufacturer. These records are usually part of the building department's file and should be requested. The annual pump test performance curves provide a comparison to the original installation test and will show any marked changes in the pump's performance over time. Both the owner and the fire protection service company should also maintain these records.

Rural Water Supply Systems

Rural areas seldom have a well-developed infrastructure for distribution of water, and generally, each home may have its own well for domestic water supply. In rural areas, there is usually little or no water distribution system for firefighting purposes. Rural fire departments therefore must develop water resources from a combination of natural water supplies, such as lakes and ponds, and hauling water with fire apparatus in the form of tankers or water tenders. The natural sources of water may include creeks, streams, ponds, lakes, wells, pools, tanks, cisterns, and rivers. NFPA 1142, *Standard on Water Supplies for Suburban and Rural Fire Fighting*[5] governs the application, development, and use of natural and manmade water sources for rural firefighting purposes.

WATER SUPPLY OFFICER

The NFPA 1142, *Standard on Water Supplies for Suburban and Rural Fire Fighting*, is a performance standard for suburban and rural water supply delivery. This means that it establishes minimum benchmarks that the local fire department must meet to achieve compliance with the standard. The methods that fire departments may employ to meet the performance criteria may vary as long as the end result meets the benchmark. NFPA 1142 requires that the fire department establish and appoint a water supply officer charged with the management and documentation of all available water resources with the jurisdiction.

Water Supply Officer Duties
A logical choice for the water supply officer is the fire inspector. This is because the inspector is familiar with the community, the types of building construction, the types of fire protection systems, and the unusual hazards in buildings, such as hazardous materials.

The rural fire inspector may be requested to be the water supply officer as part of his or her duties and would collect water source data and determine the necessary calculations for water resources. The fire inspector would be required to develop a water resource map and card system to be used by the fire department during emergencies to identify water sources. These water source maps should be distributed to all parties including the fire alarm dispatchers, the fire chief and chief officers, and all automatic and mutual aid responders. These map resources may also be computer based for easier distribution and use in fire department apparatus. There are several commercially available preplanning software tools that can make this task relatively easy.

The fire inspector as water supply officer should conduct regular inspections of critical water sources to ensure accessibility and to prevent overgrowth of vegetation. The inspections should include checking dry hydrant locations for proper signage and road stability as well as vandalism of fire protection equipment. Any resources adversely affected because of freezing, drought, or water pollution should be identified when they become unusable. Streams may also be considered when evaluating water supplies but must be capable of delivering water in capacities required by the needed fire flow for the buildings they are intended to protect.

PRIVATE WATER SOURCES

The fire inspector as water supply officer acts as a liaison between the municipal government, the fire department, and any property owners who may have private water sources that are needed for public fire protection. It is not uncommon for lakes to be privately owned by homeowner associations, who exercise control over who may use the water. An agreement between the property owners and the fire department to use a private water source must be in the form of legal agreement, or the municipality may find itself in court for property damage, especially if the fire department depletes a water source without the owner's permission. NFPA 1142, *Standard on Water Supplies for Suburban and Rural Fire Fighting*, provides samples of such agreements. Although fire departments have the right to enter properties that are on fire under exigent circumstances, they do not necessarily have the right to enter other private properties that are not on fire to secure water unless they have permission from the owner to do so or there is a legitimate state of emergency through either state or federal declaration. The fire inspector as water supply officer should establish the formal agreements with any private water source that may be required for community fire protection.

AUTOMATIC AND MUTUAL AID

Fire departments under both NFPA 1142 and the ISO public fire protection rating schedules are required to establish minimum automatic aid or mutual aid agreements to verify the capability to deliver water using a tanker shuttle delivery system. The ISO requires water supply to be established and available within 15 minutes of the initial application of water at a rate of 250 gpm from an engine company. This requires a minimum of 3,750 gallons of water on the initial response. The fire department must start the application at a rate of 250 gpm within 5 minutes of arrival on the fire scene. It is obvious that, if additional water tenders are not dispatched by an automatic aid agreement with the initial first alarm assignment, the 15-minute criterion will not be attainable. The ISO requires formal automatic aid agreements between agencies, the mutual aid fire department must be within 5 miles of the jurisdiction, and interagency training must occur between departments under the agreement.

OCCUPANCY HAZARD CLASSIFICATIONS

NFPA 1142, *Standard on Water Supplies for Suburban and Rural Fire Fighting*, further requires that a complete survey of the jurisdiction be conducted to determine the minimum amount of water that must be available to control fires based on specific risks. The building's use, occupancy, and construction type drive this analysis. The standard uses a formula to determine a building's fire flow based on a property's hazard classification, occupancy, use, and type of construction. The calculation also factors in exposed buildings that are located within 50 feet of the structure requiring a fire flow. An example of this calculation is shown in Figure 13.19.

The NFPA 1142 standard classifies buildings numerically from three (3) to seven (7) based on the expected fire severity and fire duration at the building. The lower the number, the greater the fire risk. As an example, a classification of 3 would be a severe hazard, while a 7 would be considered a light-hazard occupancy such as a single-family home. This classification is used to factor in the potential fire risk that a building may pose. The hazard classification number is based on the fire load that is expected in the building. Examples of a class 3 hazard include an oil refinery, distilleries, or lumberyards. All of these groups present a high fuel load and high fire risk. Occupancies in class 7 would

Type Construction	Classification Number
Type I Fire resistive	0.5
Type II Non-combustible	0.8
Type III Ordinary	1.0
Type IV Heavy Timber (Mill)	0.8
Type V Wood Frame	1.5

FIGURE 13.18 The construction classification along with the hazard class and potential exposures are used in NFPA 1142, *Standard on Water Supplies for Suburban and Rural Fire Fighting* to determine the fire flow for buildings. *Reprinted with permission from NFPA 1142-2012, Standard on Water Supplies for Suburban and Rural Firefighting*

include hospitals, apartments, residences, and office buildings. All of these occupancies have light fuel loads and present much smaller fire risks.

The general occupancy hazard classifications are:

- 3 – Severe-hazard occupancy – Lumberyard, refinery, distillery
- 4 – High-hazard occupancy – Department stores, barns, repair garages, warehouses
- 5 – Moderate-hazard occupancy – Dairy barns, laundries, restaurants, machine shops
- 6 – Low-hazard occupancy – Gas stations, bakeries, doctors' offices, churches
- 7 – Light-hazard occupancy – One- and two-family dwellings, fire stations, hospitals, nursing homes, schools

CONSTRUCTION CLASSIFICATIONS

The next factor to be considered is the type of building construction. Construction classification numbers are based on the five major construction types identified in all building codes. The higher the construction factor number, the greater the fire risk. It should be noted that heavy timber construction and noncombustible construction are rated equally. This is because they both perform well in fires, even though mill construction is combustible. In Figure 13.18, the types, construction, and relative fire risks are identified.

EXPOSURE HAZARDS

Once the occupancy hazard and the construction factors have been determined, any surrounding buildings also have to be considered for additional water supply. Exposure hazards are defined in NFPA 1142 as any structure more than 100 square feet in area and within 50 feet of a building under consideration. The NFPA 1142 standard takes into account that radiant heat may travel 50 feet and expose a building to damage from fire. High-occupancy hazard class buildings rated 3 or 4 are also considered exposure if they are within 50 feet of a building regardless of its size. Exposure structures require the multiplication of the water supply by a factor of 1.5, or 150 percent, to ensure adequate water for exposure protection. The minimum amount of water for any exposure shall be not less than 3,000 gallons.

Calculating Water Supply

The minimum water supply in gallons is determined by the following calculation:

$$\frac{\text{Volume of structure } (L \times W \times H) \text{ in cubic feet } \times \text{ Construction Classification Number}}{\text{Hazard Classification Number}}$$

In Figure 13.19 you see an illustrated example of the calculation for a building with a cubic volume of 34,272 square feet. The building is a light-hazard occupancy, classification 0.7, and the building is ordinary construction with a classification of 1.0. When we substitute the numbers into the formula we find that the necessary water supply for this building is 4,896 gallons. If the building had exposures, the amount of water would then be multiplied by 1.5, increasing it to 7,344 gallons of water. It should be noted that reductions in water supply are permitted when a building has a full sprinkler system in compliance with appropriate NFPA sprinkler standards.

As an example, suppose that we have a building that is 34 feet wide and 42 feet long, and the height to the attic floor is 10 feet to the ridge beam. The volume of the building will be 42 × 34 × 24 feet, which equals 34,272 cubic feet.

If the building has a hazard classification of 7, is of ordinary construction, and has no exposures, the fire flow would be:

$$\frac{34{,}272}{7} \text{ cu. ft } \times 1.0 = 4{,}896 \text{ gallons}$$

FIGURE 13.19 The minimum water supply is calculated by determining the volume of the structure, then dividing it by the hazard classification number. The result will be multiplied by the construction classification number and any exposure factor, if an exposure exists within 50 feet.

The fire department must have the capability to deliver the calculated amount of water at the flow rate specified in NFPA 1142 based on the total amount of water needed. Those flow rates are as follows:

- <2,500 gallons   The fire flow application rate is 250 gpm.
- 2,500–9,999 gallons   The fire flow application rate is 500 gpm.
- 10,000–19,000 gallons   The fire flow application rate is 750 gpm.
- 20,000 gallons   The fire flow application rate is 1,000 gpm.

The water again must be delivered by the fire department within 5 minutes of arrival at the scene.

SAMPLE PROBLEM

To better understand this calculation, let's examine the following example.

Calculate the minimum water supply for a residential dwelling of ordinary construction that is 50 feet × 24 feet with two 8-foot stories and a pitched roof attic 8 feet to the roof peak.

The first step is to determine the structure's volume.

$$\text{Volume} = \text{Length} \times \text{Width} \times \text{Height}$$

Buildings with peaked roofs or attic heights are considered to be 1/2 the height measured from the attic floor to the ridge beam for the volume calculations. In the above example, if each floor is 8 feet high, the building's total height would be 20 feet.

$$
\begin{aligned}
\text{1st floor} &= 8 \text{ feet} \\
\text{2nd floor} &= 8 \text{ feet} \\
\text{Attic} &= 1/2\,(8) \text{ or } 4 \text{ feet} \\
\text{Total:} &\quad 20 \text{ feet}
\end{aligned}
$$

We can now perform the calculation:

Step 1.   Volume: **50′ × 24′ × 20′ = 24,000 cu ft.**

Step 2.   Determine occupancy hazard and construction classification

$$\text{Hazard Class} = 0.7 \text{ and Construction Classification} = 1.0$$

Step 3.   Calculate minimum water supply:

$$24{,}000/0.7 \times 1.0 = 3{,}429 \text{ gallons}$$

Step 4.   Multiply the minimum water requirement by the exposure factor (1.5) if required.

Step 5.   Go to NFPA 1142, table 4.6.1, for minimum water delivery rate and determine the fire flow requirement. In this example the required fire flow is **250 gpm for the first 15 minutes and 500 gpm for 2 hours.**

USING STREAMS FOR FIRE PROTECTION

Natural water sources, such as streams, should be identified in the water resource plan. A stream should be capable of delivering water throughout the year in any climatic condition, such as freezing or drought, to be considered. The stream must be accessible to fire department vehicles by all-weather roads and have clear overhead passage. The water supply officer must be able to determine the flow rate of the water source for both total capacity and flow in gallons per minute. To calculate the flow rate for a lake or pond, the volume needs to be determined by measuring length × width × the average depth. The volume is then multiplied by 7.5 gallons per cubic feet to determine the total supply volume. Streams are slightly more difficult because they are flowing and cannot be effectively measured, so the fire inspector may use a cork test to determine flow.

THE CORK TEST

The use of a simple cork and the following flow formula can allow a quick determination on the fire flow available from a stream (see Figure 13.20).

$$Q = A \times V \times 450$$

The cork test should be performed in an area where both sides of the stream are perpendicular and the bottom of the stream is relatively flat. Avoid the effects of wind on the cork. The cork should be placed in the water and monitored on how long it takes it to float 10 linear feet in the stream. The 10 feet then should be divided by the number of seconds to determine the stream's velocity in feet per second.

FIGURE 13.20 The cork test can be used to calculate the flow of a stream in gallons per minute. The inspector must know the width and average depth of the stream and should measure the time it takes the cork to travel 10 feet in seconds. The formula is Q = Area × Velocity × 450. *Courtesy of J. M. Foley*

A cross-section or transect of the stream should be determined by the width and depth or average depth measured in square feet. The number 450 is a constant that resolves the measurements from cubic feet per second into gallons per minute.[6]

EXAMPLE

A creek is 4 feet wide and 12 inches deep. A cork is placed in the water and flows 10 feet in 120 seconds. What is the flow?

Step 1. Determine the area: 4 feet × 1 foot = 4 square feet.
Step 2. Determine velocity: 10 feet/120 seconds = 0.08 feet per second.
Step 3. Substitute the values in the formula and multiply by 450.

$$4 \text{ sq. ft.} \times 0.08 \text{ ft./sec.} \times 450 = 144 \text{ gallons per minute}$$

CALCULATING SWIMMING POOLS

Another possible source of rural water supply may be in-ground or aboveground swimming pools. These resources should be used only in extreme emergencies and with the owner's agreement. To determine the capacity of an in-ground swimming pool in gallons, use the same formula that is used for lakes or ponds:

Capacity in gallons = area × the average depth × 7.5 gallons/cu. ft.

If the pool has a deep end, then take the average depth of the pool to determine the actual area. The area (length × width) times the average depth is multiplied by 7.5 to determine the capacity in gallons of water.[7]

As an example, if a pool is 16 feet by 28 feet with an average depth of 4.5 feet:

16′ × 28′ × 4.5′ × 7.5 gal. = 15,120 gallons

The total gallons of water in the pool will be approximately 15,120 gallons.

Again, careful consideration must be used when considering in-ground or aboveground pools as a water resource. Pools generally are not easily accessible to fire apparatus, and property damage may occur to the pool from the vibration and movement of fire apparatus. Removing the water may also cause hydraulic pressures, which can collapse the pool. If a pool is used as a water source, small portable floating pumps may be the best solution. The pool also should not be completely emptied to avoid damage. The fire department should perform an evaluation only if there is no other possible water source available for use. A circular aboveground pool may also be considered in an emergency and is calculated in the same fashion with the following exception: The area of the round pool will be calculated at *0.8 × the pool diameter squared.*[8]

Cisterns

The fire inspector must work with the building owners and local government to make the recommendations on improvements in water capacity so that fires may be more easily controlled or suppressed. In rural areas with no water distribution system the building owner may consider the installation of a cistern with a dry hydrant connection, as seen in Figure 13.21.

Cisterns are one of the oldest methods of water storage for both firefighting and domestic use and can be traced back to the Roman Empire. Cisterns are large underground storage tanks that the fire department can draft water from in an emergency. A cistern must be of adequate capacity—at least 30,000 gallons of water (see Figure 13.21). The openings in the cisterns must be properly secured against accidental entry by children.

cistern ■ An underground tank or vat used to hold water for fire protection. Cisterns are usually 30,000 gallons or more and are equipped with dry hydrant fittings for drafting.

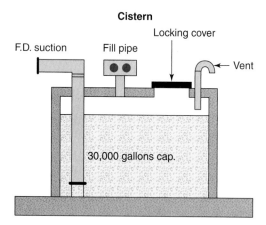

Cistern

F.D. suction   Fill pipe   Locking cover   Vent

30,000 gallons cap.

FIGURE 13.21 Cisterns are alternate water supplies when water distribution mains are not available. Cisterns should be at least 30,000 gallons in capacity.

dry hydrant ■ A dry pipe constructed of steel or plastic to convey water from a lake or pond to a fire apparatus by draft.

Cisterns will also be confined spaces and should be placarded accordingly. If water is to be drafted through a dry hydrant from the cistern, the dry hydrant should be sized to provide at least 1,000 gpm for three-quarters of the cistern's total capacity. All cisterns should be rated for highway loading so that a fire apparatus cannot cause structural damage to the tank. The use of cisterns is a reliable method of locating sufficient water supplies for firefighting purposes at important buildings such as hospitals or schools in rural areas or important or large facilities.

Dry Hydrants

The installation of dry hydrants in lakes or streams is another economical method of providing a reliable supply of water from a natural water source. Dry hydrants reduce the time that it takes to draft from a lake by conventional fire department methods. The **dry hydrant** (see Figure 13.22a and b), when installed properly, normally can be accessed from existing roadways or other all-weather surfaces installed around the fire department connection. Modern dry hydrants are constructed of schedule 40 plastic pipe and even have vandalismproof features to prevent such damage. During the review for potential locations for dry hydrants, the fire inspector needs to consider the total draft height required from the dry hydrant connection to the water supply to make sure the fire department can adequately draft from the site. The practical drafting height for fire apparatus is between 15 and 20 feet. If higher-than-normal lifts are required, technology does exist to accomplish these through the use of siphon jets to move the water up the vertical riser. The siphon jet can achieve lifts

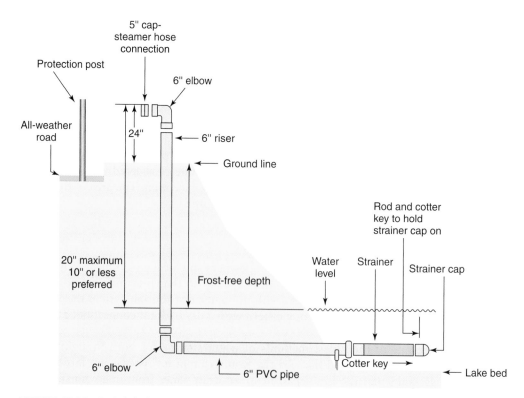

Protection post

5" cap-
steamer hose
connection

6" elbow

All-weather
road

24"

6" riser

Ground line

Rod and cotter
key to hold
strainer cap on

20" maximum
10" or less
preferred

Frost-free depth

Water
level   Strainer   Strainer cap

6" elbow

6" PVC pipe

Cotter key

Lake bed

FIGURE 13.22a Typical dry hydrant installation in a pond or lake.

of 15 to 70 feet vertically. The siphon jet is based on the principle that as water moves through a smaller pipe, it produces a higher velocity and lower pressure. The reduction in pressure is utilized to move water to the higher elevation. The fire inspector reviewing the plans for such an installation should also give considerations to requiring freeze protection, flushing fittings, identification signage, cam lever couplings, and other types of dry hydrant improvements that may be needed by the fire department.

FIGURE 13.22b Dry hydrants often are installed in lakes and ponds to provide a water supply for fire department use. *Courtesy of J. M. Foley*

Water Tender Shuttles

In rural areas, the fire departments will have to transport water to the fire using mobile water supply fire apparatus. The Insurance Service Organization, or ISO, requires that fire departments demonstrate their ability to provide water within the 15-minute requirement on initial arrival at the appropriate sustained gallons-per-minute delivery rate specified in NFPA 1142, *Standard on Water Supplies for Suburban and Rural Fire Fighting*. Mobile water supply apparatus must be designed as specified in NFPA 1901, *Standard for Automotive Fire Apparatus*, Chapter 7. Many fire departments may use converted gasoline trucks as water shuttles. The problem with these vehicles is that they are not properly designed for the weight of water nor are they properly equipped for fast unloading into portable storage tanks. The ISO also may evaluate relay water pumping with the use of large-diameter hose (3 inches or greater). The local fire department must practice these types of evolutions regularly to be proficient in its water delivery within the required fire flow time frames of the standard. The fire department must prove compliance through a demonstration of its ability to move water and maintain the minimum fire flow.

The ISO Grading schedule requires the following water shuttle procedure to be followed during the demonstration:

1. All tenders assemble 200 feet from the fire scene.
2. The time it takes to move a tender 200 feet is recorded.
3. The tender must then discharge its water into a folding storage tank and move 200 feet away.
4. The time it takes to discharge the water into the tank and move 200 feet is recorded.
5. Each tender is credited for 90 percent of its total capacity.
6. The same procedure is followed at the fill site. The time is recorded for the tender to move 200 feet, fill up, and move 200 feet away.
7. The last part of the calculation is the travel time from the supply water source to the fire location. All of these elements are recorded for one full cycle of all the required water tenders. A record is kept of the available water supply at the fire scene, and it must be equal to or greater than the minimum flow required at the application rate for 2-hour duration.[9]

The ISO evaluates the travel time from the fire station to the fire scene, and the travel times for supply engines to the water supply locations. Travel time is also calculated for any automatic aid companies on the first alarm assignment.

If a water shuttle system is to be used, at least two routes from the supply to the fire location must be calculated for travel time. Another consideration that must be determined is the total weight of the water tenders as they move through the cycle of filling, delivering, dumping, and refilling, especially on roads with small bridges. The fire department should use water tenders of the same size and weight capacity, if possible, as this eliminates the stacking up of mobile tenders at the fill or discharge locations. Additional water supply fill points should also be considered, as more tenders enter into the filling cycle to maintain system efficiency. Portable tanks should be set up in a diamond configuration to maximize the areas for water tenders to deliver their water supply. This diamond configuration reduces stacking of full water tenders at the delivery site.

The formula that the ISO uses to determine the water carrying capacity for each tender is as follows:

$$Q = \frac{V}{A + (T1 + T2) + B} - 10\%$$

V = the total capacity of the mobile water supply tanker.
A = The dump time including the 200-foot approach and return.
B = The fill time including the 200-foot approach and return.
$T1$ = Travel time from the fire scene to the water sources.
$T2$ = Travel time from the water source to the fire scene.
-10% is the amount of spillage and residual in the tankers.

Once the water carrying capacity is determined for each mobile water tender, the travel times must be calculated. From the supply to the delivery points over two routes. The formula for travel time calculation is:

$$\text{Time} = \text{Speed Constant (X)} \times \text{Distance Traveled}$$

The speed constant (X) factor is based on average speed of the water tender for each leg of the journey. In general applications, 35 mph is used by the ISO in determining maximum travel time. The fire inspector, however, needs to consider other conditions that may reduce travel time such as snow and ice, steep grades, or narrow or winding roadways. All of these conditions reduce speed, and a different constant must be used. When using the calculation, it is important to determine the average speed on each road segment. Each speed will have a different speed constant. Speeds should never be calculated above 35 mph for water tenders. Other speed constant variables are as follows:

$$\text{Time} = 1.7D \text{ at } 35 \text{ mph}$$
$$\text{Time} = 2.0D \text{ at } 30 \text{ mph}$$
$$\text{Time} = 2.4D \text{ at } 25 \text{ mph}$$
$$\text{Time} = 3.0D \text{ at } 20 \text{ mph}$$
$$\text{Time} = 4.0D \text{ at } 15 \text{ mph}$$

While these calculations may seem cumbersome, they can be calculated as part of prefire emergency plans for specific property locations. Often, the same supply and delivery routes can be used for each segment of a community. These formulas may also be placed in a spreadsheet for quicker calculations.

As an example, let's calculate the minimum continuous flow for a 1,500-gallon water tender. We will assume from previous trials that it will take this tanker 4.0 minutes to deliver its water, including the 200-foot approach and retreat, and it will take 3.0 minutes to refill the tanker at the supply site. The road distance to the supply is 2.10 miles, and the return leg to the fire scene is 1.8 miles. The first road may be traveled at 35 mph, but the return leg is only 30 mph due to traffic conditions.

$$T1 = 0.65 + 1.7D \text{ for } 35 \text{ mph}$$
$$T2 = 0.65 + 2.0D \text{ for } 30 \text{ mph}$$

$T1 = 0.65 + 1.7(2.10)$	$T2 = 0.65 + 2.0(1.80)$
$T1 = 0.65 + 3,57$	$T2 = 0.65 + 3.60$
$T1 = 4.22$ minutes	$T2 = 4.25$ minutes

Once the travel time on each leg has been calculated, we can use the next formula to determine the flow from the 1,500-gallon tender in gpm. The ISO formula is as follows, where Q = flow in gallons per minute:

$$Q = \frac{1,500 \text{ gallons}}{3.0 \text{ min} + (4.22 + 4.25)\text{min} + 4.0 \text{ min}} - 10\%$$

$$Q = \frac{1,500 \text{ gallons}}{15.47 \text{ min}} - 10\%$$

$$Q = 97 \text{ gpm} - 10\% = 87 \text{ gpm}$$

Therefore, Q = 87 gallons per minute.

To maintain a minimum fire flow of 250 gpm, at least two additional 1,500-gallon water tenders are required in the water shuttle operation on the initial alarm assignment.

History of the Fire Suppression Rating Schedule

The Insurance Service Organization evaluates communities to determine risk as it relates to fire insurance rates. The ISO generally determines the Public Protection Classification (PPC) for most communities on a 10-year basis. The classification of the community affects the fire insurance rates charged by any insurance providers to residents or businesses. While the PPC should not be used to solely determine the level of a community's fire protection, improvements should be considered because they lower costs to taxpayers who may be affected by fire protection changes. To understand the importance of the ISO rating schedule, we must first understand how the schedule was developed through some brief history, starting with water distribution systems.

Waterworks were developed initially to provide a distribution system for domestic water and sanitation purposes. Larger cities that developed water main infrastructures realized that these had an additional benefit, and that was providing water for fire suppression. By the turn of the nineteenth century, several noteworthy engineers had written papers addressing the concerns of fire protection in built-up communities and the role that the waterworks should play in fire protection. They began to develop standards on water main size and capacity as well as estimates on how many fire hose streams might be needed to suppress a large fire. Several engineering papers were presented at the Engineer's Society Meeting in the early 1900s, which began a detailed discussion concerning both the American and Canadian waterworks systems. An initial concern was to develop a method to determine the cost of water for fire protection. The engineers made estimates of the number of hose streams that would be required by fire departments in congested, densely populated cities. The engineers believed that in a rough way, the number of hose streams was related to population size.

The National Board of Fire Underwriters in its determination of the potential number of hose streams needed by population adopted the recommendations of several engineers including Factory Mutual engineer, John Freeman. All hose streams were assumed to deliver 250 gpm. Freeman took the lead in presenting progressive ideas for fire protection in water delivery in the paper he authored in 1892.[10] In that paper, he determined that domestic water supply is based on distribution, while fire protection water supply must be based on concentration. Freeman suggested that the minimum size of a water main should be 6 inches in residential districts and 8 inches in commercial districts. He suggested that the water systems be in a grid pattern to allow for better flow characteristics through the network of pipes. Freeman also suggested that fire hydrants be spaced no more than 250 feet apart in commercial areas and 500 feet apart in residential areas. In 1889, the National Board of Fire Underwriters already had begun to make fire protection surveys of municipalities with regard to water supply. This effort was intensified greatly after 1904 and the Great Baltimore City Fire, which destroyed a large portion of

15

Baltimore. John Freeman's formula for determining fire flow from his original paper was adopted and modified by the NBFU and was used until 1948, when it was changed to be more an engineering-based model than a population-based model. It was this original work by John Freeman that led to the development of the "Fire Suppression Rating Schedule" that is used to establish insurance rating today.[11]

The Insurance Service Organization Public Protection Classification Schedule

The **Insurance Service Organization** (ISO) evaluates communities with and without water supplies to provide public protection classifications for the purpose of fire insurance rating. These classifications are based on the local fire department's ability to provide water delivery to a fire scene at a minimum rate of 250 gpm for at least a 2-hour duration. The ISO has ten public protection class ratings in its insurance rating schedule. Class 1 is considered the best fire protection rating, while Class 10 is considered unprotected and the most at risk from fire. Insurance companies use this information in establishing fire insurance rates that the homeowner or business will pay in their premiums. The ISO public protection class ratings are based on three community factors: the fire department, the fire alarm communications center, and the water distribution system. As an example, the ISO considers any building more than 5 miles from a fire station to be an unprotected risk and would place it in protection Class 10. This classification means the building owner or homeowners will pay higher fire insurance premiums on their business or home to offset the additional risk to the fire insurance company. Buildings not serviced by water mains but within 5 miles of a fire station are considered semiprotected and are placed in public protection Class 9. A Class 9 is considered marginal fire protection and still has a high fire insurance rate. A community may improve the public protection classification in areas not serviced by water mains but within 5 miles of a fire station if it can demonstrate the ability to suppress a fire with a first alarm assignment by using water tender shuttles or other available methods to deliver a sustainable fire flow.

The ISO requirement for water delivery meets the requirements of NFPA 1142, *Standard on Water Supplies for Suburban and Rural Fire Fighting*, which is to achieve a minimum sustainable fire flow of 250 gpm for a 2-hour duration. Water supply under the ISO rating schedule may be delivered by use of water tender shuttles, fire hose, and relay pumping, or any combination thereof. Water supplies may be from any reliable water source provided that it meets the minimum acceptable flow requirements. Any single source of water must have a total of 30,000 gallons capacity, and all multiple sources must have at least 30,000 gallons aggregate capacity to be considered. The ISO also requires that the Department of Environmental Protection and Soil Conservation or a licensed engineer certify the capacity of any natural water source to be considered. The ISO evaluates these criteria under the fire suppression rating schedule to determine the public protection classification.

FIRE SUPPRESSION RATING SCHEDULE

The current Fire Suppression Rating Schedule (FSRS) under the ISO analyzes three areas of a community's water delivery system:

1. The supply work's capacity
2. The water main capacity
3. The distribution of fire hydrants on the system

Based on these three areas of concern, the capacity of the water system is evaluated and tested. The ISO determines the test locations by identifying buildings requiring a minimum fire flow of 3,500 gpm or more. Buildings protected by automatic sprinkler systems are not considered in the FPRS evaluation, as they fall under the commercial rating schedule.

The water supply is credited based on the total flow in gallons per minute at 20 psi residual pressure at the test location for a specific duration of time. The lowest rate of flow is used for determining the credits to be issued under water supply and are based on the following criteria:

- Needed fire flow in gpm at the test location
- Capacity of the supply system in gpm
- Capacity of the water distribution system in gpm
- Capacity and delivery of water from a fire hydrant within 1,000 feet of the property at the test location

The objective of the evaluation is to determine whether the water system is capable of supplying the needed fire flows at each selected site within a community.

SYSTEM CAPACITY

To understand the capacity of the water distribution system, you must understand the terminology that water utilities use in expressing the water being consumed.

Average Daily Demand—This is the average amount of water used every day for a 1-year period from the delivery system.

Maximum Daily Demand—This is the maximum amount of water used in any 24-hour period based on 3 years of water use data.

Peak Hourly Demand—This is the maximum amount of water used in any given hour of the day or night during the year.

Normal system demands are required to determine the effect they may have on the needed fire flows. The **needed fire flow** (NFF) is the amount of water required to control a major fire at a specific location and is expressed in gpm at 20 psi residual pressure for either 2 or 4 hours duration. The minimum NFF under ISO is 500 gpm over 2 hours, while the maximum needed fire flow is 12,000 gpm over 4 hours. This is a total water flow of 60,000 gallons and 2.8 million gallons of water, respectively, that must be available.

The needed fire flow is determined by a formula that takes into account the type of construction, occupancy hazard, exposures, and internal fire communication probability in a fashion similar to NFPA 1142, *Standard on Water Supplies for Suburban and Rural Fire Fighting*. Once the needed fire flow is established, the ISO can determine the number of engine and ladders companies necessary for the fire department. The needed fire flow calculation under the ISO differs slightly from the formula used in NFPA 1142, as it includes a communication factor.

needed fire flow ■
This is the amount of water necessary to control a major fire in a specific building expressed in gallons per minute. The minimum needed fire flow for a single building is 500 gpm for 2 hours, and the maximum needed fire flow is 12,000 gpm for 4 hours.

NEEDED FIRE FLOW FORMULA

The needed fire flow is defined as the amount of water necessary to control a major fire at a specific location in the jurisdiction. The ISO formula for determining the needed fire flow is almost the same as the formula from NFPA 1142 except that it adds a calculation for communication of the fire in an exposed building. The ISO formula is as follows:

$$NFF = (C)(O)[1 + (X + P)]$$

C = construction factor
O = occupancy factor
X = exposure factor
P = communication factors

The determination of the needed fire flow has two effects on the public protection classification rating for fire insurance purposes. First, the needed fire flow determines the basic fire flow requirements for the community. The basic fire flow is the fifth-highest needed fire flow. The basic fire flow determines the minimum number of fire apparatus, fire pump sizes, and special equipment needed for fire protection by the fire department. Second, the needed fire flow determines the amount of water available at a specific location selected by the ISO for evaluation. The maximum needed fire flow can be up to 12,000 gpm, and the minimum needed fire flow is 500 gpm. As an example, if a building had a needed fire flow of 3,500 gpm and the closest fire hydrant was 200 feet away, the credited fire flow of that hydrant will be only 1,000 gallons per minute. To meet the needed fire flow you will need an additional three hydrants, two within 300 feet and one within 600 feet of the building The needed fire low is evaluated against the supply works' water capacity, the capacity of the water distribution system in gallons per minute, and the capacity for water delivery from fire hydrants within 1,000 feet of the property. The lowest flow from these three areas will determine the percentage of points credited to the water supply distribution system.

Under the original fire suppression rating schedule, the public protection classification was measured based on deficiencies that were found; however, this terminology was changed in the mid-1970s from deficiencies to credits, which was a more positive description of the community's fire protection. The fire suppression-rating schedule examines three specific areas of fire protection delivery: the fire department, the water delivery system, and the alarm communications center. Water delivery is credited at 40 percent of the overall fire protection class rating. The water distribution rating includes a review of the water system capacity, water main distribution, hydrant installation, and hydrant inspections.

The rating schedule is divided into two parts:

Section I applies public protection class ratings for average-size buildings within a community having a needed fire flow of 3,500 gpm or less.

Section II applies to individual properties and large buildings with a needed fire flow in excess of 3,500 gpm.

The credits provided for water supply are further broken down into three subcategories for evaluation: system capacity and distribution, which accounts for 35 of the 40 possible credits; fire hydrant type and location account for 2 credits and hydrant inspection; and maintenance programs account for 3 credits. Properties being evaluated for needed fire flow must be within 1,000 feet of a fire hydrant. Any evaluated property located beyond 1,000 feet but still within 5 miles of a fire station is considered semiprotected and is placed in protection Class 9. Areas of the community greater than 5 miles from a fire station are placed in public protection class 10, which is considered unprotected. To be credited, all fire hydrants must be capable of producing at least 250 gpm at 20 psi residual pressure for a minimum 2-hour duration.

WATER SYSTEM CAPACITY

maximum daily consumption ■ This is the amount of water used by a community daily plus the amount of needed fire flow expressed in gallons per minute.

The water system must have adequate capacity for domestic consumption and fire protection. To determine this, ISO must know the **maximum daily consumption** rate for the community. The American Water Works Association projects that the average daily consumption rate for water in the United States is approximately 69.3 gallons per person per day. They also estimate that the maximum daily consumption rate is 150 percent of the average daily consumption rate.[12] You can estimate the rate by multiplying the community population by 69.3 gallons per person per day. As an example, if the population was 36,000, the consumption would be 36,000 × 69.3 = 2,494,800 gallons of water. Dividing that by 24 hours or 1,440 minutes gives a flow of 1,733 gpm. This daily consumption rate would then be multiplied by 150 percent to get the maximum daily consumption rate of 2,596 gpm. The ISO uses the highest maximum daily consumption rate to calculate the water supply capacity for the community.

To determine the supply capacity, the ISO examines the direct pump capacity plus filter capacity, emergency water reserves, and any supplemental or additional water supplies on the system. The pump capacity is credited at the effective rated capacity of the pump during normal service. The capacity of the filters is determined in gpm to credit the filter capacity. The ISO credits only potable water that meets U.S. health standards, so wells or other sources of nonportable water are not to be considered for system supply calculations.

EXAMPLE

A city has two elevated water storage tanks. The recorded average daily consumption is 175,000 gallons. The maximum fixed pump rate for each flow duration period is recorded as follows:

2 hours	3,500 gpm
3 hours	3,000 gpm
4 hours	2,500 gpm

*An emergency supply of **250 GPM** is available for **4** hours from the storage tanks.*

*The maximum daily consumption rate is **1,800 gpm**.*

How do we calculate the water supply capacity?

Step 1. To solve this problem, you first must divide the daily capacity by the appropriate flow duration to obtain a flow in gallons per minute.

Flow rate for consumption:

$$2 \ hours = 175,000/120 = 1,458 \ gpm$$
$$3 \ hours = 175,000/180 = 972 \ gpm$$
$$4 \ hours = 175,000/240 = 729 \ gpm$$

Step 2. Next, you would add the storage flow in gpm to the fixed pump rate flow, the emergency flow, and you would then subtract the *maximum daily consumption* rate. This calculation will determine the minimum system fire flow for each time duration.

Fire Flow Rate:

$$2 \ hours \ [(1,458 + 3,500 + 0 + 250) - 1,800] = 3,408 \ gpm$$
$$3 \ hours \ [(972 + 3,000 + 0 + 250) - 1,800] = 2,422 \ gpm$$
$$4 \ hours \ [(729 + 2,500 + 0 + 250) - 1,800] = 1,679 \ gpm[13]$$

FIRE HYDRANT DISTRIBUTION

Fire hydrant distribution is based on the distance of the closest fire hydrant to the building needing the fire flow. Fire hydrants must be within a maximum distance of 1,000 feet of the building. The flow of fire hydrants is calculated at 1,000 gpm if it is within 300 feet of the needed fire flow, 670 gpm if within 600 feet, and 250 gpm if within 1,000 feet. The distance is measured from the building to the fire hydrant by the length of fire hose required to make the stretch. The fire hydrants are also rated by the size and type of hydrant outlets present on each fire hydrant. If they have a steamer outlet, they are credited for 1,000 gpm. If they have only two 2-1/2-inch hose outlets, they are credited for 750 gpm, and if they have only a single 2-1/2-inch outlet, they are credited for 500 gpm

FIGURE 13.23 Fire hydrants are rated by the number of outlets provided under the ISO Fire Suppression Rating Schedule. *Courtesy of J. M. Foley*

(see Figure 13.23). What this means is that a single 2-1/2-inch outlet fire hydrant 200 feet from the building is credited only as 500 gpm and not 1,000 gpm. The needed fire flow is determined based on the *collective flows* of all the fire hydrants within 1,000 feet of the building requiring the needed fire flow.

The final factor in determining the capability of the water supply system is to evaluate the system based on the weakest link in the evaluated components. Each of the calculated factors is reviewed, including the needed fire flows, the supply capacity, the water main capacity, and the fire hydrant capacity. If all of the examined factors exceed the needed fire flow, then the NFF is credited for that test location. If one of the factors is below the needed fire flow, then that restricting element is used, the building cannot meet the needed fire flow, and credits are deducted. The fire flow for the entire community is expressed in the credit for the supply system. This determination is made by taking the total of all test locations' credited fire flows and dividing that by the total needed fire flows for all test locations. The result is then multiplied by 35 percent to determine the total water supply credit to be issued in the fire suppression rating schedule.

FIRE HYDRANT MAINTENANCE

The last part of the water distribution system that is credited by ISO is the maintenance of fire hydrants. A system of assigning points for maintenance of the hydrants is used to determine the credits to apply. All hydrants inspected every 6 months receive 100 points; if they are inspected at 1 year, they receive 80 points. The longer the inspection interval, the fewer the points that will be assigned. Hydrants over 5 years between inspections receive only 40 points. The general condition of the fire hydrants is also considered (see Figures 13.24a and 13.24b). Fire hydrants fall into three categories for condition:

FIGURE 13.24a Pictured is an example of a poorly maintained and installed fire hydrant. Notice the rust around the hydrant caps and on the bonnet. The hydrant is also too high from the curb line. *Courtesy of J. M. Foley*

FIGURE 13.24b Pictured is a red-top hydrant indicating 500 gpm or less to the fire department. Some departments require alternate hose fittings on fire hydrants, such as the Storz coupling above. *Courtesy of J. M. Foley*

standard, useable, and nonuseable. These factors are figured into a formula to determine the percentage of hydrants to be credited based on the total number of fire hydrants on the distribution system. Maintenance credits account for 3 percent of the total water distribution credit. Fire inspectors should note the condition of fire hydrants in the exterior examination of a building during regular inspections. Those fire hydrants needing maintenance should be reported to the utility for repair. Fire inspectors should also keep in mind the type of fire hydrants being installed at new construction sites or when fire hydrants are removed because they may alter the public protection class rating in the next rating period.

Summary

It is important for the fire inspector to be actively involved in the inspection, testing, and maintenance of the community water supply system. Fire inspectors must understand the proper installation and construction requirements for the water distribution infrastructure and note any deficiencies during site plan review or field inspection of the underground elements for water-based fire protection systems. It is equally important that the fire inspector understand how water systems affect fire insurance rates in a community as part of a risk reduction plan. Good planning of water distribution and fire hydrant maintenance and effective inspections of fire hydrants will ensure better fire department response in emergencies.

Review Questions

1. List and describe the five principles of water that make it a good fire suppression agent.
2. You have just conducted a flow test on a fire hydrant outlet that is 2-1/2 inches with a coefficient of 0.9. The pitot reading was 25 psi. What is the flow?
3. Determine the amount of water needed under NFPA 1142 for a building of moderate hazard and ordinary construction. The building dimensions are 50 feet wide, 100 feet long, and 16 feet high.
4. What is the capacity of a rectangular pool that is 16 feet by 30 feet and 5 feet deep?
5. Describe the three main areas of the water distribution system that are rated under the ISO Fire Suppression Rating Schedule.
6. A 2,500-gallon tanker that takes 6 minutes to fill and 5 minutes to discharge responds to a fire. The fill site is 1.5 miles each way from the fire to the lake. The tanker can only travel at 30 mph on each leg of the journey. What is the total flow in gallons per minute that the tanker can deliver to the fire scene?

Suggested Readings

Brock, Pat D. *Fire Protection Hydraulics and Water Supply Analysis.* 2nd edition. Fire Protection Publications, Oklahoma State University.

Casey, James F. *Fire Service Hydraulics.* 2nd edition. New York: Dun-Donnelley Publishing.

Hickey, Harry E. 1993. *Fire Suppression Rating Schedule Handbook.* Professional Loss Control Education Foundation.

———. 1980. *Hydraulics for Fire Protection.* Quincy, MA: NFPA.

Mahoney, Eugene E., and Hannig, Brent E. *Fire Department Hydraulics.* 3rd edition. Pearson/Brady Publishing.

National Fire Protection Association. 2008. *Fire Protection Handbook.* 20th edition. Section 15, "Water Supply for Fixed Fire Protection." Boston, MA: NFPA.

———. 1995. NFPA 24, *Standard for the Installation of Private Fire Service Mains and Their Appurtenances.* Boston, MA: NFPA.

———. 2008. NFPA 25, *Standard for the Inspection, Testing, and Maintenance of Water-Based Fire Protection Systems.* Boston, MA: NFPA.

———. 2002. NFPA 291, *Recommended Practice for Fire Flow Testing and Marking of Hydrants.* Boston, MA: NFPA.

———. 2007. NFPA 1142, *Standard on Water Supplies for Suburban and Rural Fire Fighting.* Boston, MA: NFPA.

Endnotes

1. National Fire Protection Association, *Recommended Practices for Flow Testing and Marking of Hydrants* (Boston, MA: NFPA, 2010).

2. Harry E. Hickey, *Hydraulics for Fire Protection* (Boston, MA: NFPA, 1980), 51.

3. National Fire Protection Association, *Recommended Practices for Flow Testing and Marking of Hydrants* (Boston, MA: NFPA, 2010.

4. National Fire Protection Association, *Standard for the Installation of Private Fire Service Mains and Their Appurtenances* (Boston, MA: NFPA, 2010).

5. National Fire Protection Association, *Standard on Water Supplies for Suburban and Rural Firefighting* (Boston, MA: NFPA, 2007).

6. United States Department of Interior, *Planning for Water Supply and Distribution—Operation Water* (Washington, DC: USFA), 15.

7. Ibid.

8. Ibid.

9. National Fire Protection Association, *NFPA 1142 Standard on Water Supply for Suburban and Rural Firefighting* (Quincy, MA: NFPA, 30.

10. John R. Freeman, "The Arrangement of Hydrants and Water Pipes for the Protection of a City Against Fire," *Journal of the New England Water Works Association* 7 (September 1892).

11. Ibid.

12. http://www.drinktap.org/consumerdnn/Home/WaterInformation/Conservation/WaterUseStatistics/tabid/85/Default.aspx

13. Harry E. Hickey, "Fire Suppression Rating Schedule Handbook," *Professional Loss Control Education Foundation* (1993), 180.

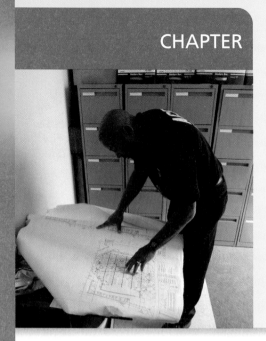

Courtesy of J. M. Foley

KEY TERMS

OBJECTIVES

After reading this chapter, the reader should be able to:

- Understand the plan review process.
- Understand the application and review of site plans using local planning and zoning boards.
- Know the fire protection items to review on site plans that affect community fire protection.
- Understand the application and plan submission process including the types of drawings that must be submitted for review.
- Know the key fire protection elements that the reviewer needs to examine in fire alarms, automatic sprinklers, standpipes, and other types of fire suppression systems.
- Know the plan review of passive fire protection features and identification of key inspection elements.

Professional Levels of Job Performance for Fire Inspectors as Cited in NFPA 1031 and NFPA 1037

- NFPA 1031 Fire Inspector I *Obj. 4.3.11 Inspect emergency access for site*
- NFPA 1031 Fire Inspector II *Obj. 5.3.11 Verify compliance with construction documents*

- NFPA 1031 Fire Inspector II *Obj. 5.3.12 Code compliance of heating, ventilating, and air conditioning*
- NFPA 1031 Fire Inspector II *Obj. 5.4.4 Review the installation of fire protection systems*
- NFPA 1031 Fire Inspector II *Obj. 5.4.6 Verify the construction type of a building and portions thereof*
- NFPA 1031 Plan Reviewer I *Obj. 7.2.2 Facilitate the resolution of deficiencies identified during plan review*
- NFPA 1031 Plan Reviewer I *Obj. 7.2.3 Process plan review documents*
- NFPA 1031 Plan Reviewer I *Obj. 7.2.4 Determine applicable codes and standards*
- NFPA 1031 Plan Reviewer I *Obj. 7.3.1 Determine requirements for fire protection or life safety equipment*
- NFPA 1031 Plan Reviewer I *Obj. 7.3.2 Determine classification of the occupancy*
- NFPA 1031 Plan Reviewer I *Obj. 7.3.7 Evaluate code compliance of emergency vehicle access*
- NFPA 1031 Plan Reviewer I *Obj. 7.3.8 Recommend policies and procedures for the delivery of plan review services*

Introduction

It is important to ensure that new construction and buildings undergoing renovations comply with the fire safety requirements of the building and fire code. Municipal, state, and federal code authorities may enforce the building codes depending on the type of industry or building under construction. Many construction projects may be reserved to specific state or federal inspection agencies because of the nature of the industry and its impact on the surrounding population. For example, to build a nuclear power plant, all regulations must go through the nuclear regulatory agency, as nuclear generating plants are an interstate utility. Large entertainment complexes or industrial plants such as water treatment facilities, refineries, hospitals, or sports stadiums may require state plan reviews and approvals. In most areas of the country, commercial and residential building plan reviews are conducted at the local level of government and involve a review by the fire inspector. The preparation of building plans may include many design professionals, including architects, engineers, and fire protection specialists as part of the approval process. Most jurisdictions have a fire inspector assigned or detailed to the local building department to ensure that the needs of the fire department are considered during the **plan review** process. A fire inspector relegated to this duty requires additional training in the building code and may also require additional licensing or certifications under local or state building code legislation. In some states, the fire inspector must hold licenses for specific types of structures in order to review the fire protection plans. These licensing levels generally are determined by the complexity of the building. The lowest license is residential commercial specialist, or RCS. Generally, fire inspectors with these licenses can conduct plan review on one- and two-family dwellings and small, light commercial buildings. The next license level is the industrial commercial specialist, or ICS. The fire inspector at the ICS level can review plans and conduct fire protection inspections in all of the RCS structures

plan review ■ The systematic review of building and equipment plans and specifications to verify compliance with the building code.

and light- to middle-sized industrial and commercial structures that usually are below 75 feet in height and do not contain any high-hazard occupancies. The highest license level the fire inspector can achieve is high hazard specialist, or HHS fire protection inspector. The fire inspector at the high hazard level can review and conduct inspections in high-rise buildings and more technically complex industrial structures containing high-hazard occupancies. A fire inspector may also be required to have a specific amount of field experience and specific college-level training in building construction and code application to attain this license. It should also be pointed out that in many states, fire inspectors may also be required to pass a national certification competency exam on the building code. These examinations are provided by code-writing agencies such as NFPA or the International Code Council. A fire inspector who is assigned administrative duties under the building code may also require an administrative license to perform that function. In some states, this is referred to as a **fire protection subcode official** license, and their administrative duties are needed for plan approvals. Generally, states requiring licenses also require the fire inspector to maintain the license by attending additional continuing education training over the licensing period. In some jurisdictions, design or fire protection professionals may be hired by the building or fire department to conduct the fire protection plan reviews. These professionals may perform the review and may direct fire inspectors in the testing and inspection aspects of final approval and commissioning of fire protection systems.

The fire inspector is only one-fourth of the team necessary to review and approve building plans. The building codes are intended to prevent the many perils in the construction of a building from fire, natural, disasters, gravity, and weather events and from other building systems that have safety issues, such as electrical and plumbing systems. The building codes also require building, electrical, plumbing, mechanical, and special inspectors to review specific portions of the plans related to their field of expertise. Special inspectors may review elevator or escalator operation and installation and smoke control systems.

While the building codes generally address natural hazards such as fire, snow, hurricanes, earthquakes, and floods, they generally do not address technological disasters such as terrorism, bombings, or transportation accidents. After the World Trade Center disaster on September 11, 2001, the National Institute of Standards and Technology made many suggestions to improve building safety from such technological events. The new editions of the building codes are beginning to address these concerns by improving the application of spray on fireproofing, hardening the egress systems, using elevators as part of the building evacuation process, and changing fire alarms to mass evacuation systems. The fire inspector conducting plan review must stay abreast of these changes in the building codes.

Fire inspectors also have the responsibility of keeping the building construction site safe from fire hazards brought to the site. Fire inspectors must ensure that fire safety permits are obtained for fuels, welding, cutting, explosives, and other materials regulated by the fire prevention code and that these materials are properly managed on the construction site.

Fire inspectors must have an understanding of the building's use and occupancy as well as the type of construction-required fire-resistance rating and the requirements for means of egress, fire alarm, and automatic sprinkler system design. Fire inspectors may also assist building inspectors in inspection of ventilation systems and smoke control management systems, elevator recall operations, and special electrical systems, such as emergency generators, as well as in many other facets of system inspection and testing. While compliance with the building and fire codes is ultimately the responsibility of the design professional, the fire inspector's role in plan review is to ensure that all of the minimum building and fire code requirements related to fire protection have been adequately addressed in the building design and that all critical fire protection systems are properly tested before occupancy of the building.

fire protection subcode official ■ The administrator of the fire code component of the building code. The fire subcode official oversees fire protection plans review, conducts inspections, and issues certificates of acceptance on fire protection systems.

The Plan Review Process

A major portion of all building codes is oriented to both passive and active fire protection. The plan review process generally defines the plan review and inspection responsibilities for each inspection discipline. The plan review process begins with the application for a building permit and the submission of signed and sealed plans from a design professional. The application process is defined in the building code's administrative chapter. The building code includes provisions for minor work or building repairs that do not require the submission of signed and sealed plans. Architects and engineers, however, must provide signed and sealed plans for large renovations, new construction, or an increase in an existing building's size or height requirements. The building permit application generally specifies the minimum number of plan sets that must be submitted for permit review purposes. Usually, at least three or four sets are required for permit application approval.

The building code may identify the appropriate design professionals who possess knowledge of the particular system design, and the design professionals must prepare and seal the plans with their professional license. This is why fire inspectors may be required to meet with multiple design groups on large construction projects. Generally, architects prepare the building design, structural, and foundation components and plans, while professional engineers provide the plans for electrical, fire protection, fire alarm, and HVAC systems. Some states may require **fire protection engineers** to prepare all of the plans for fire protection, including the automatic fire alarm and all fire suppression equipment. The municipal construction official usually provides guidance to the designers on the necessary information that must be submitted with the application and plans to ensure a proper plan review by all agencies.

fire protection engineer ■ A professional engineering degree in fire protection engineering science that specializes in the design, analysis, and testing of fire protection systems.

The plan review process is generally time-driven from the submission of the building permit application to the denying or approval of the plans. It is important for the fire inspector to understand and adhere to these time constraints. Most building codes establish the review process at 20–30 days maximum, at which time the plans must either be approved or denied. The building code identifies each area of joint plan review and areas where a single agency may perform the review. For example, the means of egress is usually a joint responsibility of both the building and fire inspectors, and all comments on egress should be coordinated between the building and fire inspectors. The installation of the automatic sprinkler system falls strictly under the fire inspector's review, and he or she would have approval responsibility. The building department establishes a set of plan review forms that identify the comments of each plan reviewer and the applicable building code violations that need to be addressed. In some cases, comments may be just for additional information from the designer to ensure code compliance such as the referencing of an applicable standard or a UL design number. The review of plans for new construction generally begins with a planning and zoning application by the owner to ensure proper land use and zoning approval. The fire inspector usually sees these plans before the building permit application. The planning and zoning board will approve a site plan that is included in the building plan submission for a permit. Many states may have additional prior approvals that must be satisfied before the application for a building permit is accepted, and these must be included in the permit application for the plan review to be completed.

Application of Planning and Zoning Regulations

Planning and zoning regulations are established in local jurisdictions to regulate community growth and development. Communities are divided into specific land use categories that determine what type of buildings can be constructed in that particular area of the community. These zoning categories usually include low-, medium-, and high-density residential housing use, commercial or business use, industrial or factory use, open spaces, agricultural farms, and other types of growth-restricted areas such as watersheds or green acres preservation areas. Only buildings of the permitted use may be constructed within each specific designated zone. The purpose of zoning regulation is to control development and prevent situations where a factory is located within a residential neighborhood. These situations have happened in the past when there were few or no zoning restrictions on building construction. As an example, in many older cities like Philadelphia or New York, commercial and industrial uses sometimes are mixed into the residential neighborhoods. Consequently, many serious conflagrations in industrial buildings have caused the evacuations of surrounding residences.

In most parts of the country, anyone wishing to construct a new building or altering an existing one must file a zoning application with the zoning and planning board for approval. A locally appointed zoning and planning board then reviews the proposal and makes comments or grants approval for the structure building use or may grant a variance to a specific zoning requirement. Frequently, these applications must also go before local government-elected bodies for additional approval. The zoning and planning board's approval extends only to the permitted use of the building in the particular zone and is considered a preapproval for a building permit. The **zoning and planning board** may also issue variances from certain zoning requirements or restrictions such as a building's height or the setbacks of the building from the interior property lines. The zoning and planning board usually has a member of the fire department on the board or at least will forward review copies of proposed plans and variances to the fire chief or fire official for comment. This is an opportunity for the fire inspector to raise any concerns related to water supply, site access, road improvements, or other fire safety–related issues identified in the fire prevention code such as outside control areas for hazardous materials or location of additional fire hydrants.

SITE PLANS

The **site plan** is a two-dimensional drawing of the building located on the proposed building lot. Site plans usually include a locator map that shows the location of the tax map block and lot within the jurisdiction as well as street or cross street locations of the property. This information is critical to the fire inspector to ascertain proper site location to determine fire department access and the available water supply. The site plan details show the relationship of the building to the street frontage and the interior lot lines of any adjoining properties (see Figure 14.1). These lot line distances are critical in determining the permissible height and area as well as the type of construction that must be used for the building. The distance to lot lines also establishes the fire-resistance ratings of the building's exterior walls. Generally, if the building is located 30 feet or less from the interior lot lines, a fire-resistance rating or firewalls are required by the building code.

VEHICLE ACCESS

Another important factor in community fire protection is to ensure that the fire department has vehicle access to fight a fire. The building codes and fire prevention codes usually establish the minimum size of a fire lane. In most codes, a fire lane must be at least 20 feet in width. Under the building code, fire lane access may allow an increase in the building's area or height based on access to open perimeter around the building. While the fire code establishes a fire lane's width, other local zoning codes may also address road access within the community. As an example, in San Jose, California, the zoning code requires roads to extend to within 150 feet of a building, and any roads longer than 1,000 feet in length must have two access points and cannot be dead-ended because of the potential for wild land fires.

Santa Clara, California, establishes the minimum width of a street at 36 feet if cars are parked on both sides of a street and 28 feet if they park on one side, and no parking is

---

zoning and planning board ▪ An appointed or elected group of citizens, professionals, and government officials who review proposed constructions for conformance with local community planning and zoning laws and regulations.

site plan ▪ The site plan is a diagram of the building's position in relation to property lines. The site plans show access and inner structure details, including utility and water-line locations and elevation changes.

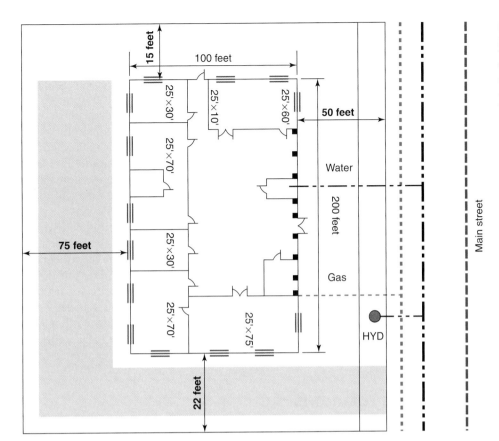

permitted on streets less than 28 feet in width. Santa Clara also has additional requirements for aerial ladder access in certain parts of the community. The fire inspector should be aware of the size of local fire apparatus and the required clearances necessary for access or turning. As a general consideration, fire engines are generally between 8 feet 6 inches and 9 feet 6 inches in width. When considering an aerial apparatus, an additional 6 feet must be provided for outrigger placement. If fire apparatus must pass each other on the roadway, at least 26 feet in width is required and an additional 8 feet should be provided next to fire hydrants to allow vehicle passage.

Another access issue is the dead-end streets or **cul de sacs**, which must have sufficient radius for all fire apparatus turning. The fire inspector must ensure that the fire department apparatus has an adequate amount of area necessary to make safe turns. The average fire apparatus turning radius is 20 feet inside and 40 feet outside, but this may not be enough for certain aerial tower apparatus with long wheelbases. Fire inspectors should consult with apparatus manufacturers to learn the correct turning radius of their equipment. Another concern that the fire inspector must consider is the placement of any architectural features or utility on the curb lines that can obstruct fire apparatus turning. All single-access roadways should require a "T" at the end so apparatus may turn around safely. It is unsafe to back up a large fire apparatus over a long distance, and there are many cases of firefighters that have been killed by apparatus backing up. Dead-end roads should be avoided, if at all possible. The fire inspector should also look for any overhead obstructions during the site plan review. The average fire apparatus height clearance is 13 feet 6 inches to 14 feet. In many cities, it is common for walkways and buildings to be built across city streets. The fire inspector should look for such obstructions and ensure that the height clearances meet the acceptable height of fire department equipment.

cul de sac ■ A cul de sac is a circular turnaround at the end of a dead-end street. Cul de sacs must have a sufficient turning radius for fire apparatus.

GRADES AND VEHICLE WEIGHT

The road grade and construction of the road surface also are important considerations to fire department vehicle access that should be noted in the fire inspector's site plan review. An aerial apparatus can weigh up to 60,000 pounds, and a pumper can weigh 45,000 pounds. Road surfaces must be capable of supporting at least 75,000 pounds of gross vehicle weight for fire department equipment. Roads that are not adequately constructed can lead to the sinking of fire apparatus or the collapse of aerial apparatus during fire operations. The road's grade or incline also must be considered in the review. Fire apparatus on asphalt roads can take about an 8 to 10 percent incline in grade. Fire apparatus may be capable of steeper grades up to a 15 percent incline but requires concrete instead of asphalt for more support and vehicle traction. Grades above the 15 percent incline will require specially designed fire apparatus. As an example, in San Francisco, which can have 25 percent street grade inclines, the fire apparatus must be equipped with special transmissions and engines. These engines are also constructed to be lighter in overall weight to negotiate the steep climbs. The fire inspector must examine the topography and grade on all access driveways to the construction site, especially in areas at flood plain. These grade inclines may be too short or too steep, which can cause fire apparatus to bottom out or scrape the tailboards, damaging the undercarriage of the vehicle. As a recent example, President Obama's armored limousine ran into such a problem in England when it became stuck on a security speed bump upon leaving Buckingham Palace. Any sharp or steep incline at an entrance or exit of a property must be carefully examined for proper ground clearance.

FIRE HYDRANT LOCATIONS

The locations of fire hydrants as well as other utilities in the surrounding streets are usually labeled on site plans. Typically, fire hydrants may be marked to be installed, removed, or relocated at the building site. The fire hydrant spacing may vary based on the specific fire demand zone where the building is located (residential or commercial zone). Typically, fire hydrants should be no more than 500 to 800 feet apart in residential areas and 300 to 500 feet apart in commercial areas.

In industrial areas because of the higher potential for fire, the fire hydrants should be no more than 300 feet apart. The Insurance Service Organization, or ISO, establishes a maximum fire hydrant distance of 1,000 feet with a fire flow credit of 250 gpm for the hydrant.[1] The 1,000-foot distance assumes a hose load on 1,200 feet of 2-1/2-inch or large-diameter fire hose. To be credited at a fire flow of 1,000 gpm, which is typically needed in industrial areas, the fire hydrant must be within 300 feet of the building and must be capable of the 1,000 gpm fire flow. Often, fire hydrant spacing is a local issue, and spacing rules do vary across the country. Fire inspectors need to check their local jurisdiction on fire hydrant spacing requirements. Zoning requirements may also establish rules on the installation and spacing of private fire hydrant systems per NFPA 24, *Standard for the Installation of Private Fire Service Mains and Appurtenances.*

STANDPIPE AND AUTOMATIC SPRINKLERS

The installation of fire suppression systems must also be examined as part of the site plan submission if the building requires their installation. The fire inspector should identify the location of the water supply for the building site and the location of the fire department connection to ensure that these are in accessible locations. While these systems are subject to a more stringent plan review in the building permit process, the general location and accessibility of the connections for fire apparatus should be identified in the site plan. Keep in mind that the location of the fire department connections is where the fire department—not the designer or fire protection contractor—chooses. It is also important to identify the water main size available for the fire protection systems. The fire inspector may have to request changes in water supply connections to accommodate proper fire flows for sprinkler systems and fire department outside hose allowances. It can be embarrassing to approve a site plan and find out later that insufficient water exists at the site.

If the fire marshal, fire official, or fire inspector is not an active participant in the zoning review process, critical elements of community fire protection can be overlooked. It is difficult to rectify these deficient conditions once the building is completed.

Understanding Fire Protection Drawings

The fire inspector must have the ability to read and understand building and fire protection plans. Fire inspectors delegated to conduct plan review should take a course in blueprint reading or should work with a more experienced inspector who has performed plan reviews to learn how to properly examine plans. Let's examine the basic skills and the process of conducting the plan review.

THE PERMIT APPLICATION PROCESS

Every municipality has a building permit application and plan review process for contractors to follow when applying for a construction permit. Often, these processes are established by state regulations, which specify the information that must be included in the application. The application process identifies the time elements for the review and how any disputes will be settled in the process. The process identifies prior approvals such as zoning, environmental protection, flood plain, or other state or federal permits approvals necessary to build or operate the structure. For example, a local building department could not approve a building permit for the construction of a water treatment plant without prior approvals from other state regulatory agencies such as the Department of Environmental Protection. In fact, the local department most likely will not have jurisdiction at all. The permit application process specifies the type of drawings that must be submitted and how many sets are required. The building department must review the plans and either approve or deny the permit application within the time frame required by the statute. In many cases, a partial plan release may be made while other detailed plans are still being reviewed. These partial plan releases may be for erection of structural steel, pouring foundations, or utility improvements that must be made before building construction. When a construction official issues a partial release of any plans, the plans are usually subject to any corrections that may be required when the complete set of drawings is reviewed by the appropriate discipline. This partial release policy allows the contractor to proceed with construction at some risk for future changes that may come up during complete review.

Fire inspectors may have sole or joint plan review responsibility for specific building code review areas. The fire inspector may have joint plan review responsibilities with the building and/or the electrical inspectors. The fire inspector usually has sole responsibility for review of the automatic sprinkler and standpipe system and fire alarm system and may have joint responsibility with the building or electrical inspectors for review of fire-resistance ratings, means of egress, exit and emergency lighting, elevator operation, and installation of HVAC mechanical systems. The fire inspector must understand the scope of responsibility assigned to him or her and must work cooperatively with other officials in the examination of the plans. Fire, building, electrical, and plumbing inspectors usually employ some uniform plan review report or checklist system where each official signs off on what he or she has reviewed and attaches any correction comments, lists, code deviations, or discrepancies to the application for correction and/or resubmission of the plans. The fire inspector may request a conference with the designer or the building owner to discuss any code deficiencies or variations identified during review. Code conferences are helpful sometimes to review design concepts on systems operations or the applicable

NFPA standard requirements. The **construction code official** will gather all of the plan review reports and comments and decide whether the permit should be approved, partially approved, or denied, or whether the plans require a resubmission. The construction official then will contact the owner or design professional to pick up the plans and will discuss the needed corrections required before resubmitting them for further review and final approval. In most cases, the construction official will permit building construction to begin with the agreement that the designer will address all of the comments identified in the plan review. If the designer or contractor fails to comply, the construction official generally has authority to stop construction until the proper compliance is achieved. If there are disputes, they generally are resolved in a construction board of appeals hearing similar to the appeals board for fire code violations.

Architectural Drawings and Specifications

Architectural drawings tell the story of how a million pieces of construction materials will be assembled into a useful building. In order to do this, the plans and specifications must be organized in a fashion so that every building trade can work on its piece of the puzzle, while not having to review all of the construction documents in their totality. As you can imagine, the coordination of large complex buildings takes tremendous cooperation and control by professional designers, construction managers, and construction workers from the drawing board to the ribbon cutting.

In small buildings, architectural drawings may be limited to foundation, floor plan, and electrical, plumbing, and mechanical plans. For the purpose of plan review by the fire inspector, these reviews are relatively simple, and all specifications generally are identified directly on the drawings. In large buildings, the plans are organized much differently using the Construction Specification Institute's Master Specification document. This breaks construction documents down into 16 divisions of material and their associated subsystems, similar to book chapters. The fire inspector will have to review the plans and check the specification book to identify important components for each system. The Construction Specification Institute's Master Specification identifies the following construction elements:

Division 01 — General Requirements
Division 02 — Site Construction
Division 03 — Concrete
Division 04 — Masonry
Division 05 — Metals
Division 06 — Wood and Plastics
Division 07 — Thermal and Moisture Protection
Division 08 — Doors and Windows
Division 09 — Finishes
Division 10 — Specialties
Division 11 — Equipment
Division 12 — Furnishings
Division 13 — Special Construction
Division 14 — Conveying Systems
Division 15 — Mechanical
Division 16 — Electrical

The fire inspector must examine the applicable specification for each element of construction that they review. For example, automatic sprinkler systems and fire alarms are listed under Division 13, Special Construction, but the fire inspector may also have to review sprinkler specifications in Division 4 for dry pipe valve enclosures or Division 2 for the installation of the underground piping. Understanding where the system specification is located can assist the fire inspector in determining building code compliance.[2]

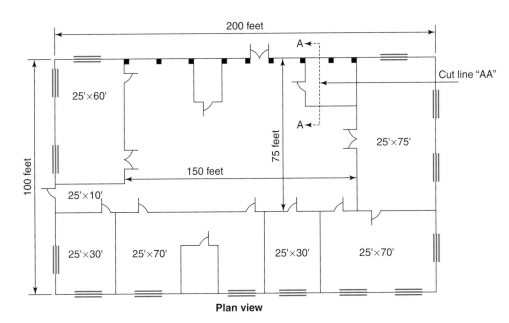

FIGURE 14.2 The plan view drawing represents a two-dimensional aspect of the building and details interior wall and door locations as well as building dimensions.

Blueprint reading can be difficult unless you understand the basic format used by architects and engineers in preparing building documents. The blueprints are a two-dimensional representation of a three-dimensional object; therefore, the third dimension of height will be found in different areas of the drawings. Architectural drawings will consist of four drawing elements or views: the plan view, the elevation views, the section views, and the detail views. The fire inspector must be capable of identifying each view and how it applies to the three-dimensional building component. Let's examine each type of drawing and the information it provides to the fire inspector.

PLAN VIEW DRAWINGS

The plan view drawing is a two-dimensional drawing that represents the length and width of the building. The fire inspector can think of this view as being above the building and looking downward on it without the roof in place. The plan view provides all of the interior room dimensions and partition wall locations and also shows where doors and window openings are located (see Figure 14.2). In most plans, multiple plan view drawings will be provided for different building systems such as electrical, plumbing, or mechanical systems. The fire inspector may find the fire alarm equipment on the electrical plan, or it may be a separate fire alarm system plan. The same holds true for automatic sprinklers, which may be found on a reflected ceiling plan, a plumbing plan, or a separate fire protection system plan. The complexity of the building usually is what decides where these systems are drawn. As an example, a single-family home could be a 5-page drawing or a 25-page drawing depending on the building's size and complexity. The plan view drawing also will identify the elevation view drawings by the application of a symbol indicating the elevation drawing number and the viewing direction (see Figure 14.3). Additionally, sometimes the building floor plan may not be able to be drawn on a single drawing sheet. In these cases there will be match lines provided. The match lines usually are identified by column numbers or letters that help identify to the fire inspector where drawings join together. The plan view drawings

Architectural Symbols

Elevation 2
On page A-4 the arrow shows direction of view

Detail 3 on page A-5

FIGURE 14.3 Symbols, such as those pictured, direct the plan reviewer to the third height dimension of the building and associated design details of walls. The symbol shows the number of the elevation or detail, the arrow indicates viewing direction, and the lower number indicates the page location in the plan set.

FIGURE 14.4 NFPA 170 provides standardized symbols for architectural, electrical, and fire protection drawings. Reproduced with permission from NFPA 170-2012, *Standard for Fire Safety and Emergency Symbols*

Common NFPA 170 Symbols
Fire-resistant walls
S
Smoke barrier
1 hour wall
2 hour wall
3 hour wall
½ hour wall
¾ hour wall
Fire alarms
Horn
Horn/Strobe
P
Smoke detector
Photoelectric
F
Heat detector
Fixed temperature
Duct smoke
detector
Fire sprinklers
Upright sprinkler
Upright sprinkler
Nipple up
Pendant sprinkler
Pendant sprinkler
Nipple down
Sidewall sprinkler

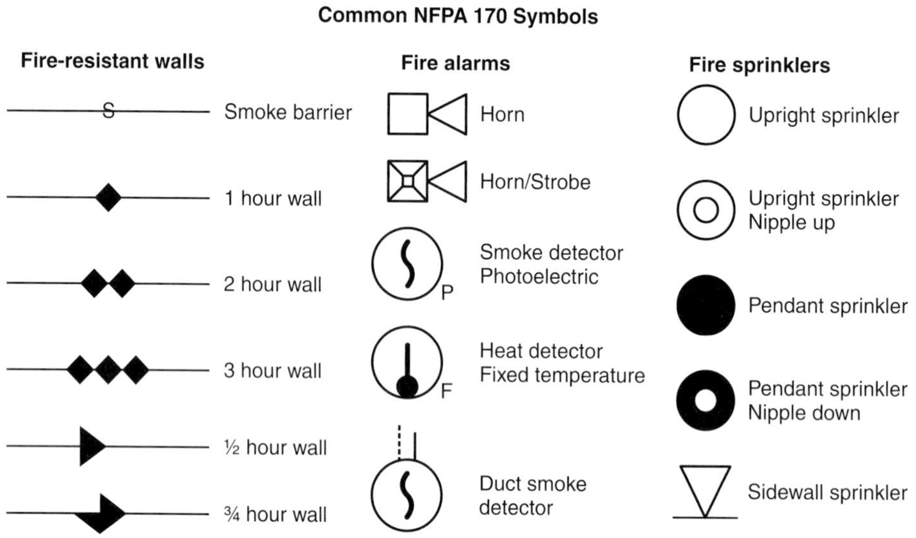

may also have "cut lines" through the plan that identify a sectional view of the building's interior (see Figure 14.2). The cut lines normally have an arrow at each end of the cut to show the direction of the view for the section drawing. Sectional drawings may also have arrows that identify specific construction details on a section component. Details usually demonstrate how a wall or connection is to be assembled. The use of these specific drawings and symbols allows the plan reviewer to identify the location of the third dimension of height as well as any critical assembly and construction details and demonstrates where they are located in the drawings (see Figure 14.3). Typical plan view drawings show the rooms and spaces, with overlay drawings for electrical wiring, plumbing fixtures, reflected ceilings, fire alarm, sprinklers, and mechanical HVAC systems. Each drawing is provided with a symbol legend describing the types of symbols the designer has used to represent different building components on the drawing. NFPA 170, *Standard for Fire Safety and Emergency Symbols* provides standardized symbols for all fire protection and fire alarm systems and often is used by designers in preparing their drawings for plan review (see Figure 14.4).

ELEVATION VIEW DRAWINGS

The elevation drawing represents a view of the exterior sides of the building or interior wall sections as viewed from ground level facing the wall. Elevation drawings provide the height dimension and generally illustrate the finish materials of the exterior or interior walls. The elevations show the egress openings, duct and vent heights, interior ceiling heights, and exterior elevation changes of the building. Elevation drawings may be exterior street views of the structure or cuts of specific parts of the building. A directional arrow usually represents a building cut line with a letter set such as "A-A" as shown in Figure 14.5. The cut lines will also identify the drawing page and location so the fire inspector can find the correct drawing.

DETAIL VIEW DRAWINGS

The detail drawing is a sectional cutaway designated by a symbol through either the plan view or the elevation view drawings. The symbol identifying the detail page and detail number for the fire inspector is shown in Figure 14.5. The actual detail shows the building system connection methods and the fire-resistance rating of the particular wall or ceiling assembly (see Figure 14.6). Details identify the UL or FM design number for fire-resistant

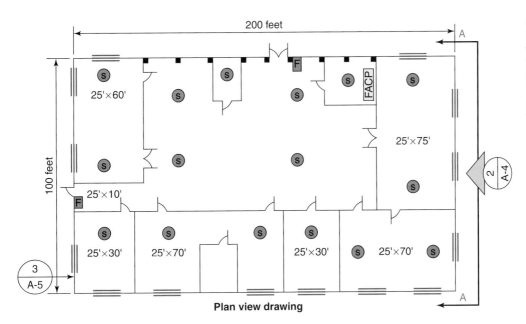

Plan view drawing

FIGURE 14.5 The cut line at location "A-A" would direct the reviewer to the elevation showing the interior side of the 100-foot wall and corridor partitions.

assembly or the design calculation for cement masonry unit walls. The detail also lists the components that are to be used in the assembly.

NOTES

The architect or engineer often places note flags on the plans to identify additional information about specific aspects of the drawings. Notes include the edition of the building code that was used, specifications on the sprinkler or fire alarm design, hourly fire-resistance rating designs, and other useful information to clarify the drawings. A note usually is symbolized by a small triangle with a corresponding number inside for the specific note. This triangle is placed on the appropriate drawing corresponding to the note. The architectural symbols that each designer uses will be located in the symbol legend. On plan revisions, the architect often places a note on the revision and draws a cloud around the revised components to identify the change on the drawing to the fire inspector or contractor constructing the building.

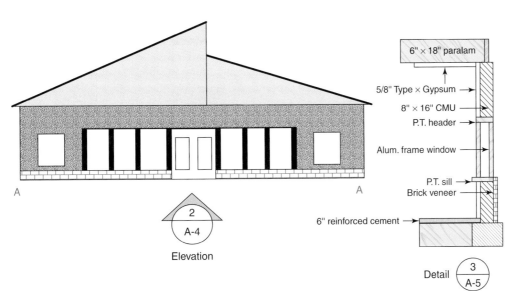

FIGURE 14.6 The elevation drawing demonstrates the exterior appearance of the structure, and the detail drawing demonstrates the assembly of a particular building component.

DETAIL SCHEDULES

The professional designer also provides a detail schedule for all doors, windows, fixtures, and interior finishes as part of the plans. The detail schedule identifies the door style, fire rating, opening size, hardware, and the finishes of doors, windows, and other elements of the construction. The fire inspector can use the detail schedule to identify the different symbols on the plan and then can go to the schedule to obtain any specific detail for the construction element.

Electrical Drawings

The electrical system drawings may be illustrated on the architectural plan view drawing for small buildings or may be a separate plan overlay for larger buildings. The review of electrical plans is a shared review responsibility between the building, electrical, and fire inspectors. The building and fire inspectors' role in the electrical plan review is to identify the proper locations of emergency lighting units and exit signs (see Figure 14.7). The fire alarm and sprinkler system also has electrical system connections that should be referenced on this electrical plan. The fire alarm system's connection to the building's electrical power supply should be identified on the circuit box panel diagram and should have its own circuit breaker. The fire inspector should note the wiring method for the system to determine whether it is a power-limited or non-power-limited fire alarm system.

EXIT AND EMERGENCY LIGHTING

Egress and emergency lighting systems must be connected to both the primary and emergency power supplies. These emergency power supplies may be a storage battery or a connection to an emergency electrical generator. Emergency lighting is required in all paths of exit travel when the building is occupied. The emergency lighting system must provide at least one foot-candle of light along the path of exit travel and must be independent of any dwelling unit electrical systems. In the lighting plan, ceiling fixtures should be identified as connected to the emergency electrical circuits. The fire inspector also should note whether or not the exit discharges have adequate emergency lighting. Battery-operated emergency lighting units must have sufficient battery power to remain operational for 90 minutes. All rooms and spaces that have multiple exits also require emergency lighting

FIGURE 14.7 The electrical plan will show the locations of emergency lighting, exit signs, and associated fire alarm and sprinkler system electrician connections.

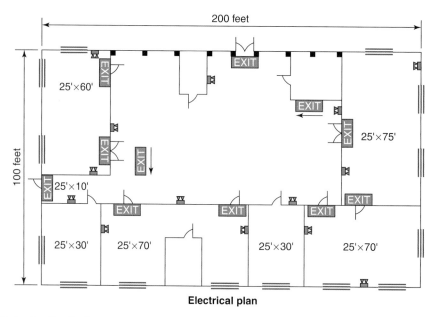

Electrical plan

to illuminate both paths of egress. The building code also requires that all exit signs be connected to the emergency electrical system. Electrical exit signs are listed to UL 924 *Emergency Lighting and Power Equipment*, and should be identified in the electrical equipment schedule even if they are the self-luminous signs. The International Building Code and NFPA also require low-level, self-luminous exit marking systems in high-rise stairway applications. These low-level lighting systems must comply with the requirements of UL 1994, *Luminous Egress Path Marking Systems*, for placement and photoluminous capability. All elements of egress will also require normal ambient lighting at all times that the building is occupied; this is especially important in buildings with self-luminous signs that gather their energy from the ambient lighting.

FIRE ALARM SYSTEMS

Fire alarm systems may appear on the electrical plan for small buildings, or they may have a separate fire alarm drawing for larger buildings. The fire inspector should first determine whether the system is power-limited or non-power-limited. Power-limited systems are 12–24 volts and are the predominant fire alarms systems employed today. Non-power-limited fire alarm systems are 110-volt systems used predominantly in single-family dwellings and small apartment buildings. Non-power-limited systems use single-station smoke alarms listed under UL 217, *Single- and Multiple-Station Smoke Alarms*, that have interconnected audible circuits so that activation of one smoke alarm will sound all of the interconnected smoke alarms on the system. Non-power-limited smoke alarm systems are usually limited to no more than 12 interconnected smoke alarms according to their UL listing. Power-limited fire alarms generally are used in larger buildings and are low-voltage systems that use UL 268-listed smoke detectors and UL 521-listed heat detectors connected to a UL 864-listed fire alarm control panel. The fire inspector needs to identify the type of system power in order to identify the correct listing for the equipment being installed. The fire inspector should request the fire alarm specification and all of the manufacturer's equipment specification sheets for review. The examination of the fire alarm plans begins with the fire inspector identifying the location of each alarm-initiating device and determining whether it is within its correct spacing. To make this determination, the fire inspector should refer to NFPA 72, *National Fire Alarm and Signaling Code*, spacing rules. The NFPA 72 standard requires all smoke detectors to be spaced within their "circle of protection" (see Figure 14.8), which is defined by their UL-listed spacing. Standard smoke detectors are listed at 900-square-foot spacing per smoke detector. Underwriters Laboratories has determined this spacing by a fire test. The test consists of placing a pan of burning heptane in the center of a room. Automatic sprinklers are spaced at 10 feet by 10 feet around the pan fire. The smoke detectors are placed at 30 feet by 30 feet away from the fire in the center of the room. The test requires that the smoke detectors detect the fire and activate before the sprinkler heads fuse and operate, suppressing the fire. This test established the smoke detector spacing at 30 feet by 30 feet, or 900 square feet. The NFPA 72 "circle of protection" uses the test data information in reverse; the theory is that if the smoke detector is in the middle of the square, it can detect a fire 30 feet out in all directions or can detect anything within 21 feet, which is the detector's radius. This is also referred to as the *0.7X spacing rule*.

$$0.7 \times 30 \text{ ft.} = 21\text{-foot radius}$$

The NFPA 72 standard requires the first smoke detector to be placed at half of its listed spacing from a wall, and each additional smoke detector is placed at full spacing. Typically the first smoke detector would be installed at 15 feet, and each additional detector would be 30 feet apart (see Figure 14.8). If the designer uses the 0.7R method, then the first smoke detector can be spaced at 20.5 feet from the wall and the next detector at 41 feet. This eliminates the need for one smoke detector in the corridor because the corridor

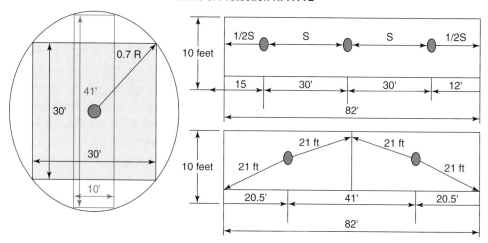

Circle of Protection NFPA 72

FIGURE 14.8 The circle of protection takes into account that smoke detectors function in 360 degrees; therefore, the rectangle that is 10' × 41' fits in the circle and can be protected by a single smoke detector. This rule allows fewer smoke detectors to protect the same area, thus reducing equipment and false alarm potential.
Reprinted with permission from NFPA 72®-2013, 2010, 1999 *National Fire Alarm and Signaling Code*

width fits within the circle of protection. The fire inspector must verify that all smoke detectors are within their listed spacing or fit inside the circle of protection by making sure they are within the 20.5-foot radius.

The fire inspector must also verify the placement of manual pull fire alarm stations at the exits and in the path of exit travel. The building codes require that manual pull stations be located within 5 feet of the exit door and should be spaced so that they are 200 feet apart. The fire inspector should review the location, spacing, and type of each manual alarm-initiating appliance. Manual pull stations should not be placed in locations where they are prone to false alarms such as elevator lobbies. The fire inspector should determine whether the manual pull stations are single-, double-, or triple-action devices by reviewing the manufacturer's specification. The specific location of the manual pull station often determines the proper type of device that should be employed based on the probability of false alarms. Manual pull stations in locations where they may be vandalized or falsely activated should have either double- or triple-action boxes. Next, the fire inspector needs to examine the locations of alarm-indicating appliances, including the horns, speakers, and visual strobe lights. The building code will identify specific speaker termination locations within buildings where speakers must be installed. Additional speakers will be required to meet the minimum acceptable sound output required by NFPA 72. Speaker placement is based on the watt output of each speaker appliance and the decibels produced at the minimum hearing distance from the speaker specified in UL 464, *Audible Signal Appliances*. Most indicating appliances are rated at 10 feet for decibel output. The fire inspector must review the manufacturer's specification for rated output based on the wire size used in the system. Typically, speakers use 18 awg wire, which has a voltage drop of approximately 6.4 volts per 1,000 feet of wire. The fire inspector must know the wire gauge and appliance voltage drop to verify the voltage drop calculations required to be submitted in the plan review process. The voltage drop is important because it determines whether all of the speakers in a circuit will operate properly. On a typical review of voltage drop, the fire inspector should make sure the starting voltage takes into account the secondary power supply requirements of 24 hours of operation and 5 minutes in alarm. The minimum starting voltage should be 21 volts, and the voltage at

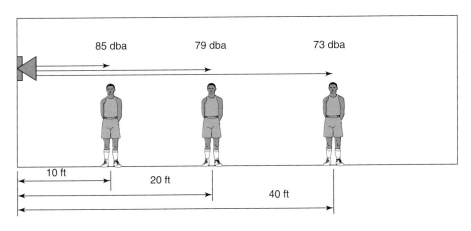

FIGURE 14.9 Sound pressure decreases inversely with distance. As the distance is doubled, the sound level decreases by 6 decibels. Minimum sound pressure is 15 dba above the ambient noise level but not less than 75 dba.

the last audible appliance should be no less than 16 volts for the system to operate properly. The calculation for voltage drop should identify the operating voltage drop of each appliance and the voltage drop in the gauge of the amount of wire used in the circuit.

In emergency mass evacuation systems, the speakers also must comply with voice intelligibility requirements. The manufacturer's specifications for the specific audible equipment provide this information in the specification sheets for the appliance spacing requirements. As a general rule, audible appliances are usually spaced at 50 feet in corridors and one speaker every four floors in stairways. This may differ slightly depending on the equipment manufacturer. Sound travels by the inverse square law, which states that about 6 decibels of sound is lost every time the distance to the speaker is doubled (see Figure 14.9). The fire inspector can use this formula in the testing phase of the fire alarm by using a sound meter to measure the audible output of the speakers at different distances.

Visual fire alarm indicators are also required to be installed in buildings to alert the hearing impaired. Visual alarm indicators are required by both the building codes and the American with Disabilities Act (ADA). Typically, visual indicators are required to be spaced so that they can be seen regardless of the viewer's orientation in a room and at a maximum spacing of 100 feet. If two visual indicators are within the field of view, they must be separated by 55 feet and shall flash in a synchronized fashion. Visual indicators may be integrated within speakers or may be of a standalone variety. The flash rates for a visual alarm must be between 1 to 3 hertz. At 1 hertz, the device produces 60 flashes per minute and at 3 hertz it will produce 180 flashes per minute. Visual alarm indicators should be installed no higher than 96 inches nor lower than 80 inches from the floor.

The architectural /electrical plans may not always provide all of the necessary information for proper location of the fire alarm system devices. The fire inspector may have to request additional specification or shop drawings for the fire alarm system. In the shop drawings, several additional drawings will be provided, including a fire alarm system riser diagram that illustrates the number of initiating and indicating appliances on a circuit and also indicates the annunciator zone circuits per floor level. The shop drawings provide a point-to-point diagram of the fire alarm control panel showing the attachment of each circuit to the fire alarm circuit board (see Figures 14.10, 14.11, and 14.12). Additionally, the fire inspector needs to review the **alarm matrix**. The matrix describes the sequence of operation of each initiating device and what actions, such as sounding an alarm, grounding the elevators, or releasing a fire door, should occur when the device is activated. The fire inspector should also review the specifications for secondary power battery calculations. It is important that standby batteries have sufficient power to keep the alarm system operational for 24 hours and at the end, be capable of sounding the alarm for at least 5 minutes. All secondary power supplies according to NFPA and UL must provide at least

alarm matrix ■ The alarm matrix describes the operation of every fire alarm-initiating appliance and describes all the functions that should occur upon its activation, including ancillary functions for systems attached to the fire alarm such as elevator recall, smoke control, door closers, and other devices.

FIGURE 14.10 Typical fire alarm system plan. Note legend of symbols for different alarm system appliances.

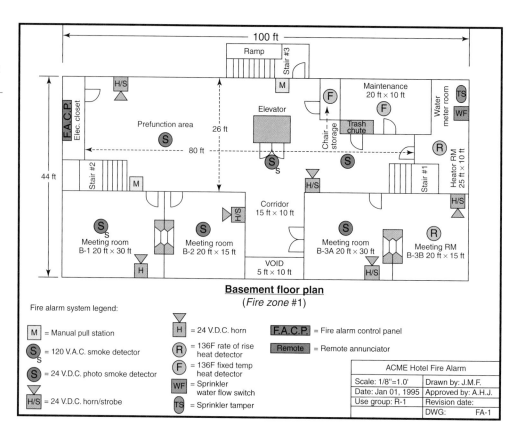

Basement floor plan
(*Fire zone #1*)

Fire alarm system legend:

M = Manual pull station

S/S = 120 V.A.C. smoke detector

S = 24 V.D.C. photo smoke detector

H/S = 24 V.D.C. horn/strobe

H = 24 V.D.C. horn

R = 136F rate of rise heat detector

F = 136F fixed temp heat detector

WF = Sprinkler water flow switch

TS = Sprinkler tamper

F.A.C.P. = Fire alarm control panel

Remote = Remote annunciator

ACME Hotel Fire Alarm	
Scale: 1/8"=1.0'	Drawn by: J.M.F.
Date: Jan 01, 1995	Approved by: A.H.J.
Use group: R-1	Revision date:
	DWG: FA-1

FIGURE 14.11 The riser diagram shows each zone circuit and the number of devices attached to it. Note that initiating circuits are on one side, and indicating appliances are on the other.

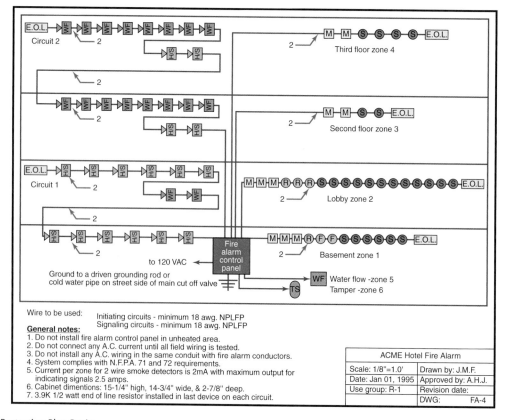

Wire to be used: Initiating circuits - minimum 18 awg. NPLFP
Signaling circuits - minimum 18 awg. NPLFP

General notes:
1. Do not install fire alarm control panel in unheated area.
2. Do not connect any A.C. current until all field wiring is tested.
3. Do not install any A.C. wiring in the same conduit with fire alarm conductors.
4. System complies with N.F.P.A. 71 and 72 requirements.
5. Current per zone for 2 wire smoke detectors is 2mA with maximum output for indicating signals 2.5 amps.
6. Cabinet dimentions: 15-1/4" high, 14-3/4" wide, & 2-7/8" deep.
7. 3.9K 1/2 watt end of line resistor installed in last device on each circuit.

ACME Hotel Fire Alarm	
Scale: 1/8"=1.0'	Drawn by: J.M.F.
Date: Jan 01, 1995	Approved by: A.H.J.
Use group: R-1	Revision date:
	DWG: FA-4

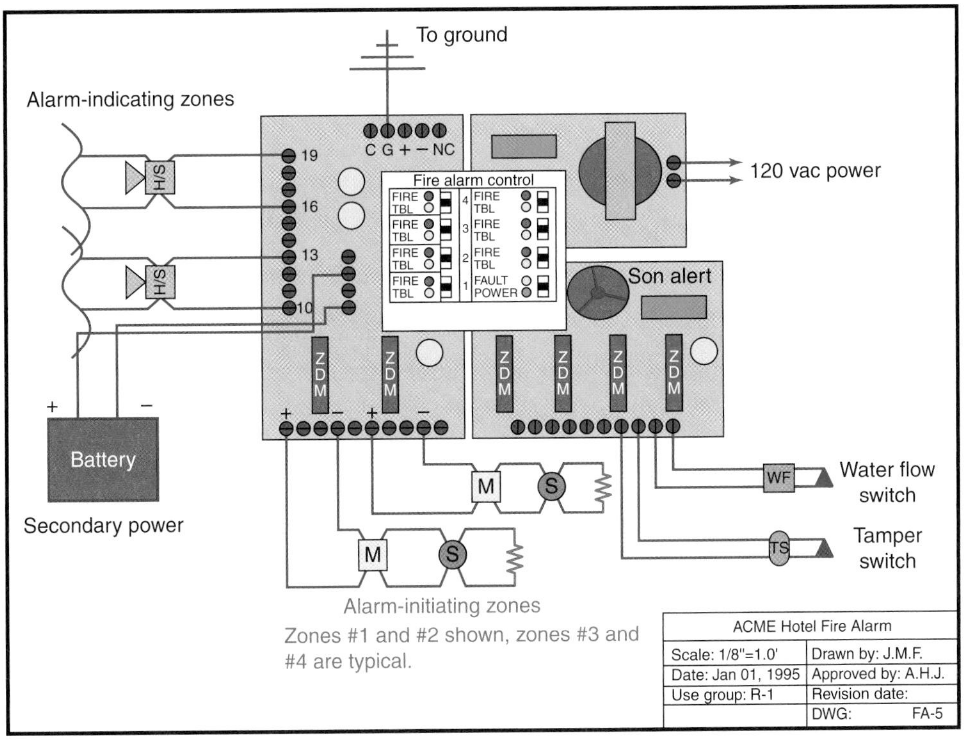

FIGURE 14.12 The point-to-point diagram illustrates each circuit connection to the fire alarm control panel.

85 percent of the power supply voltage. In a 24-volt fire alarm system, the minimum starting voltage for determining the battery calculation is 20.4 volts or 85 percent. The battery calculation should indicate the total current for standby and alarm conditions in amp/hours multiplied by 1.2, which is a safety factor to determine the minimum battery capacity in amp/hours. The fire inspector should also request a copy of the fire alarm operation manual and system warranty information that further explains maintenance and operation of the fire alarm system. This is especially important in approval of central station signaling systems.

Mechanical Drawings/Special Construction Reviews

The mechanical systems to be examined during plan review include HVAC systems, smoke control systems, stairwell pressurization systems, kitchen exhaust ducts, and appliance chimneys and vents. The HVAC system often has heat and smoke detectors installed as required by NFPA 90A, *Standard for the Installation of Air-Conditioning and Ventilating Systems*, and the ICC mechanical code. Smoke or heat detectors may be placed in the both the air returns and air supply ducts of the HVAC system. These detectors are required when HVAC systems serve multiple floor levels or provide more than 15,000 cubic feet per minute of air. These specialized heat or duct smoke detectors must shut down fans to avoid spreading products of combustion throughout the building via the HVAC system. Duct smoke detectors are usually placed near the fan supply motor and before the air filter rack in the return air duct. These detectors are generally for system control purposes and may or may not activate the fire alarm system depending on the local building code requirements.

HVAC

The heating, ventilation, and air conditioning, or HVAC, system is usually a joint plan review by both the fire inspector and the building inspector. The fire inspector's review is for both the interconnection of duct detectors to the fire alarm system as well as the locations of air supply and return diffusers that can affect area smoke detector or sprinkler locations. Smoke detectors must be located at least 3 feet from the air supply diffusers and should be located to favor the air returns in the space. The fire inspector also should determine whether the return air system is ducted or a variable air volume plenum return system. All fire alarm system wiring installed in the return air plenums must be fire-resistant for plenum application and have a Class 1 flame spread of 0–25 in an ASTM E-84 Steiner tunnel test. The fire inspector must also determine whether the building requires a smoke control system. Typically, smoke control is required in building atriums that connect two or more stories. Smoke control is also required in covered malls with atriums and underground buildings. High-rise buildings also have smoke removal systems to eliminate smoke after a fire by either natural ventilation or mechanical means. The plan review of smoke control systems is complicated at best and requires special inspection analysis by design engineers to verify code compliance beyond the normal system operational testing. Information that must be provided in the plan submission includes a complete overview of the system's operational scheme and a design manual on how the system is to be operated. When smoke control systems are installed in a building, the system design must include manual controls for the fire department's use to start or stop fans. This manual operating control panel must be located in the building's fire command center according to the building code. An operational matrix also must be provided so that the fire inspector can determine the sequence of operation for each initiating device. The operational matrix describes which fans start and stop and which mechanical dampers will open or close during system activation. The system designer shall provide the documentation on the proper number of air changes based on the cubic volume of the space to be protected. Smoke control systems are joint plan review responsibilities between both building and fire inspectors.

Another important mechanical system to be reviewed is the stairwell pressurization system. The building and fire inspector must ensure that the stairwell pressurization system is properly connected to the building's fire alarm system. During the plan review process, the fire inspector should beware of any notes calling for "connection by others" as sometimes these interconnections can be missed between different building trades. An example would be that a fire alarm speaker and a firefighter's communications connection must be placed inside each elevator car; but who makes that connection to the fire alarm panel—the elevator technician, the fire alarm technician, or the electrician? These points should be specified and defined to ensure proper installation of the system. This confusion of responsibility can occur in many systems including magnetic door closers, elevator lobby smoke detectors, or fire protection system shunt devices and smoke control supply and return air dampers.

AUTOMATIC SPRINKLERS

Automatic sprinkler system plan review also requires shop drawings from the fire protection installer's engineers. Typically, the architect may show sprinkler heads on the reflected ceiling plan and may have notes on compliance with NFPA sprinkler standards. While this architectural plan shows the installation of the sprinkler heads, it does not provide sufficient information for the fire inspector to conduct an adequate system plan review. The fire inspector needs the engineered **shop drawings** and the supporting equipment specification sheets for each device attached to the sprinkler system as well as the hydraulic calculation to determine proper system hazard design. The fire inspector must determine compliance with NFPA 13, *Standard for the Installation of Sprinkler Systems*; NFPA 13R, *Standard for the Installation of Sprinkler Systems in Low-Rise Residential Occupancies*; or NFPA 13D, *Standard for the Installation of Sprinkler Systems in One- and*

shop drawings ■ Drawings prepared for the installation of specific systems, including fire protection systems. Shop drawings provide additional detail on system assembly.

Two-Family Dwellings and Manufactured Homes, as well as other fire sprinkler standards for specialized system performance characteristics such as warehouses, high rack storage, or hazardous materials. The applicant should provide the following types of information in the plan submittal.

The fire sprinkler plans should specify which NFPA standards are being employed and should identify the use of the building. The fire inspector must know the type of ceiling being installed in the building: Is it a smooth ceiling, or will there be obstructions of beams, posts, open bar joists, or open ceiling grids? The type of ceiling alters the sprinkler system installation patterns. The plans should specify the hazard classification, the system design density, and the total design area of sprinkler operation. The water supply and flow test information also should be identified on the hydraulic calculations that are submitted with the plans. The sprinkler designer must specify the pipe schedule to be used and identify the coefficient of pipe roughness or the pipe's "C" factors. The plan drawings identify the types of sprinkler head to be used including the count of upright, sidewall, or pendant heads as well as the symbols used in the plan legend. The fire inspector should also review the specification sheets to determine sprinkler orifice size and the "K" factors and sprinkler temperature rating. The shop plans should include a riser diagram to verify control valve locations, the type and size of pipes and fittings, and the location and type of pipe hangers including the earthquake sway bracing and hanger styles. The shop drawings will also specify the type of sprinkler design such as wet pipe, dry pipe, preaction, or deluge. The riser detail should include the valve arrangement and all the ancillary connections to the fire alarm system for pressure switches or water flow and tamper switches.

SPRINKLER PLAN REVIEW

The fire inspector should examine each floor sprinkler plan sheet and determine the correctness of the following information:

1. Count all sprinkler heads and ensure that each is within the specified design coverage for the square footage of the floor area based on the hazard classification (*Light hazard, Ordinary Group 1 & 2 hazard, Extra Group 1 & 2 hazard*).
2. Identify the type of sprinkler heads and temperature requirements (*Upright sprinklers, pendant sprinklers, sidewall sprinklers, ESFR sprinklers*).
3. Identify the location of the inspector's test connections and drain lines.
4. Identify the control valves, pressure reducing valves, water flow switch, and tamper switch locations and how they are attached to the fire alarm.
5. Identify the type of sprinkler pipe and the size of risers, mains, cross mains, and branch lines.
6. Identify the remote area of operation and the method of sprinkler calculation (*area method or room design method or pipe schedule*).
7. Is the permissible fire area for the hazard group exceeded in square feet? (*Light and ordinary hazard – 52,000 sq. ft. Extra hazard pipe schedule – 25,000 sq. ft., hydraulically calculated – 40,000 sq. ft.*).
8. Do the design reference points match up with the node tags on the hydraulic calculation sheet?
9. Have elevation changes been properly identified and referenced?
10. Are the proper number of pipe hangers and earthquake braces demonstrated on the drawing?
11. Are there combustible concealed spaces requiring additional sprinkler protection?
12. Have all rooms and spaces been sprinklered as required by the building code for the specific building use group?
13. Have the lengths of sprinkler risers and drops from the branch lines been included in the calculated design?

HYDRAULIC CALCULATION REVIEW

The fire inspector will be required to verify the information on the hydraulic calculation sheets against the information provided on the plans. The fire inspector is not expected to be an engineer, although at times, that could be very helpful. The inspector's role is to ensure that the minimum requirements of the appropriate NFPA sprinkler standards have been complied with.

The examination of the hydraulic calculations for the sprinkler system can be an intimidating part of the plan review. The fire inspector must realize that his or her role is not to find mathematical errors in the calculations, but to verify that the correct design information was used. Engineers that design sprinkler systems use computer programs to calculate the hydraulics of the sprinkler systems. Usually, the computer analysis is accurate in the calculations, provided that the correct information was entered into the program. The challenge for the fire inspector is to ensure that it is not "garbage in, garbage out" by verifying the basic entry data information found on the sprinkler design summary sheet. The fire inspector should answer the following questions to determine whether the information entered was correct:

1. Is the contractor's information, building location, use, type of construction, ceiling height, and occupancy information provided on the hydraulic calculation summary sheet?
2. Has the type of system been specified including the inside and outside hose allowances? (See Figure 14.13.)
3. Are the system design and calculations summaries correct for total gpm required to the point of connection? Has any new underground pipe added from the point of connection to the water main not been calculated?
4. Is the flow test data current and correct? Does it take into account seasonal low water pressures?

FIGURE 14.13 The plan checker must review the basic system design information for correctness.

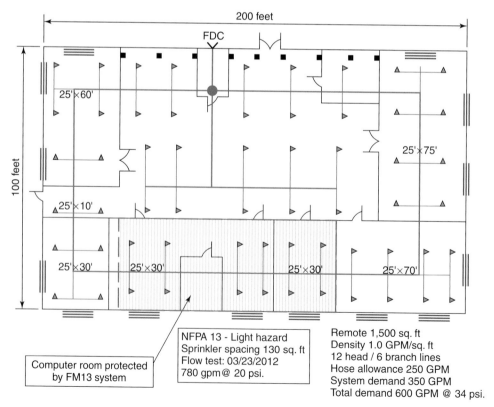

200 feet

FDC

100 feet

25'×60'

25'×75'

25'×10'

25'×30'   25'×30'   25'×30'   25'×70'

Computer room protected by FM13 system

NFPA 13 - Light hazard
Sprinkler spacing 130 sq. ft
Flow test: 03/23/2012
780 gpm @ 20 psi.

Remote 1,500 sq. ft
Density 1.0 GPM/sq. ft
12 head / 6 branch lines
Hose allowance 250 GPM
System demand 350 GPM
Total demand 600 GPM @ 34 psi.

5. Are commodities present that require application of additional NFPA sprinkler standards that affect system design such as racks, aerosols, or pallets?
6. Has the remote area of sprinkler operation been properly calculated using the following formula:
   a. Step 1: The number of sprinklers required to operate is found by dividing the total operating area by the sprinkler head coverage.
      i. *Example 1,500 sq. ft./130 sq. ft. = 11.5, or 12 sprinklers must operate*
   b. Step 2: The remote area must be 1.2 × square root of the operating area divided by the number of operating sprinklers.
      i. *Example 1.2 × Square root of 1,500 = 38.73/12 = 2.98 or 3 automatic sprinklers per line must be calculated*
   c. Step 3: Examine the most remote area of operation. Is the long leg of the rectangle parallel to the branch lines, and are there 12 sprinklers and 3 branch lines calculated?
7. If a dry pipe valve supplies the sprinkler system, has an additional 30 percent increase been added to the remote area of operation? Was the correct design density curve used for the hazard classification in accordance with NFPA 13, *Standard for the Installation of Sprinkler Systems*?
8. Are the correct sprinkler K-factors being used for the type of sprinkler head calculated in the remote area of operation?
9. Was the correct coefficient of roughness or C-factor used in calculating the friction loss in the system piping?
10. Does the calculation identify the minimum starting pressure and flow at the most remote sprinkler head:
    a. $Q = A \times D$ or Flow = Area of sprinkler coverage × density in gallons per minute.
       i. *Example: Q = 130 × 0.2 = 20 gpm at first sprinkler head.*
    b. To calculate the first head starting pressure you would use the formula pressure = (flow/sprinkler K-factor)^2
       i. *Example: P = (20 gpm/5.6)^2 = 12.75 psi*
11. Were the sprinkler pipe drops or sprigs considered in the sprinkler calculations?
    a. *Sprinkler drops are the pipes from the branchline to the finished ceiling.*
    b. *Sprigs are the pipes from the branchline up to the finished ceiling or roof.*
12. Are the sprinklers properly spaced on each branch line and are the branch lines properly spaced according to the hazard classification in NFPA 13?
13. Was the correct water demand used (*number of sprinklers × minimum flow per head plus hose allowance = total minimum flow*)?
    a. Example: 15 sprinklers × 20 gpm = 300 gpm + 250 gpm hose allowance = 550 gpm
14. Were couplings and elbows calculated into equivalent pipe lengths for friction loss?
15. Do the calculations reflect all the changes in elevation of pipe around ducts or other building obstacles?

   While this is a brief overview of the types of questions that must be asked, the purpose of the fire inspector's plan review is to determine that the correct system information was applied in the hydraulic calculation of the system and the correct water supply is available.

STANDPIPES

Standpipes are required in buildings once they exceed three stories in height. During the plan review, the fire inspector should check for the following standpipe information:

1. Has the type of standpipe been identified (Class I, II, III, or combination) by the designer?
2. Are control valves located at the base of each standpipe risers?
3. Are standpipes located in the exit stairs out of the path of egress travel?
4. Are standpipes at the proper spacing to the most remote location of the floor areas?

5. Did the designer use the proper standpipe flow calculation based on whether the building has automatic fire sprinklers:
   a. 500 gpm for initial standpipe and 250 gpm for each additional standpipe up to 1,250 gpm max in nonsprinklered buildings?
   b. 250 gpm for initial standpipe and 250 gpm for each additional standpipe up to 750 gpm max in sprinklered buildings?
6. What types of hose valves are being installed? Are there pressure-reducing devices or pressure-reducing valves on the system and were the specifications of the valves provided?
7. Are the standpipes and sprinklers systems cross-connected?
8. Has the fire department connection been properly identified and is it in the correct location?
9. If the building has split zone risers, are they identified properly at the fire department connection?
10. Are the standpipe risers of the proper minimum size based on the building's height?

Fire inspectors should develop a checklist of all the standpipe information necessary to evaluate the plans for proper installation.

Other Fire Protection Systems

The fire inspector may also be required to examine plans for other types of fire protection systems including kitchen suppression systems, total flooding CO_2 systems, clean agent fire protection systems, or special hazard systems such as water mist, explosion control, and firefighting foam application systems. All of these different types of fire protection systems are addressed by specific NFPA standards specified by the building code:

- Commercial Kitchen Systems Mechanical Code or NFPA 96, *Standard for Ventilation Control and Fire Protection of Commercial Cooking Operations*
- Wet Chemical Systems NFPA 17A, *Standard for Wet Chemical Extinguishing Systems*
- Dry Chemical Systems NFPA 17, *Standard for Dry Chemical Extinguishing Systems*
- Foam Fire Protection NFPA 11, *Standard for Low-, Medium-, and High-Expansion Foam* and NFPA 16, *Standard for the Installation of Foam-Water Sprinkler and Foam-Water Spray Systems*
- Carbon Dioxide Systems NFPA 12, *Standard on Carbon Dioxide Extinguishing Systems*
- Halon Systems NFPA 12A, *Standard for Halon 1301 Fire Extinguishing Systems*
- Clean Agents NFPA 2001, *Standard on Clean Agent Fire Extinguishing Systems*

While each standard has unique requirements based on the type of extinguishing agent being used, there are many commonalties in the building code standards for these types of systems. These commonalities include a storage vessel for the extinguishing agent, a detection method and control head to fire the systems, a dispersion system of piping and application nozzles, and a pull station for manual operation. The building code establishes the initial installation and acceptance testing, and the fire code will provide the service, maintenance, and annual requirements for the life of the system.

KITCHEN RANGE HOODS (WET OR DRY CHEMICAL)

While the fire suppression agents in each type of system may vary, surface and plenum fire systems will have the following components that should be identified in the building plans (see Figure 14.14):

1. The location of agent storage system. The storage system must be in an acceptable location for inspection and servicing and should not be above the ceiling or hidden.

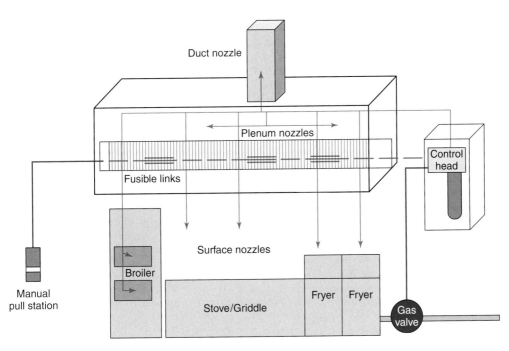

FIGURE 14.14 Basic system design for a kitchen fire suppression system.

2. These systems must be monitored by the building's fire alarm system.
3. A manual pull station must be installed, usually within 10 to 20 feet of the appliance and within the path of egress travel.
4. The system must activate an automatic gas valve to shut the gas supply down to the appliances. The valve must be manually reset.
5. Electrical equipment under the hood must shut down when the system operates.
6. The systems must operate the makeup supply air and exhaust fan in accordance with the manufacturer's specifications.
7. The specification should identify a portable K-class fire extinguisher and its location.
8. The plan should identify the number of nozzles, the orifice type, and the position over the appliances.
9. The plan should describe the method of fire detection by either fusible link or automatic rate compensation detectors.
10. The specification of a range hood system should identify that it is compliant with UL 300, *Fire Testing of Fire Extinguishing Systems for Protection of Commercial Cooking Equipment.*
11. All of the ductwork construction should be specified, including that it is liquid-tight welding; the gauge of metal; and that all system cleanouts, motors, fans, and attachments comply with the ICC mechanical code or NFPA 96, *Standard for Ventilation Control and Fire Protection of Commercial Cooking Operations.* If the vent penetrates more than one floor, it must be in a fire-resistant shaft, and the hourly rating should be identified.
12. The plan should also specify the method of firestopping or protection of combustible construction around the duct shaft and the rear wall of the cooking line.

TOTAL FLOODING FIRE PROTECTION SYSTEMS

Total flooding fire protection systems may use carbon dioxide gas, water mist, inert gas agents such as halocarbons, dry chemicals, and clean agents as fire suppressants. While the agents are different, many of the basic system components are the same or

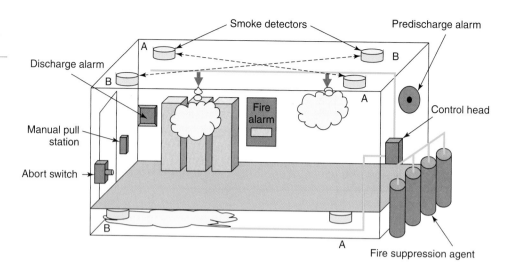

FIGURE 14.15 Typical components for gaseous fire suppression systems.

similar (see Figure 14.15). The fire inspector should examine the plans for the following criteria:

1. All total flooding systems are operated by an interconnected fire detection system. The fire inspector should determine the zoning areas of each fire detector. In most systems, the detection components are cross zoned on two circuits. The activation of a single detector provides a predischarge alarm warning to the room occupants. If the second detector on the cross-zone operates, then the system discharges. The reason for the cross zoning is to provide discharge warning and to reduce the possibility of accidental discharge.
2. The fire inspector must ensure that the total flooding system is connected to the building fire alarm system for notification.
3. The system installer must provide a description of the sequence of system operation for each type of system.
4. The design drawing must specify the type and amount of fire suppressing agent required as well as the type of storage cylinders to be installed.
5. Redundant agent supplies must also be provided for maintenance and system or refilling purposes in Halon replacement, CO_2, and water mist systems.
6. The manual pull station should be located in the path of egress travel and at the correct height.
7. Total flooding systems require an abort or **dead man switch**, which must be located at the exit for the space. The abort switch, when pushed in, delays the system's discharge until it is released to provide additional egress time for the room occupants. Releasing the switch causes an immediate discharge once the system is in alarm.
8. The designer must provide the calculations on the cubic volume of the space and the concentration of suppressant agent being used. Each type of system will have specific discharge time requirements as well as concentration levels, so the fire inspector should refer to the appropriate NFPA standard.
9. The plan must indicate the locations of both audible and visual alarm signaling appliances.
10. The plan must indicate the size of discharge piping and the number of nozzles to be installed as well as any sectional valves in the system.
11. A critical aspect of gaseous total flooding systems is the construction of the room enclosure including door seals, HVAC supply and damper locations, and sealing of the compartment. Most of these gaseous fire suppression agents are heavier than air and will run out of the enclosure if it is not properly sealed.

dead man switch ■ This is a constantly-on, spring-loaded switch at the exit from spaces provided with total flooding fire suppression systems. Upon discharge alarm activation, this switch can be held in to abort the system's activation until the area is clear of personnel. If the switch is released, the system immediately discharges.

On acceptance testing, the fire inspector must pay careful attention to any penetration of the enclosure; all penetrating items, such as door openings, must have automatic door closers and door sweeps. Vents for air conditioning or heating must have automatic air dampers to close the openings on system discharge. If a drop ceiling is installed within the space, it must be secured with special ceiling tile clips to prevent tile blowouts or possible ceiling collapse during the system discharge. Gaseous total flooding systems will also require special testing and inspection to pressurize the room to detect leaks.

FOAM FIREFIGHTING SYSTEMS

Foam firefighting systems are installed on helipads in aircraft hangers, flammable liquid storage rooms, and on high-voltage electrical transformers. These types of fixed fire protection systems are installed in accordance with NFPA 11A, *Standard for Medium- and High-Expansion Foam Systems*; NFPA 16, *Standard for the Installation of Foam-Water Sprinkler and Foam-Water Spray Systems*; or NFPA 409, *Standard on Aircraft Hangars*, for foam systems in aircraft hangers. The key elements to consider in the plan review of any fixed foam system are the placement of automatic monitor nozzles to ensure proper area coverage and the location of the manual pull activation station. Foam fire suppression systems for heliports must have at least a 10-minute minimum firefighting foam supply. These systems are frequently installed on high-rise buildings and hospital trauma centers (see Figure 14.16).

SMOKE CONTROL AND SMOKE EVACUATION SYSTEMS

The control of smoke in a building is a very difficult engineering task, at best, and has been a component of the building codes since the 1970s. The problem in designing smoke control systems is that the atmospheric conditions constantly change both inside and outside the buildings, caused by the **stack effect**. The stack effect is the movement of air along a pressure gradient created by different temperatures between the inside air and outside air of the building. These temperature differences cause air to move into the building below the neutral pressure plane and move out of the building above the neutral pressure plane. This effect also causes wind around high-rise buildings or the whistling of air

stack effect ■ The air movement through tall buildings caused by differences in air pressure due to temperature change.

FIGURE 14.16 Firefighting automatic remote monitor nozzle with 10-minute foam concentrate supply. *Courtesy of J. M. Foley*

around elevator shafts. It is this constant air movement that makes smoke control systems difficult to balance to remove smoke effectively. The general theory of smoke management falls into two categories. The first category is to create a negative pressure on the fire floor by exhausting 100 percent of the air volume from the floor. The surrounding floors above and below the fire floor are then pressurized with 100 percent air supply, creating a positive air pressure around the fire floor. This balance between higher and lower pressure should cause the smoke to migrate in the direction of the low pressure toward the exhaust fans. The second category is to create a smoke-collecting area, or smoke bank, such as at the top of an atrium. The smoke bank is provided with roof exhaust to relieve the smoke, causing a negative pressure below and making smoke continue to migrate into the smoke bank area. This type of system often is used in covered malls and enclosed arenas. The smoke control system supplies air to the floors around the atrium and exhausts smoke at the top of the atrium. Smoke control systems may be dedicated systems or may be part of the normal HVAC system. The most common method used is to integrate smoke control with the normal HVAC to avoid having two systems and additional construction and maintenance costs. The issue with integrated HVAC/smoke control systems is air balancing. To remove smoke, high fan speeds are required, and this usually has a negative effect on normal heating and cooling, which is the predominant purpose of the HVAC system. Smoke control systems are required to undergo a performance test for smoke removal at the system's commissioning. Depending on how well the system works, the fan speeds usually must be increased to effectively remove smoke. After the acceptance testing, the system usually is rebalanced for normal HVAC operation. This rebalancing often renders the reliability of the smoke removal system suspect at best. Another significant feature of smoke control systems is that each system is unique and cannot be compared to other existing systems because of differences in design concepts and HVAC equipment. The building code requires that designers of smoke control systems conform to NFPA 90A, *Standard for the Installation of Air-Conditioning and Ventilating Systems*, and NFPA 90B, *Standard for the Installation of Warm Air Heating and Air-Conditioning Systems*, as well as the standards of the American Society of Heating, Refrigeration, and Air Conditioning Engineers, or ASHRAE. The design engineer must provide a complete overview of the conceptual smoke removal design, and also must provide an operations manual and matrix for the system. The smoke control system requires acceptance after a special inspection by a certified engineering agency to verify the operability and design of the system. The fire inspector during plan review should examine the following elements:

1. The overall concept of design for the system in the operations manual.
2. The location of each air supply fans and return air dampers should be identified on the system drawings.
3. The sequence of system operation for each initiating device should be explained and represented in the system matrix.
4. The duct smoke detector should be indicated on the plans as to their locations with regard to fan motors and filter racks as well as the cubic feet per minute that each fan provides.
5. The location of fire and smoke barriers should be identified and the designer should have an explanation of how multiple systems are to be managed across fire and smoke barriers.
6. The perimeters of each of the smoke evacuation zones should be identified.
7. The firefighter smoke control panel and operational diagram should be provided. The smoke control panel must be located in the building fire command center (see Figure 14.17).
8. The interface of the smoke control system with the fire alarm system should be indicated.

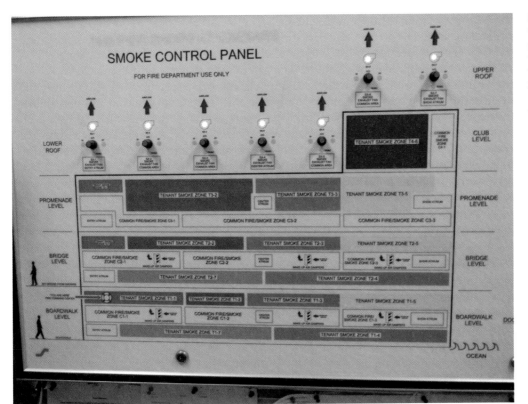

FIGURE 14.17 Fire-fighter smoke control panel. Each switch controls a fan to start or stop the system. Each smoke zone is in a different color.
Courtesy of J. M. Foley

FIRE-RESISTANT CONSTRUCTION FEATURES

The examinations of fire-resistant construction as well as the means of egress are typically a joint plan review responsibility between building inspectors and fire inspectors. The key elements of the review include examining the building's firewalls, fire separation walls, and fire-rated ceiling and floor assembly, which are found in the architectural floor plans. The architectural floor plan also references the fire door schedule and the associated fire door hardware. The fire-resistance ratings for vertical shafts may appear on both the architectural and the mechanical plans when the electrical, mechanical, and plumbing shaft systems penetrate multiple floors of the building. The fire inspector should identify the UL design fire-resistance rating of all items penetrating the vertical shaft walls. It is also key for the fire inspector to identify the boundaries of each fire area within the building in order to determine the operational zones for the fire alarm and the fire suppression system's alarm-indicating appliances. In large buildings with open floor areas, each fire area may contain many fire alarm annunciator zones and multiple sprinkler systems. The fire zone is essentially the area within defined firewalls, fire separation walls, or fire barriers and between the rated floor ceiling assemblies that form a compartment. Fire alarm audible devices must sound throughout the entire fire area. The fire alarm annunciator zones are defined as specific alarm activation zones within a fire area that identify the operation of a particular alarm-initiating device. Annunciator zones are established by floor level, square footage of area, and travel distances for each floor. Each floor must have a minimum of one fire alarm annunciation zone, and the largest that a single zone may be is 20,000 square feet. The longest dimension of an alarm annunciator zone is 300 feet in any direction. While a fire area may have multiple annunciation zones, the evacuation signal must go throughout the entire fire area. These items often are confused in design.

The fire inspector must review and identify all of the fire-resistant wall and ceiling assemblies and their UL design listings for the proper fire-resistant wall construction. The fire inspector should also examine the mechanical and electrical plans and identify the method to be used to firestop any penetrations through fire-resistant assemblies, including the placement of fire and smoke dampers and the associated service access openings. This is especially important in floor and ceiling assemblies, which can become pathways for smoke and fire travel if not properly firestopped. The fire inspector must also identify the locations and fire resistance ratings of all stairwells, vertical shafts, and fire and smoke barriers. A method that can be used to identify fire-resistant features is to highlight each wall, door, damper, or other component with a different color highlighter based on the required hourly fire-resistance rating. This helps make these elements jump out of the blue-line drawings and assists the fire inspector in ensuring that all of the required fire-resistant components are present within each fire area. It should also be pointed out that the marking of a plan set should be performed only on the fire department's copy and not on the copy being returned to the owner or professional designer as approved plans. Always request an extra set of plans for markup and review.

MEANS OF EGRESS

In smaller buildings, the egress system and occupant loads may be identified on the architectural floor plans. In large buildings, the designer prepares a separate egress drawing to identify the occupant load of each room or space and the paths of travel to the exits. The egress plan should include the total travel distances to the closest exit, the capacity of each egress component, and the total floor occupant load. The fire inspector should cross reference the egress plan with the electrical and fire alarm plans for placement of exit signs, emergency lighting, and fire alarm manual pull stations. The fire inspector's goal for the means of egress review is to ensure that all minimum exit requirements have been addressed and that the exit capacity is sufficient for the intended occupant load of the building. The design professional should identify how each occupant load was determined and should also identify the minimum designed occupant load and any increased occupant loads in any rooms or spaces that may apply.

As-Built Drawings

An important aspect of the plan review process is to obtain a complete set of as-built drawings before the final *certificate of occupancy* is issued. The certificate of occupancy issued by the construction official identifies that the building is code compliant and can be occupied and used. The as-built drawings represent the actual installation of the fire protection equipment or reflect any changes in design that have occurred over the course of the building's construction. In some cases, the as-built plans differ significantly from those that were originally reviewed because of unforeseen circumstances encountered between building trades, system locations, or owner's changes in the physical design. All plan changes must be reviewed by the authority having jurisdiction before making the change, but the reality is that often, a number of small changes do occur that are not reviewed. The construction code official must decide at what point it is imperative for the building or fire inspectors to review a change in the design. The as-built plans most accurately reflect the final building configuration and the way it was constructed.

Authority and Responsibility of Plan Reviewers

The authority and responsibility for each element of a plan review process is specified in the adoptive legislation of the local or state building code. Fire inspectors must understand the responsibilities delegated to them specifically, and those responsibilities shared by themselves and the building, electrical, and plumbing inspectors. The fire inspector's responsibilities fall into two specific areas during a building's construction. These

The fire at the Beverly Hills Supper Club in 1977, where 167 persons lost their lives, was one of the most tragic fires to occur in a place of public assembly since the Cocoanut Grove fire in 1942. A leading and contributing factor to loss of life was a combination of failures to comply with state building and fire codes and the lack of identifying deficiencies in the early construction stages. The state of Kentucky at the time had a mandatory building code but also allowed local jurisdictions to adopt local codes provided they were as stringent as the state code. The Beverly Hills Supper Club was originally constructed in 1937 as a country club. In 1970 it underwent a $170,000,000 renovation under building permits issued by the city of Southgate. The building suffered a fire during reconstruction, and its opening was delayed for a year. The club reopened in 1971 even though there were questions with regard to fire code compliance with the state's fire marshal's office and the resolution of ten code violations. In 1974 and 1975 the owner performed additional renovations and additions with permits from the city of Southgate, adding the Cabaret and the Zebra rooms to the existing facility. The Zebra room was determined to be the area of origin in the deadly 1977 fire.

Due to the large loss of life, the Kentucky State Police, the NFPA, and the National Bureau of Standards participated in the fire investigation. It was evident that numerous, serious building and fire code deviations existed in the Beverly Hills Supper Club that contributed to the loss of life. Most of these code deviations should have been identified in the plan review and approval process or during building inspections to verify code compliance. Some of these deviations may have been the result of piecemeal construction over the years, with each area complying to different versions of building codes, or just ineffective plan review processes. As each new addition to the existing building was made, exit capacity for the entire building was reduced. In addition, the building's square footage was being increased, and the area eventually exceeded the permitted construction classifications at the time of the building code. The construction type at the time of the fire was determined to be unprotected combustible construction, which would not be permitted in a building of that square footage under any building code in existence in 1977. The NFPA determined that the exit systems were grossly inadequate for the occupant loads permitted. The state fire marshal's office established the occupant load at 2,349 occupants, and the NFPA calculated the occupant load to be 2,735. According to the NFPA, this would require at least 27.5 units of egress width within proper travel distances for safe exiting. The entire facility at the time of the fire had only 16.5 units of egress width, providing a calculated exit capacity of 1,511 persons. The building had many other building and fire code defects including unprotected wiring systems, lack of proper firestopping, lack of firewalls, improper interior finish materials, no fire alarm systems, and no fire suppression systems. All of these defects are code items that should have been identified in the plan review process and corrected before occupancy. The best time to address code defects is when they are just pencil lines on a piece of paper. Proper plan review alleviates these types of issues and corrects them long before construction begins, ultimately saving lives in the process.

responsibilities include conducting of the plan review and the final inspection and acceptance testing of fire protection systems. As an example, the electrical and fire inspector may have joint plan review responsibility on the fire alarm system; however, the functional testing of the alarm system is usually the fire inspector's responsibility. The construction of any building takes a great deal of cooperation and communication between the inspection agencies and the owner and contractors. The fire inspector must realize that he or she cannot operate in a vacuum; rather, he or she needs to work cohesively with the other building inspectors to complete the assigned tasks. Fire inspectors should note any deficiencies in construction and refer those deficiencies to the appropriate inspector for determination and review. Fire inspectors should be aware that often the building trades contractors may pass on information concerning items that they believe are not up to code. That information should be passed on to the appropriate AHJ for further investigation. Fire inspectors assigned to plan review units must have the capacity to work well with nonfire personnel and must be good communicators and cooperative participants in the plan review process.

Summary

Fire inspectors and fire prevention activities begin before the building is constructed and occupied. The fire prevention inspector who can participate in reviewing plans can ensure that fire safety systems and fire prevention measures are adequately addressed in the planning stages of building construction. Fire inspectors assigned to plan review duties must develop their professional knowledge in the fire safety aspects of the building code. The fire inspector must be well versed in the applicable NFPA standards, especially those related to fire protection systems and automatic fire alarms. Fire inspectors must be able to work cooperatively with outside agencies and to communicate on a professional level with architects and engineers when addressing concerns during the plan review. The fire inspector must have a good working knowledge of the plan review process and be diligent in following the time constraints and inspection requirements specified under these regulations.

Review Questions

1. Examine the site plan review process of your jurisdiction. Describe what agencies are involved in the process and what their responsibility is in examining the site plan.
2. What areas of your jurisdiction's building code are relegated to the fire department for plan review? What are the areas requiring joint plan review? Who has inspection authority in these joint review areas? How are disputes or disagreements between reviewing authorities addressed in the regulation?
3. Review your local zoning and planning regulations. Describe the fire department's level of involvement in the process and whether it is sufficient to secure proper community fire protection.
4. What fire protection features should be examined in a site plan review?
5. What type of drawings should be submitted by an applicant to conduct a thorough plan review?
6. Describe the plan submittal process in your jurisdiction. What time constraints are placed on the review for each agency's approval or denial?
7. List and describe the key elements of performing a plan review on an automatic sprinkler system.
8. What standards are employed in your building code for the review of fire protection systems? What are the key standards that the reviewer must be well versed in?
9. List the performance characteristics of a smoke control system. How does your jurisdiction inspect these systems?
10. What are the key inspection points for the plan review of passive fire protection features? List the elements that should be examined and the applicable tests or standards that would apply.

Endnotes

1. http://www.isopropertyresources.com/Landing-Pages/On-Location/LOCATION-GeoTRIVIA-December-2006.html

2. Kenneth E. Isman, *Layout, Detail and Calculation of Sprinkler Systems* (National Fire Sprinkler Association, 2007), 39.

INDEX

NFPA 92A *(Standard for Smoke Control Systems Utilizing Barriers and Pressure Differences)*, 274

NFPA 92A *(Standard for Smoke-Control Systems Utilizing Barriers and Pressure Differences)*, 210

NFPA 30B *(Code for the Manufacture and Storage of Aerosol Products)*, 186

NFPA 51B *(Standard for Fire Prevention During Welding, Cutting, and Other Hot Work)*, 142

NFPA 13D *(Standard for the Installation of Sprinkler Systems in One- and Two-Family Dwellings and Manufactured Homes)*, 298

NFPA 13E *(Recommended Practice for Fire Department Operations in Properties Protected by Sprinkler and Standpipe Systems)*, 309

NFPA 13R *(Standard for the Installation of Sprinkler Systems in Low-Rise Residential Occupancies)*, 297

for water-based fire protection system, 283

Standard Sprinkler Pendent (SSP), 299

Standard Sprinkler Upright (SSU), 299

Standard Tank Institute (STI), 178

Standard time/temperature curve, 224, 225

Standpipes, 311–316, 356–357
class I, 311–312
class II, 312
class III, 313
defined, 311
and fire inspection, 314–316
operational categories, 311
plan review, 371–372

State inspection agencies, fire safety, 21

Station nightclub fire tragedy, Warwick, 12, 142

Statistical analysis, of data, 42–45

Status quo, 53, 62

Steiner tunnel test (ASTM E84), 234–235

STI. *See* Standard Tank Institute (STI)

Storage, buildings
arrangement, 147–148
hazardous materials, 176
of flammable and combustible liquids, 179

Storage cabinets, hazardous materials, 175–176

Storage tanks, water
elevated, 329
pressure, 329–330

rooftop, 329–330

Streams, as natural water source, 336
transect of, 337

Strobe lights, 258–259

Superfund, 161, 162

Superfund Amendment and Reauthorization Act (SARA), 162, 163
reporting requirements of, 165–166

Sustained, objection, 127

Swimming pools, as water resource, 337

Systematic inspection pattern, 95
ceiling inspection, 95–96
equipment and storage, 96–97
fire protection system inspection, 96
wall inspection, 96

Tamper switches, 273, 305–306

Task force, 86, 88

Tax records, for data collection, 37–38

Technology
fire alarm system, 272–273
and organizational change, 71–72
and record keeping, 108–109

Teflon® tape, 182–183

Templer, John, 199

Tenth Amendment, 90

Theatrical fires, 141

Theory X, 61

Theory Y, 61

Three "E's," fire prevention, 80

Threshold planning quantities (TPQs), 165

Tier I/II report, SARA, 165

Time march multiplexing systems, 273

Tire recycling fire, Philadelphia, 84

Torches, as open flame device, 141, 142–143

Total flooding fire protection system, 373–375

TPQs. *See* Threshold planning quantities (TPQs)

Transponders, 272

Travel distance
defined, 202
exit access, 204–205

Treadles, defined, 152

Triangle Shirtwaist Company fire tragedy, New York City, 9

Trouble alarm relay, 259, 269–270

UL-300, 149

Ultraviolet (UV) detectors, 255–256

Underwriters Laboratories (UL), 19
fire alarm detection equipment listings, 248

Uniform Building Code and the *Uniform Fire Code* (ICBO), 24

Uninterrupted power supply (UPS) battery storage systems, 157

Union Carbide disaster, Bhopal, 11–12, 161–162

Union Fire Company, 3–4

Unit of egress width model, 198

Unprotected risk, 53

Unsafe structure order, 118

Upright sprinkler, 299

Urban water supply systems, 326–332
elevated water storage tanks, 329
fire booster pumps, 331–332
fire hydrants, 327–329
pressure/rooftop storage tanks, 329–330

U.S. Coast Guard, 20

U.S. Fire Administration (USFA), 20, 284

U.S. Fish and Wildlife Service, 21

U.S. Forestry Service, 21

USDA. *See* Department of Agriculture (USDA)

Vacant buildings, 144–146
securing of, 118

Values, of leaders, 64–65

Valves, 330
alarm check, 290
automatic sprinkler systems, 304
closed, 309
dry pipe, 291
inspector's test, 290, 291
low differential, 291
OS&Y, 289
pressure-reducing, 312, 313–314

Variations, in fire prevention code, 116–117

Vehicle access, fire department, 354–355

Vehicle impact protection, 146–147

Ventilation fans, inspection of, 148

Verbal orders, 110

Vertical shafts, 242

Very early smoke detection aspirator (VESDA), 256–257

Vigiles, 2

Violations, fire prevention code, 240–241, 308–309. *See also* Notice of violation (NOV)

Vision, fire department, 63–65

Visual fire alarm indicators, 365

Voice alarm communication systems, 261–262

Walls
fire-resistant, 228–229
fire separation, 229
fire/smoke barriers, 229
firewall, 228–229
inspection of, 96